For a listing of recent titles in the
Artech House Antennas and Propagation Library,
turn to the back of this book.

Wearable Antennas an

Wearable Antennas and Electronics

Asimina Kiourti
John L. Volakis
Editors

ARTECH
HOUSE
BOSTON | LONDON
artechhouse.com

Library of Congress Cataloging-in-Publication Data
A catalog record for this book is available from the U.S. Library of Congress.

British Library Cataloguing in Publication Data
A catalog record for this book is available from the British Library.

ISBN-13: 978-1-63081-821-0

Cover design by Charlene Stevens

685 Canton Street
Norwood, MA 02062

10 9 8 7 6 5 4 3 2 1

Contents

CHAPTER 1
Introduction 1
1.1 History of Wearables 1
1.2 Applications of Wearables 2
1.3 The Future of Wearables 5
1.4 Book Overview 6
References 7

CHAPTER 2
Basic Approaches for Printing and Weaving Wearables 9
2.1 Introduction 9
2.2 Basics of Embroidery 11
2.2.1 Operating Principle 11
2.2.2 Types of Conductive Threads 14
2.2.3 Substrates Used for Embroidered Prototypes 15
2.2.4 Nonconductive Threads 16
2.3 Advanced Aspects of Embroidery 17
2.3.1 Improving Precision 17
2.3.2 Grading the Embroidery Density for Foldable Prototypes 18
2.3.3 Colorful Prototypes 19
2.4 Polymer Integration 21
2.4.1 Polymer Substrates 21
2.4.2 Stretchable Prototypes Embedded in Polymer 21
2.4.3 Magneto-Actuated Prototypes 23
2.5 Performance 24
2.5.1 Radio-Frequency Performance 24
2.5.2 Mechanical Performance 28
2.5.3 Launderability 31
2.6 Example Applications 31
2.6.1 Textile-Based Antennas 32
2.6.2 Electromagnetic and Circuit Components 33
2.6.3 Sensors and Actuators 36
References 36

CHAPTER 3
Wearable Electronics with Flexible, Transferable, and Remateable Components 41
3.1 Technology Drivers 41
3.2 Functional Building Blocks 43
3.2.1 System Architecture and Components 43
3.2.2 Power and Data Telemetry 46
3.2.3 Energy Storage: Batteries and Supercapacitors 48
3.3 Technology Building Blocks for Heterogeneous Component Integration 50
3.3.1 Thin Substrates 51
3.3.2 Circuit Formation: Metallization, Photopatterning, or Additive Deposition 55
3.3.3 Device and Component Assembly 60
3.3.4 Encapsulation 66
3.4 Transferable On-Skin Electronics 66
3.4.1 Laser or Thermal-Assisted Release 67
3.4.2 Transfer with an Elastomeric Stamp 68
3.4.3 Transfer with a Water-Soluble Tape 68
3.4.4 Direct Flex Transfer onto Skin: Cut, Paste, Peel, and Release 69
3.4.5 Flex Substrate Embedding into E-Textiles 70
3.5 Biosignal Interfaces: Electrode and Photonic Interfaces 71
3.5.1 Ag/AgCl Electrodes 74
3.5.2 Dry Electrodes 75
3.5.3 Carbon- or Conducting Polymer-Based Electrodes 76
3.5.4 Fractal Gold Electrodes 77
3.5.5 Electrochemical Electrodes 78
3.6 Remateable Connectors 78
3.6.1 Pin-Socket Connectors 80
3.6.2 Flat Connectors 81
3.6.3 Reworkable Adhesives 84
3.7 Conclusion 85
References 86

CHAPTER 4
Wearable Antennas 91
4.1 Introduction 91
4.2 Embroidered Antennas 92
4.2.1 Design and Construction 94
4.3 Screen-Printed Antennas 95
4.3.1 Design and Construction 99
4.4 Inkjet-Printed Antennas 100
4.4.1 Design and Construction 101
4.5 Material Considerations: Fabrics, Inks, and Threads 102
4.5.1 Fabrics 102

4.5.2 Conductive Fibers 103
4.5.3 Conductive Inks 104
4.6 Applications 104
References 110

CHAPTER 5
Wearable Sensors 115
5.1 Sensing with Wearables 115
5.2 Wearable Electronics for Biomarker Extraction 116
5.3 Wound Monitoring RFID Bandage on Textile Surface 117
5.4 Textile Based Voltage-Controlled Oscillator 118
5.5 Wound Assessment Using Data Modulation 119
5.6 Smart Bandage Integration for Practical Measurements 123
5.7 Wireless Power Telemetry Link 124
5.7.1 Near Field Power Transfer Using a Corrugated Crossed-Dipole Antenna 124
5.7.2 Textile-Based Rectifier 124
5.8 Measurement Setup Realized to Emulate In Vivo Electrochemical Sensing and Monitoring Scenarios 127
5.9 Conclusion 129
References 130

CHAPTER 6
Wearable RF Harvesting 133
6.1 Part 1: Far-Field Integrated Power Transfer and Harvesting for Wearable Applications 133
6.1.1 Introduction 133
6.1.2 Conductive Thread Embroidery-Based Fabrication of Patch Antenna 134
6.1.3 Textile-Based Single-Diode Rectifier in Wearable Applications 138
6.1.4 Design and Optimization of Textile Rectenna Array 140
6.1.5 RF-Power Availability Tests 140
6.1.6 Power Harvesting Using Textile Rectenna Arrays 143
6.2 Part 2: Near-Field Integrated Power Transfer and Harvesting for Wearable Applications 145
6.2.1 Introduction 145
6.2.2 Anchor-Shaped Antenna: Fundamentals 146
6.2.3 Textile-Integration of an Anchor-Shaped Antenna and Its Ergonomic Applications 163
6.2.4 RF-to DC Rectifier Design and Optimization 167
6.2.5 System Design and Tests Using RF Rectifier and Anchor-Shaped Antenna 172
6.3 Conclusion 175
References 176

CHAPTER 7
Radiofrequency Finger Augmentation Devices for the Tactile Internet 181
7.1 Introduction 181
7.2 Communication Models for the Fingertip-Wrist Backscattering Link and Its Variability 183
7.3 Constrained Design of R-FADs 186
7.4 R-FAD Manufacturing 189
7.5 R-FAD Applications to Aid Sensorially Impaired People 191
7.5.1 Sensing an Item's Temperature 192
7.5.2 Discrimination of Materials 194
7.6 Application to Cognitive Remapping 197
7.7 Conclusion 197
7.8 Acknowledgments 199
References 199

CHAPTER 8
Wearable Imaging Techniques 203
8.1 Wearable Imaging Algorithms 205
8.1.1 Radar-Based RF and THz Imaging 205
8.2 Ultrasound Imaging 223
8.3 Optical Tomography 225
8.4 Photoacoustics Imaging 229
References 235

CHAPTER 9
Wearable Wireless Power Transfer Systems 241
9.1 Introduction 241
9.2 WPT Methods 242
9.2.1 Inductive Power Transfer 244
9.2.2 Resonant Inductive Coupling 245
9.2.3 Strongly Coupled Magnetic Resonance 245
9.3 CSCMR Systems for Wearable Applications 249
9.3.1 CSCMR System Design 249
9.3.2 Performance of CSCMR System on the Human Body 251
9.3.3 Magnetic Field Distributions 254
9.3.4 Specific Absorption Rate 255
9.4 CSCMR Systems for Implantable Applications 258
9.5 Conclusion 260
References 263

About the Editors 267
About the Contributors 271
Index 275

CHAPTER 1

Introduction

Asimina Kiourti and John L. Volakis

1.1 History of Wearables

Wearables are defined as electronic technologies incorporated into accessories or garments that can be worn comfortably on the body. This approach has gained much interest in recent years with a goal to incorporate hands-free wireless devices into garments for communications and sensing, including information gathering for the Internet of Things (IoT) and medical applications. Notably, the concept of wearables dates back to the thirteenth century (see Figure 1.1) [1] with renewed interest in smart wears starting in the 2000s. Spectacles were the first widely used wearable, followed by the first pocket mechanical watch in the sixteenth century, and abacus rings in the seventeenth century, mostly used by Chinese traders as counting tools for quick calculations. Other instances of wearables include the first watch designed to be worn on the wrist for the Queen of Naples in 1810, and timing devices hidden in shoes to cheat in roulette games in the 1960s and 1970s [2].

Wearable computers were first introduced in the 1980s, and in the 1990s wearable systems were considered, including Global Positioning System (GPS) and private eye displays [3]. The first commercially available wearable computing hardware was developed in the same decade by Rockwell International. In the 2000s, progress on wearables was relatively slow, but in the 2010s efforts on wearables with sophisticated (or smart) functionalities was intensified by the dramatic growth and ubiquitous usage of wireless devices in every aspect of daily life [4].

The first wearable devices that attracted consumer interest at a reasonable price and performance were fitness trackers. Developed by companies in the sports sector (e.g., Fitbit, Jawbone, and Nike), these fitness trackers are still in use today and have set the stage for a new realm of possibilities among wearables. More recently, wearable products have expanded to include smart watches, biomonitoring clothes and straps, smart eyegear, and navigational shoes (Figure 1.2). Concurrently, there is growing research activity in making such devices more comfortable, washable, reliable, and adaptable to a variety of applications and human needs.

Needless to say, the market of wearables is growing rapidly, as seen in Figure 1.3, with body wear research requiring more attention in the years to come.

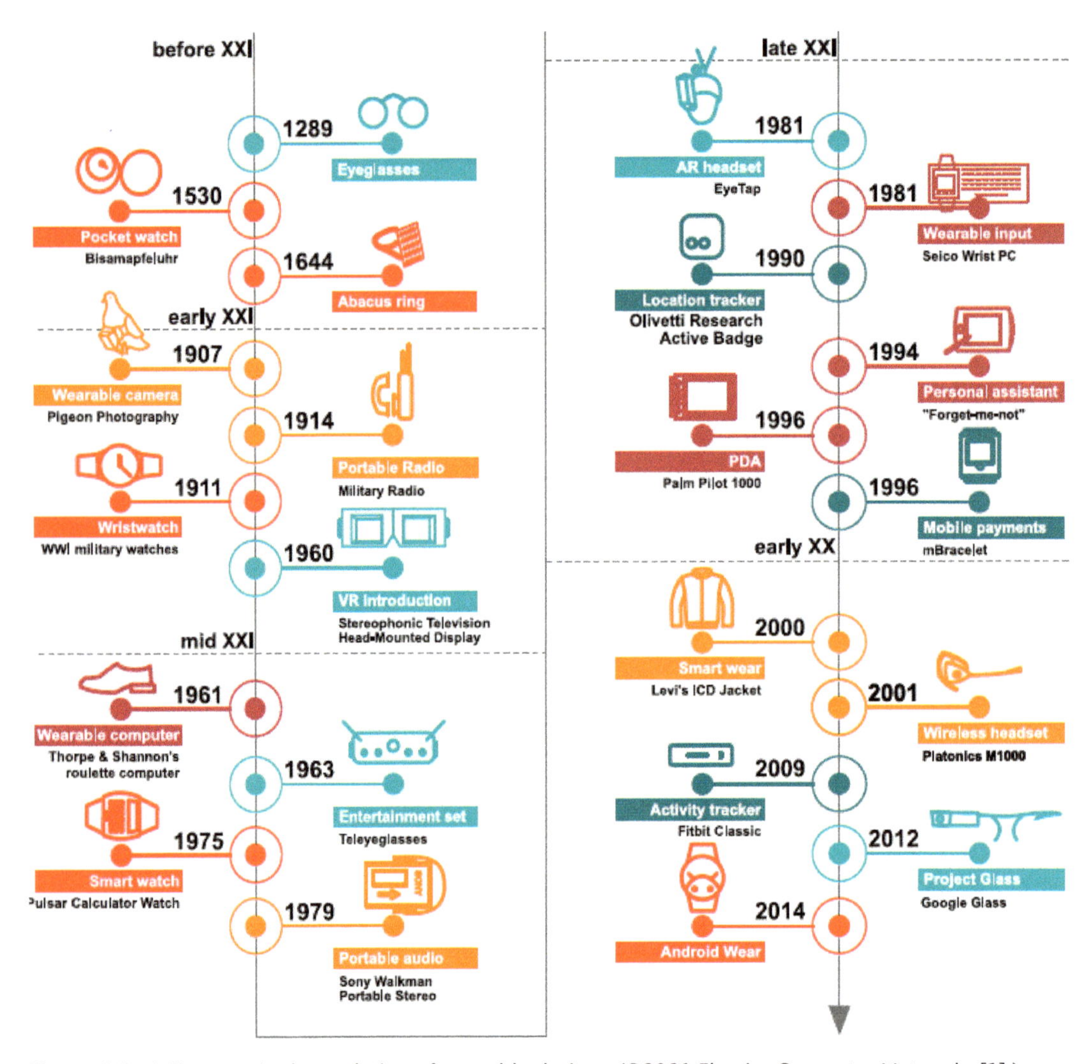

Figure 1.1 Milestones in the evolution of wearable devices. (©2021 Elsevier Computer Networks [1].)

Notably, shipments of smart wearables stood at 266.3 million units in 2020 globally, and is projected to reach 776.23 million units by 2026, registering a compound annual growth rate (CAGR) of 19.48% during the period of 2021–2026 [5].

1.2 Applications of Wearables

To date, the vast majority of commercially available wearables have focused on fitness and consumer electronics applications. The main reason for this focus is because they avoid the manufacturing challenges associated with textile integration. The latter is a continuing challenge and is the focus of this book. Some indicative examples of the former include:

- Fitness bands that help people understand their sleep cycles and how to move more and eat better (JawBone UP);

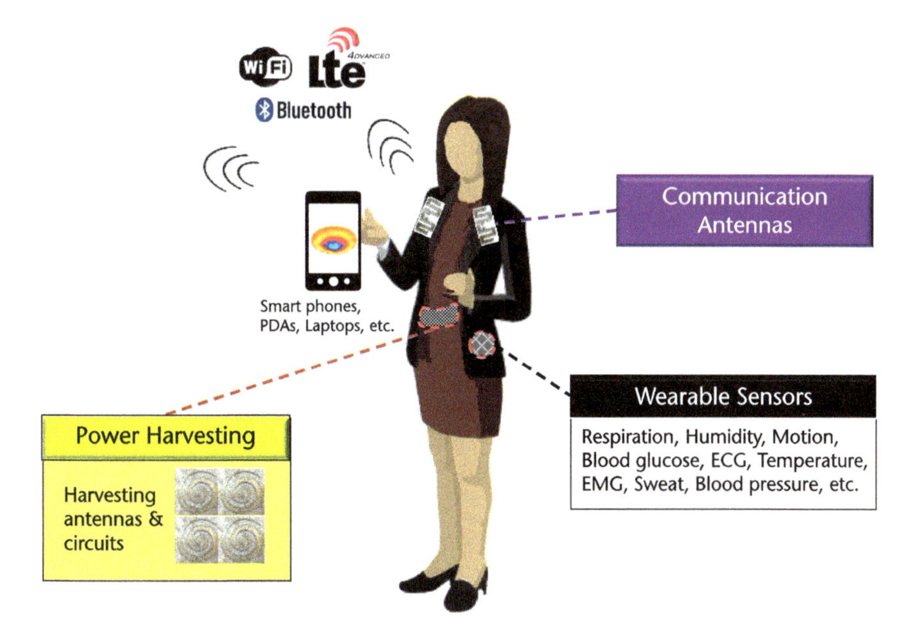

Figure 1.2 Examples of wearable sensors with integrated communication antennas for remote monitoring.

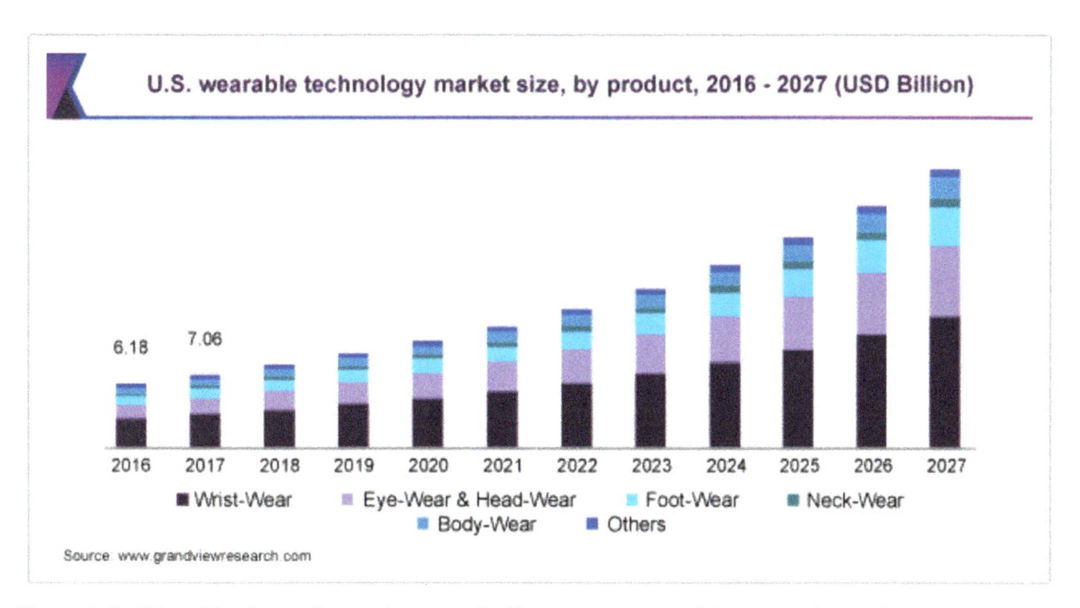

Figure 1.3 Wearable electronics market growth. (Source: www.grandviewresearch.com.)

- Chest straps with heart-sensing fabric technology for sports performance monitoring (NuMetrex);
- Glasses that help wearers that access the internet via language commands (Google Glass);
- Shoes that employ haptic feedback to notify users which direction to take (LECHAL GPS shoes);

- Rings for gesture-based control of devices (Fin Ring);
- Watches with biomonitoring capabilities (Apple watch);
- Shirts that monitor heart rate variability, anaerobic thresholds, as well as fitness and stress levels (Ambiotex);
- Socks for monitoring oxygen levels and heart rates while people (babies in particular) sleep (Owlet).

With the increasing use of wireless devices in our daily lives, there is concurrent research to allow them to be used in many areas, including

- *Medicine:* Wearables are envisioned to provide around-the-clock monitoring for medical condition prevention, diagnosis, and therapeutic applications. The ultimate goals is to shift the focus from reactive to proactive healthcare and, eventually, to generic therapeutic models for personalized healthcare. Example wearables devices include sensors for deep-tissue imaging [6], garments for motion capture in real-world environments [7], navigation systems for the visually impaired [8], monitors for the elderly [9], eye trackers for mental health monitoring [10], glucose monitors for individuals with diabetes [11], remote fetal monitors during pregnancy [12], and robotic devices for stroke rehabilitation [13].
- *Defense:* Wearables are envisioned to update traditional uniforms to include smart functionalities that sense vitals, alert for damage/control events, enable wireless connectivity, and harvest energy [14]. Notably, the large surface area associated with garments enables large antennas to be woven into fabrics, hence boosting communication range and/or enabling larger amounts of power to be captured in such challenging environments. Wearables reported in this regard include gloves for chemical sensing [15], antennas for off-body communication [16, 17] wearable biofuel cells for bioenergy harvesting [18], and glasses for training based on augmented reality [19].
- *Emergency:* The emergency services sector can also benefit by wearables that are seamlessly embedded in personnel uniforms. Firefighters, policemen, and emergency medical service providers, among others, can improve the quality of their work and maintain safer lifestyles through next-generation uniforms functionalized with sensing and communication abilities. Wearables reported in this regard include vests with photoplethysmography capabilities [20], garments for assessing an individual's physiologic state [21], systems that collect environmental information and predict danger [22], electronic triage devices [23], and cognitive assistant systems [24].
- *Space:* Healthcare provisioning for space applications is sparse, with only a few missions having dedicated medical professionals. But the vast majority of missions rely on live communication links with Earth-based physicians. Concurrently, the unique challenges that astronauts will face for long-term spacecraft and during extravehicular activity increase chances in-flight injuries. Preflight training and postflight reconditioning also require extensive vitals monitoring to improve performance and personalize medical monitoring. With the above in mind, research on wearables for space applications

has gained significant traction. Wearables reported in this regard include situational awareness terminals [25], augmented reality devices for training [26], visual odometry systems for navigation [27], biosuits for enhanced locomotion [28], and haptic feedback sensors to prevent disorientation [29].

1.3 The Future of Wearables

The term wearables has often been synonymous with accessory-like devices, such as bracelets, watches, shoes, or straps. Nevertheless, such accessories are bulky and obtrusive, and cannot be integrated and weaved into textiles. For example, a smart blouse in today's market is typically a regular fabric with a set of electronics housed in a casing and attached to the fabric. These electronics must be removed prior to washing. They are also rigid and aesthetically unpleasant, requiring frequent recharging, making them unsuitable for traditional clothing. However, future wearables promise to be seamlessly embedded into fabrics. In these products—often referred to as electronic textiles or e-textiles—conductors (such as antennas, transmission lines, and sensing elements) will be stitched using conductive threads, batteries will be replaced by textile-based power harvesters, and electronics will be seamlessly embedded into the fabric itself. That is, smart garments of the future will look much like the regular garments that we traditionally wear. Indeed, as rapid advances in electromagnetics, sensors, powering, materials, and electronics are being introduced and deployed, electronic textiles are becoming a greater part of the industry's focus. Indeed, as indicated in Figure 1.4, the benefits of improved connectivity using weaved antennas and wearable radio frequency (RF) electronics are significant, making the case for e-textile products. In the figure, a traditional copper-based antenna used for Wi-Fi applications is compared to an e-textile antenna with a much larger footprint as enabled by its integration into the fabric.

As would be expected, several challenges remain before the vision of seamlessly integrated wearables can become a reality. Conductive textile surfaces should feel

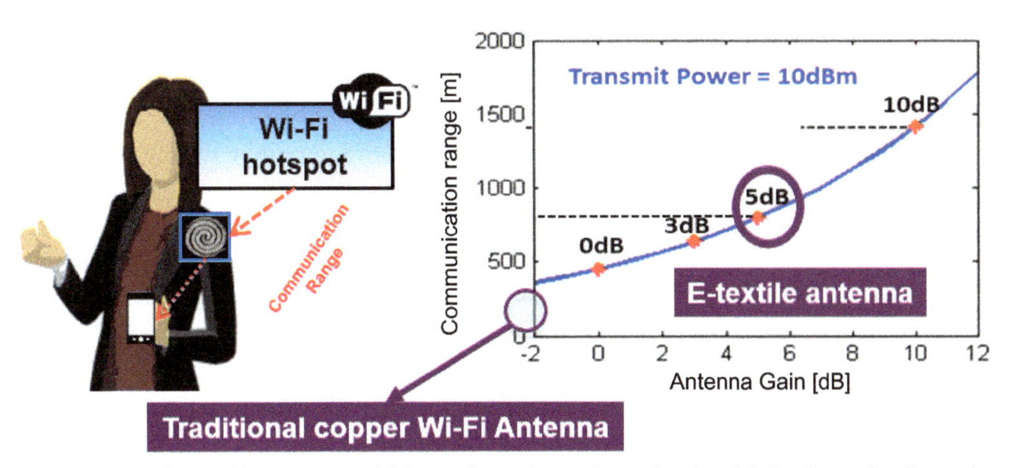

Figure 1.4 E-textiles enable antennas with larger footprints to be realized on fabrics, hence leading to larger antenna gains. In turn, higher gain e-textile antennas increase the maximum communication distance. A simple calculation using the Friis equation shows that a 3-dB increase in antenna gain with 10 dBm of transmitted RF power can increase communication range by ~40% (~200 m).

and behave much like regular fabrics. Concurrently, fabrics should exhibit high conductivity and electromagnetic performance similar to their copper counterparts. Connectors and electronics should be miniaturized, implemented on flexible materials, and seamlessly fused with flexible conductive surfaces. Also, fabric-based power harvesters are required to collect the abundant RF, solar, and thermal energy from the surrounding environment to eliminate batteries altogether. Certainly, RF noise associated with operation in-the-wild should be eliminated such that wearable devices are reliable for critical applications. Further, privacy and security concerns must be addressed by embedding suitable algorithms and hardware within the wearable device. Finally, personalization and interoperability represent two cornerstones of wearable systems. In summary, much remains to be developed in up-and-coming wearables with the future prospects being highly encouraging in realizing fascinating devices that will be part of our lives.

1.4 Book Overview

This book presents state-of-the-art developments on wearable antennas and RF electronics with applications to sensing, imaging, communications, power harvesting, medical data gathering, and sports applications, to mention a few. The book chapters are organized as follows.

Chapter 2 focuses on the embroidery of conductive electronic threads or e-threads as one of the most prominent implementations for functionalized textiles. The chapter discusses the basics of embroidery, selection of e-threads weaved into fabrics, and their performance for RF applications.

Chapter 3 addresses some of the challenges of component integration into wearable electronics. This chapter introduces the key system components for realizing flexible electronics into garments. It also describes processes for integration into flexible substrates, and how to create comfortable integration of lumped and wearable devices via remateable technologies.

Chapter 4 introduces fabrication methods for wearable fabric antennas, including challenges, advantages, and applications. The chapter discusses current research and progress made on diverse types of flexible materials in constructing these antennas.

Chapter 5 presents several sensing applications of e-textile wearables, particularly smart biochemical bandages that are textile integrated and self-powered. It is demonstrated that robust and reliable assessment of chronic wounds is feasible via reduced electronics and textile-based voltage-controlled oscillators.

Chapter 6 focuses on fabric-integrated power transfer and harvesting, emphasizing the advantage of clothing surfaces for larger power collection due to the available larger surface area on garments. The chapter considers integration of embroidered antenna arrays and rectifiers into fabrics for far-field power collection.

Chapter 7 reviews state-of-the-art RF) finger augmentation devices (RFADs) for tactile internet applications. The chapter provides design methods and flexible implementations for sensory/therapeutic applications.

Chapter 8 provides a review of the most commonly used wearable medical imaging techniques. The chapter discusses microwave tomography, ultrasound imaging, optical topography, photoacoustic hybrid imaging, and photoacoustic microscopy, as well as their importance in medical imaging.

Chapter 9 focuses on wireless power transfer systems for wearable and implantable applications. The chapter discusses related operating principles and system designs. Efficiency and specific absorption rate (SAR) performance are addressed.

References

[1] Ometov, A., V. Shubina, L. Klus, et al., "A Survey on Wearable Technology: History, State-of-the-Art and Current Challenges," *Elsevier Computer Networks,* Vol. 193, No. 5, 2021.

[2] Popat, K., and P. Sharma, "Wearable Computer Applications a Future Perspective," *International Journal of Engineering and Innovative Technology,* Vol. 3, No. 1, 2013.

[3] Sultan N., "Reflective Thoughts on the Potential and Challenges of Wearable Technology for Healthcare Provision and Medical Education," *International Journal of Information Management,* Vol. 35, No. 5, 2015.

[4] https://www.nytimes.com/2011/10/25/science/25shirt.html; https://www.newswise.com/articles/computers-in-your-clothes-a-milestone-for-wearable-electronics; https://www.embs.org/pulse/articles/e-textiles-for-health-monitoring-off-to-a-slow-start-but-coming-soon/; https://www.e-fermat.org/files/communication/Kiourti-MUL-LAPC2015-2016-Vol14-Mar._Apr.-009%20Wearable%20Antennas%20Sensors%20and%20.pdf.

[5] *BusinessWire,* "Global Smart Wearable Market," https://www.businesswire.com/news/home/20210208005342/en/Global-Smart-Wearable-Market—Market-to-Grow-by-19.48-from-2021–2026—ResearchAndMarkets.com.

[6] Salman S., L. Lee, and J. L.Volakis, "A Wearable Wrap-Around Sensor for Monitoring Deep Tissue Electric Properties," *IEEE Sensors Journal Special Issue Antenna Design and Integration in Smart Sensors,* August 2014, pp. 2447–2451.

[7] Mishra, V., and A. Kiourti, "Wearable Electrically Small Loop Antennas for Monitoring Joint Flexion and Rotation," *IEEE Transactions on Antennas and Propagation,* Vol. 68, No. 1, January 2020, pp. 134–141.

[8] Bai, J., S. Lian, Z. Liu, et al., "Virtual-Blind-Road Following-Based Wearable Navigation Device for Blind People," *IEEE Transactions on Consumer Electronics,* Vol. 64, No. 1, 2018.

[9] Kekade, S., C. Hseieh, M. Islam, et al., "The Usefulness and Actual Use of Wearable Devices Among the Elderly Population," *Computer Methods and Programs in Biomedicine,* Vol. 153, 2018.

[10] Vidal, M., J. Turner, A. Bulling, et al., "Wearable Eye Tracking for Mental Health Monitoring," *Computer Communications,* Vol. 35, No. 11, 2012.

[11] Cappon, G., G. Acciaroli, M. Vettoretti, et al., "Wearable Continuous Glucose Monitoring Sensors: A Revolution in Diabetes Treatment," *MDPI Electronics,* Vol. 6, No. 3, 2017.

[12] Signorini, M., G. Lanzola, E. Torti, et al., "Antepartum Fetal Monitoring through a Wearable System and a Mobile Application," Vol. 6, No. 2, 2018.

[13] Ren, Y., Y. Wu, C. Yang, et al., "Developing a Wearable Ankle Rehabilitation Robotic Device for in-Bed Acute Stroke Rehabilitation," *IEEE Transactions on Neural Systems and Rehabilitation Engineering,* Vol. 25, No. 6, 2017.

[14] Vital, D., P. Gaire, S. Bhardwaj, and J. L. Volakis, "An Ergonomic Wireless Charging System for Integration with Daily Life Activities," *IEEE Trans. Microwave Theory and Techn.,* Vol. 69, No. 1, 2021, pp. 947–954.

[15] Hubble, L., and J. Wang, "Sensing at Your Fingertips: Glove-based Wearable Chemical Sensors," *Electroanalysis,* Vol. 31, No. 3, 2018.

[16] Reddy, R. S., and A. Kumar, "Tri-band Semicircular Slit Based Wearable Antenna for Defense Applications," in *Innovations in Infrastructure. Advances in Intelligent Systems and Computing,* D. Deb, V. Balas, and R. Dey (eds.), Vol. 757, Singapore: Springer, 2019.

[17] Wang, Z., L. Z. Lee, D. Psychoudakis, and J. L. Volakis, "Embroidered Multiband Body-Worn Antennas for GSM/PCS/WLAN Communications," *IEEE Trans. Antenna Propagat.,* Vol 62, No. 6, June 2014, pp. 3321–3329.

[18] Jeerapan, I., J. Sempionatto, and J. Wang, "On-Body Bioelectronics: Wearable Biofuel Cells for Bioenergy Harvesting and Self-Powered Biosensing," *Advanced Functional Materials,* Vol. 30, No. 29, 2020.

[19] Barfield, W., and T. Caudell, *Military Applications of Wearable Computers and Augmented Reality,* CRC Press, 2001.

[20] Coca, A., R. Roberge, W. Williams, et al., "Physiological Monitoring in Firefighter Ensembles: Wearable Plethysmographic Sensor Vest versus Standard Equipment," *Journal of Occupational and Environmental Hygiene,* Vol. 7, No. 2, 2009.

[21] Tartare, G., X. Zeng, and L. Koehl, "Development of a Wearable System for Monitoring The Firefighter's Physiological State," *Proc. IEEE Industrial Cyber-Physical Systems,* 2018.

[22] Bu, Y., W. Wu, X. Zeng, et al., "A Wearable Intelligent System for Real Time Monitoring Firefighter's Physiological State and Predicting Dangers," *Proc. IEEE International Conference on Communication Technology,* 2015.

[23] Gao, T., and D. White, "A Next Generation Electronic Triage to Aid Mass Casualty Emergency Medical Response," *Proc. IEEE International Conference of the Engineering in Medicine and Biology Society,* 2006.

[24] Preum, S., S. Shu, M. Hotaki, et al., "CognitiveEMS: A Cognitive Assistant System for Emergency Medical Services," *ACM SIGBED Review,* Vol. 16, No. 2, 2019.

[25] Carr, C., S. Schwartz, and I. Rosenberg, "A Wearable Computer for Support of Astronaut Extravehicular Activity," *Proc. International Symposium on Wearable Computers,* 2002.

[26] Vizzi, C., K. Helin, and J. Karjalainen, "Exploitation of Augmented Reality for Astronaut Training," in *Proc. European Association for Virtual Reality and Augmented Reality,* 2017.

[27] Wu, K., K. Di, X. Sun, et al., "Enhanced Monocular Visual Odometry Integrated with Laser Distance Meter for Astronaut Navigation," *MDPI Sensors,* Vol. 14, No. 3, 2014.

[28] Newman, D., M. Canina, and G. Trotti, "Revolutionary Design for Astronaut Exploration—Beyond the Bio-Suit System," *AIP Conference Proceedings,* Vol. 880, No. 1, 2007.

[29] Bernard, T., A. Gonzalez, V. Miale, et al., "Haptic Feedback Astronaut Suit for mitigating Extra-Vehicular Activity Spatial Disorientation," *Proc. AIAA Space and Astronautics Forum and Exposition,* 2017.

CHAPTER 2

Basic Approaches for Printing and Weaving Wearables

Balaji Dontha and Asimina Kiourti

2.1 Introduction

Traditional copper-based electronics and electromagnetic components are rigid and typically nonconformal to curved surfaces. Upon flexing or deformation, they tend to crack, break, and eventually deteriorate from their targeted operational performance. Over the years, various advances have been made toward realizing antennas and electronics that are flexible, conformal, lightweight, and mechanically robust. These devices rely upon printing on flexible substrates and are capable of conforming upon various complex surfaces, such as the human body. Table 2.1 compares various fabrication approaches reported in the literature, as discussed in further detail in [1]. Figure 2.1 shows related fabricated prototypes with example applications including smart watches, radio frequency identification (RFID) sensing, augmented reality (AR), and virtual reality (VR) devices [1–4].

In particular, a branch of flexible electronics involves their realization on fabrics, leading to what is called textile electronics. With an increasing demand for wearable devices in sectors as diverse as healthcare, military, smart homes, and the entertainment industry, textile electronics are becoming highly popular. Techniques used to create such textile electronics include copper sheets adhered to fabric textures, electrotextiles, screen printing, inkjet printing, and embroidery of conductive threads. Table 2.2 compares these approaches and Figure 2.2 shows example prototypes, with further details reported in [1].

In this chapter, we focus on embroidery as one of the most prominent implementations for functionalized textiles. The approach relies on conductive threads (known as e-threads) and home-style machines that support digitized embroidery, and includes multiple variables such as the type of e-threads, stitching density, thread orientation while stitching, and fabric material. Advantages of fabrics integrated with embroidered antennas and electronics entail high printing resolution (order of 0.1 mm), RF performance similar to that of their rigid copper counterpart, repeatable fabrication, high endurance to thermal and mechanical changes,

Table 2.1 Techniques to Fabricate Flexible Nonfabric Sensors

Type	*Definition*	*Advantages*	*Disadvantages*
Screen printing	Additive stencil-based process that involves printing of conductive inks through a patterned fabric screen [5]	Simple and cost-effective [6]	Limited control over the thickness of the metal deposits
Photolithography	Patterning process in which a photosensitive polymer is selectively exposed to light through a patterned mask [7, 33]	Easy and low-cost fabrication process in laboratory settings	Sensitive to extensive flexing and mechanical stress
Inkjet printing	Additive, direct-write technology that employs printers to deposit highly conductive inks upon flexible substrates [22]	Simple, fast, environmentally friendly, and high-resolution printing	Limited achievable thickness of ink layer resulting in poor conductivity
Manual patterning of conductive polymers	Conductive polymers are used to create thick conductive surfaces	Low cost, can create complex geometries, enables thicker and biocompatible films	Limited resolution, chemical and thermal instability
Graphene printing	Graphene, the allotrope of carbon nanotube, is used in printing to improve conductivity [59]	Highly conductive, flexible, lightweight, and low cost	Binder-free method has relatively low conductivity for RF applications, and graphene with binder such as ethyl cellulose (EC) is unsuitable for heat sensitive substrates
Injection of liquid metal alloys in microstructured channels	Conductive layers are created as a grid of liquid-metal-filled channels inside a flexible substrate [8]	Low cost, high performance, mechanically robust, and thermally stable	Limited to simple geometries

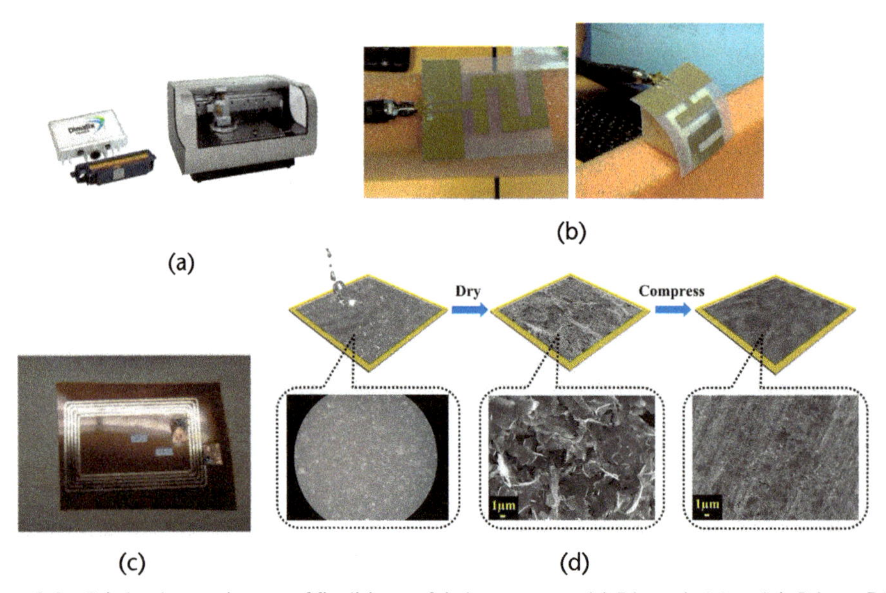

Figure 2.1 Fabrication and types of flexible nonfabric antennas: (a) Dimatrix Materials Printer DMP-2831 and DMC-11610 cartridge. (©2012 Int. J. Antennas Propag. [22].) (b) Inkjet-printed Z shape antenna on PET substrate. (©2018 Int. J. Antennas Propag. [63].) (c) Manual patterning of conductive polymers to create a flexible antenna (UWB antenna created via 158-μm-thick PPy conductive polymer). (©2012 Int. J. Antennas Propag. [22].) (d) Graphene printing on substrate. (©2015 Scientif. Rep. [23].)

Table 2.2 Techniques to Fabricate Flexible Fabric Sensors

Type	*Definition*	*Advantages*	*Disadvantages*
Copper sheets adhered to fabric textures	Utilizes adhesion of flexible copper sheets to fabric substrates	Low cost, easy, fast, and reliable	Delamination, poor accuracy, cannot withstand high temperatures and humidity
Electrotextiles	Conductive fabrics that are obtained by interleaving normal fabrics with conductive metal/polymer threads	Easy to perform cutting and sewing operations; can withstand washing, drying, ironing, and multiple deformations	Sensitive to extensive flexing and mechanical stress, less accurate when manually cut, can cause delamination during extreme deformations and skin irritation in nickel-based electrotextiles
Screen printing on fabrics	A screen is used for injecting pressurized conductive inks onto fabrics; techniques include direct to single-layer garment (DTG) printing [54] as well as several layers of screen-printing techniques [55]	Easy to fabricate, low cost, suitable for thick-film printing	Accuracy depends on surface geometry and roughness
Inkjet printing on fabrics	Uses silver or gold nanoparticles to create conductive lines on fabrics [57, 58]	Fast, low cost, and easily adaptable to various designs	Difficult to create highly conductive continuous ink tracks
Embroidery	Uses a computer-aided embroidery machine to print desired pattern using conductive threads	High-precision embroidery, launderability, tolerant to extreme temperatures and mechanical stress	High cost, limited operating range of frequencies

launderability, and easy integration with diverse types of polymers. Notably, embroidery fabrications are not just restricted to garments, but may readily expand to other wearable accessories (e.g., shoes, caps, belts, backpacks), as well as fabrics other than wearables (e.g., curtains, artwork) [1].

The rest of the chapter discusses basics of embroidery, e-thread and fabric materials that can be employed during the process, advanced embroidery techniques that achieve superior performance, polymer integration, RF performance of the resulting embroidered antennas and electronics under diverse conditions, and real-world applications.

2.2 Basics of Embroidery

2.2.1 Operating Principle

E-textile embroidery is a process that uses a computerized embroidery machine and conductive threads to embroider antennas and electronics upon fabrics in an automated manner [9, 53]. The embroidery process requires a series of steps as outlined below and further shown in Figure 2.3. We note that the description relates to e-textile antennas, yet the process can be readily expanded to other electromagnetic components (e.g., transmission lines) and/or circuits and electronics that would be conventionally printed on copper.

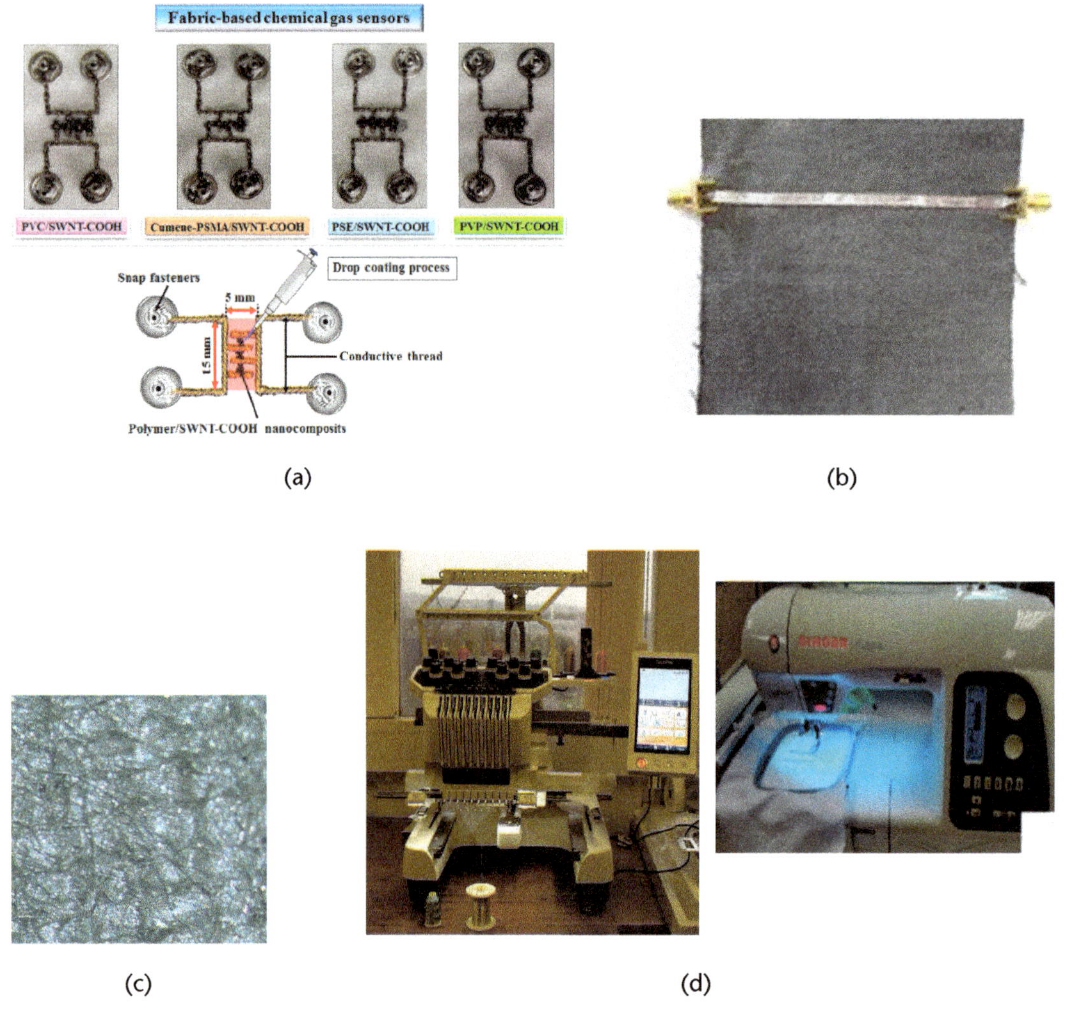

Figure 2.2 Types of fabric-based sensors. (a) Fabric-based chemical gas sensors. (©2015 Sensors [27].) (b) Microstrip line on denim substrate. (©2012 Int. J. Antennas Propag. [19].) (c) Details of conductive surface of a RFID antenna screen-printed on cotton fabric. (©2012 Int. J. Antennas Propag. [25].) (d) Automated machines used for embroidery. (©2014 MDPI Electronics [3], ©2020 Micromachines [26].)

2.2.1.1 Antenna Design

Antenna design is performed first so that it satisfies predefined performance criteria such as gain, bandwidth, size, and operational frequency [9]. This can be done via any electromagnetic computational solver tool. An important parameter that needs to be controlled while designing the antenna is to avoid sharp edges and details of higher than 0.1 mm in resolution, as these will likely be challenging in implementing during embroidery.

2.2.1.2 Digitization

Digitization implies identification of the path that the needle will follow during embroidery, stitch by stitch [10]. It involves conversion of a computer-aided design (CAD) based antenna design into an embroidery machine readable format.

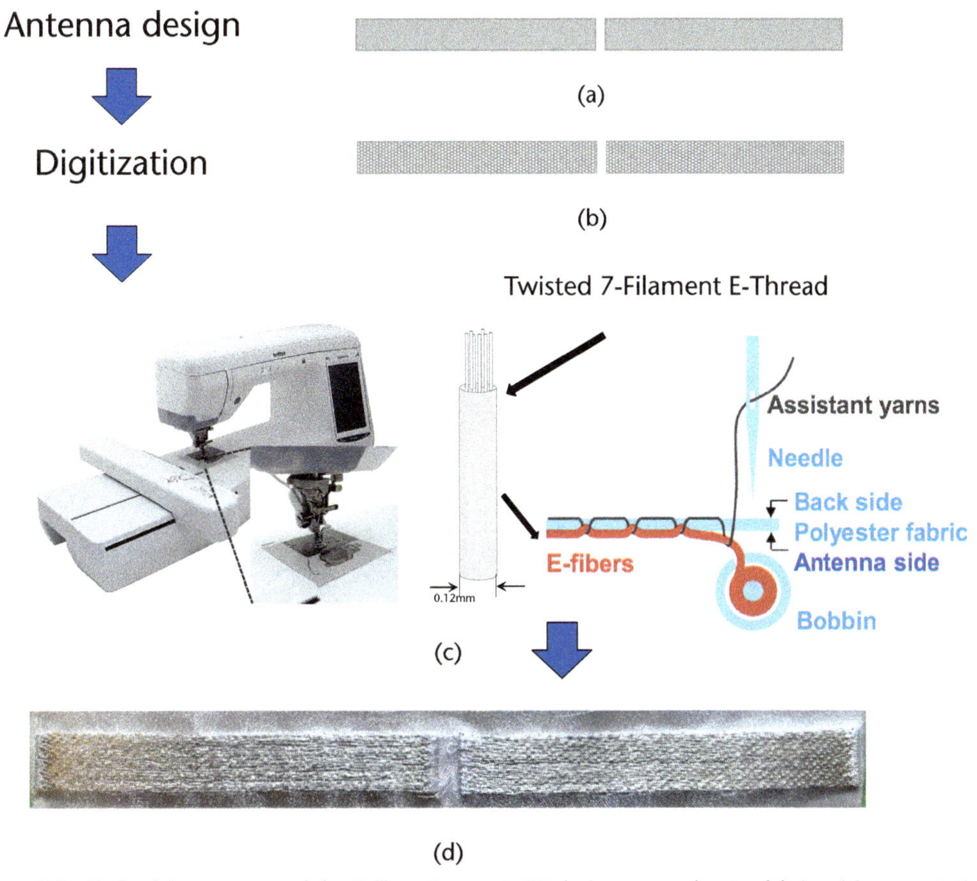

Figure 2.3 Embroidery process of the E-fibers to create RF designs on polyester fabrics: (a) computational software model, (b) digitized format in personal embroidery design software, (c) computerized embroidery machine, and (d) final embroidered antenna on a fabric material.

To achieve this, the antenna designed in any preferred computational software is converted from a Drawing Interchange File (DXF) to a Windows Metafile Format (WMF) and then imported as a vector file into the software that accompanies the embroidery machine. For example, a popular computerized embroidery machine is the Brother 4500D, which uses the PE Design-Next software tool.

2.2.1.3 Embroidery

E-textile embroidery involves: (a) the nonconductive fabric that the design is printed on, (b) a spool of conductive thread (e-thread) that is placed in the bobbin case, and (c) a spool of nonconductive thread that is threaded through the needle. As such, the nonconductive thread prints at the top of the fabric, and the conductive thread prints at the bottom of the fabric. Different e-threads and fabric materials can be used, which will be discussed in later sections of the chapter. Figure 2.3 shows an example antenna embroidered on a polyester fabric using silver-coated Amberstrand e-threads. In other cases, nonconductive threads of different colors can be alternated to create colorful prototypes. Further, diverse settings can be tuned on the embroidery machine, such as tension (e.g., Brother 4500D has 1-to-10 tension scale to quantify the pulling stress), e-thread density (e.g., varying from 1 e-thread/mm to 7 e-threads/mm for the Brother 4500D) [10], and direction of stitching.

2.2.1.4 Testing and Validation

Once the e-textile antenna is finalized, it can readily be tested. Its performance should ideally match the simulation results and/or measurement results acquired using gold-standard copper counterparts. Should any discrepancies occur, one may need to fine-tune the design so that it accommodates potential limitations of the embroidery process. Alternatively, parameters of the embroidery can also be modified so that the embroidered surface accurately depicts the simulated design and achieves a surface conductivity close to copper.

2.2.2 Types of Conductive Threads

A wide range of conductive threads (e-threads) are commercially available, although in-house options are also possible. It is worth noting that the vast majority of off-the-shelf e-threads are targeting shielding applications rather than embroidery of antennas and other pertinent electromagnetic components. However, with appropriate fine-tuning of the embroidery parameters reported in Section 2.1, such e-threads can readily be employed toward this end. Example e-threads available in the market are summarized next.

2.2.2.1 X-Steel

X-steel entails stainless-steel filaments covered with an outer metal layer, such as nickel, copper, or silver. The standard filament count is 16 (i.e., 16 filaments are twisted together to form one e-thread). The end thread has a diameter of 0.19 mm. Other filament counts are also possible. Overall, X-steel exhibits excellent thermal stability and tailored electrical conductivity. These e-threads can be connected to diverse other components by soldering or via other forms of connectors [12].

2.2.2.2 Liberator

Like X-steel, Liberator is a multifilament metal-clad e-thread. There are Liberator-20/40/80 e-threads available where the succeeding number indicates the number of filaments that are twisted together to form a single e-thread. Each filament is composed of ~23 μm thick liquid crystal polymer (LCP) Vectran fiber core and is coated with two metal layers such as copper (inner layer) and silver (outer layer). Depending on the twist per inch (TPI) used to form the twisted e-thread, the end diameter may vary. For example, the literature has reported Liberator-20 and Liberator-40 e-threads twisted at a TPI of 4.5 that exhibit diameters of 0.22 mm and 0.27 mm, respectively. These e-threads are flexible, lightweight, exhibit excellent thermal stability, and have good mechanical strength [11, 12].

2.2.2.3 Amberstrand

As with X-steel and Liberator, Amerstrand is a multifilament metal-clad e-thread. Amberstrand 166/332/664 e-threads are available where the succeeding number indicates the number of filaments. Each filament is composed of Toyobo Zylon core and is coated with copper, nickel, or silver outer metal layers. Like Liberator, the end diameter of the e-thread varies depending on the filament number and the TPI. For example, Amberstrand 166 at a TPI of 4.5 has been reported to exhibit a 0.24-mm diameter. These e-threads are flexible and lightweight, exhibit excellent thermal stability, and have good mechanical strength [11, 12, 16]. Their diameters, however, are large as compared to those of Liberator, making them suitable only for designs that do not require fine printing precision.

2.2.2.4 Elektrisola

Elektrisola e-threads are ultrafine enameled conductive wires. The core of these e-threads entails pure metals, alloys, or metal-plated wires. The outer layer consists of an enamel coating that works as insulation for the bare wire. Colored enamels may be used to provide different colored e-threads. Monofilaments are usually fine and smooth but less resistive to abrasive forces. Multifilament e-threads are more flexible than monofilaments and offer high strength and resistance to stretching [34]. Depending on the number of filaments, the diameter of the Elektrisola e-threads typically ranges from 0.010 mm (56 AWG) to 0.500 mm (24 AWG). Choice of different core and outer materials allows the control of various parameters, such as corrosion resistance as well as electrical, thermal, and mechanical properties [14].

2.2.3 Substrates Used for Embroidered Prototypes

Embroidery is always performed on a (nonconductive) fabric substrate. Depending on the application, diverse types of fabrics may be used and the embroidery settings may need to get optimized accordingly (e.g., embroidery density or tension).

One of the most commonly used fabrics in e-textile embroidery is organza. Besides being flexible and lightweight, a major advantage of organza is that it exhibits dielectric properties close to air [9]. By doing so, the organza substrate does not interfere with the electromagnetic performance of the e-textile component. Authors in [18] used organza to print tightly coupled dipole arrays on conductive textiles, while authors in [10] embroidered origami dipole antennas with graded embroidery on organza fabric material.

However, several other fabrics besides organza have also been reported for e-textile embroidery. For example, Kevlar is well known for its high strength, robustness, and ability to withstand extreme conditions. Authors in [17] embroidered a textile spiral antenna on Kevlar fabric for airframe applications intended to withstand extreme temperatures and mechanical fatigue. In other cases, felt has been used as a low loss and low dielectric permittivity material. For example, a dual-band E-shaped textile-based wearable patch antenna has been reported on felt fabric [52], while felt-based patches with coplanar waveguide (CPW) feeding on artificial magnetic conductors (AMC) are also available [19].

An added advantage to e-textile prototypes embroidered on fabrics is that they can be readily integrated with flexible substrates of varying dielectric properties. A common option is to use a polymer-based substrate and adhere it to the e-textile during the curing process. For example, polydimethylsiloxane (PDMS) is a low loss and highly flexible polymer substrate that has been frequently reported as an e-textile substrate. It exhibits tunable dielectric permittivity, room temperature fabrication, inherent chemical stability, and water resistance. The dielectric constant can be altered by controlling the dispersion of ceramic powders into the PDMS matrix. Authors in [11] employed PDMS (Sylgard 184, Dow Corning Corp.) as a substrate of an embroidered patch antenna. PDMS was used in [15] to protect electrotextile antennas during washing cycles. Shao and others incorporated PDMS into RFID tags to enable repetitive deformation and stretching while also preserving the e-threads from corrosion [16]. In other cases, shape memory polymers (SMPs) can be used that convert into preprogrammed structures under certain stimuli. These polymers are durable and insensitive to environmental variations such as temperature, and humidity. However, a drawback of SMPs is their limited ability to adhere to conductors [10]. Figure 2.4 shows example antenna and electromagnetic component prototypes implemented on nonconductive substrates.

2.2.4 Nonconductive Threads

Popular nonconductive threads used in e-textile embroidery include cotton, polyester, nylon, and silk. Besides the material of these threads, their diameter may also

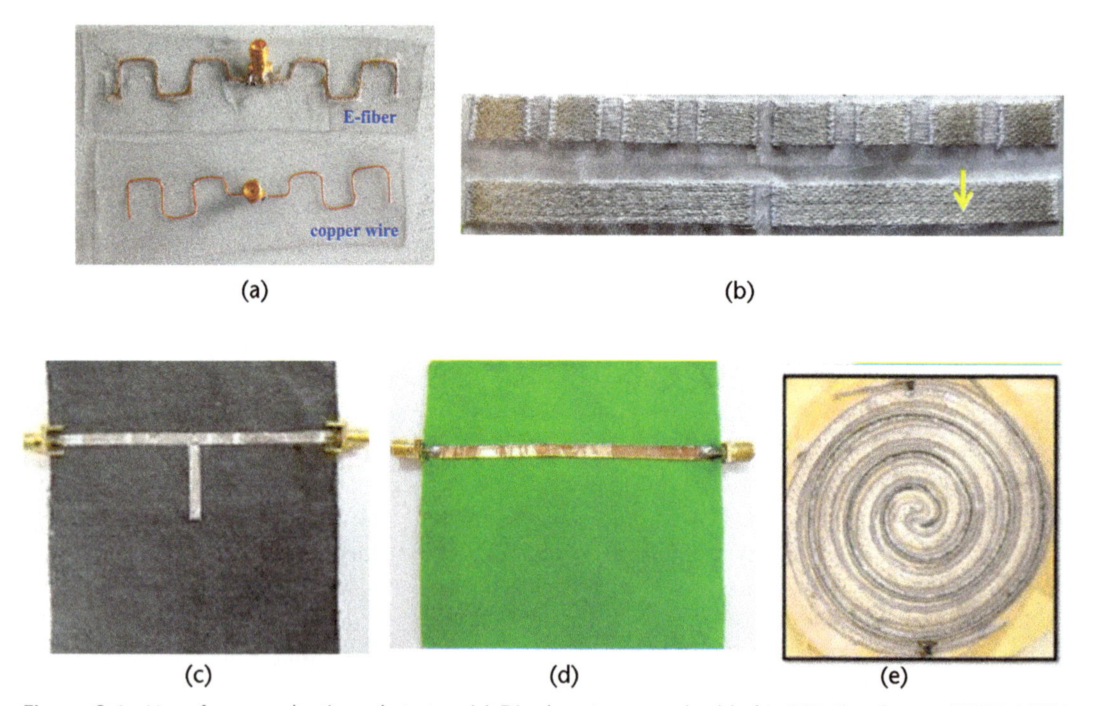

Figure 2.4 Use of nonconductive substrates. (a) Dipole antenna embedded in PDMS polymer. (©2014 IEEE [39].) (b) Graded dipole antenna using organza substrate. (©2018 IEEE [10].) (c) Microstrip line on denim. (©2012 Int. J. Antennas Propag. [19].) (d) Microstrip line on felt. (© 2012 Int. J. Antennas Propag. [19].) (e) Spiral antenna using Kevlar substrate. (©2017 IEEE [17].)

vary. Tensile, frictional, and bending properties are essential and the embroidery machine needs to be adjusted in order to obtain high-quality seamless stitching [20]. Colorful nonconductive threads can be used to provide color to the corresponding embroidered side of the fabric.

2.3 Advanced Aspects of Embroidery

2.3.1 Improving Precision

Electromagnetic components embroidered on e-threads need to be produced with high precision so that they exhibit excellent performance. Gold-standard milling machines print copper at a resolution of 0.1 mm, and a similar resolution is desired for embroidered prototypes as well. To this end, parameters affecting the embroidery precision are the thickness of the e-thread, the density of e-threads (i.e., number of e-threads printed per mm), the TPI characteristic of the e-thread, and the embroidery machine tension. These parameters can be individually or concurrently controlled to achieve the desired printing resolution per application scenario.

Some of the first embroidered prototypes exhibited a printing resolution of 0.5 mm. As an example, the embroidered multiband body-worn antenna developed by the authors in [50] for Global System for Mobile Communications/pulse-code modulation/wireless local area network (GSM/PCS/WLAN) communications used 332-filament silver-coated Amberstrand e-threads and demonstrated a printing resolution of 0.5 mm, which was capable of accurately realizing the antenna slots and arms. Figure 2.5(a) shows the use of e-threads to implement this particular antenna with a slot width of 1.5 mm. Advantages of using thicker e-threads in this case included increased conductivity and reduced stitching density that accelerated the overall time needed to perform the embroidery. However, drawbacks of using thicker e-threads included limited printing resolution and inability to print sharp corners [50].

To overcome the above, work by the researchers in [13] demonstrated a new technology of embroidering e-threads with resolution as high as 0.3 mm. The approach uses 40- and 20-filament Liberator e-threads having diameters of 0.27 mm and 0.22 mm, respectively. Figure 2.5(b) shows the resolution improvement enabled by Liberator e-threads with high TPI (~4.5) printed at a density of 7 e-threads/mm. Nevertheless, a limitation in this case is the lower conductivity of Liberator compared to Amberstrand. Specifically, the DC resistance of Amberstrand-664 e-threads is 2.3 Ω/m and that of Liberator-40 and Liberator-20 is 3.3 Ω/m and 6.6 Ω/m, respectively. Along the same lines, antenna prototypes created using Liberator-20 have shown to exhibit 0.6-dB lower gain as compared to Liberator-40. This can be addressed by practicing denser embroidery and/or by stitching two conductive layers one on top of another. Of course, such an approach increases embroidery time and may lead to breakage of the embroidery needle. Thus, it is important to optimize the embroidery density for high conductivity and geometrical accuracy per need case [13].

Further developments by the authors of [29] ultimately led to an improvement of precision as high as 0.1 mm. The process employed seven-filament Elektrisola e-threads having a very thin diameter of 0.12 mm. These e-threads exhibit very

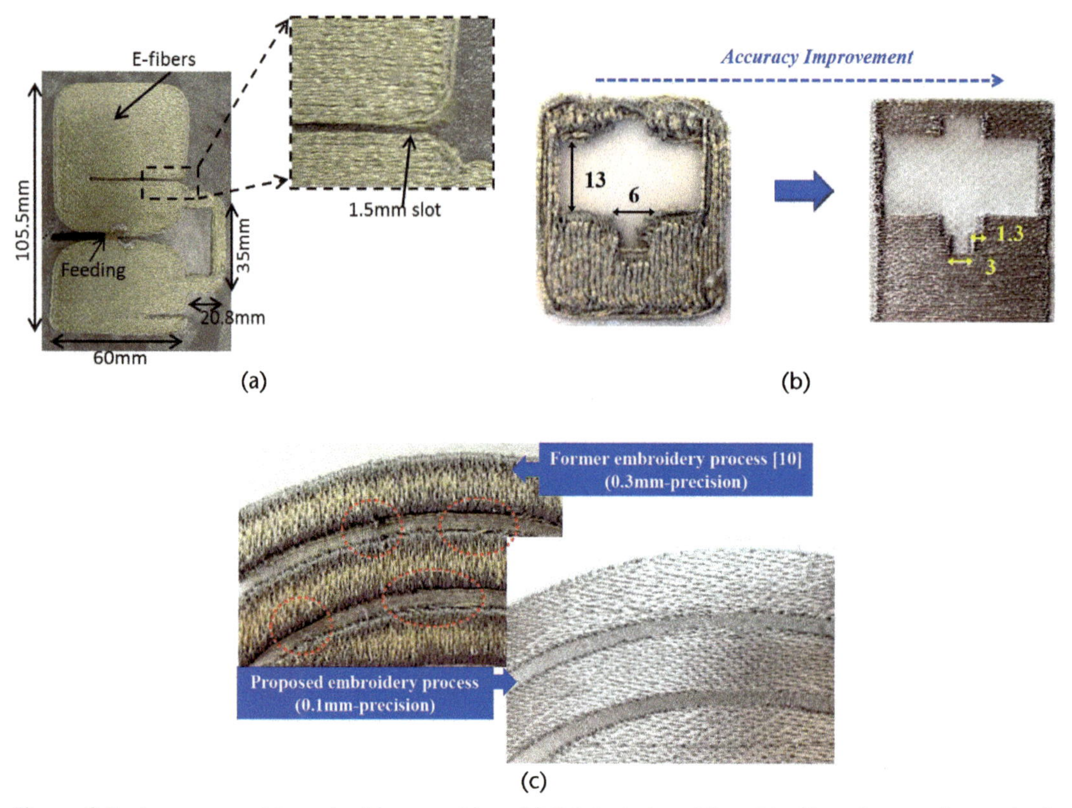

Figure 2.5 Improvement in embroidery precision. (a) Fabricated multiband textile antenna using embroidered e-threads. (©2017 IEEE [50].) (b) Precision improvement from 0.5 mm to 0.3 mm. (©2015 IEEE [13].) (c) Precision improvement from 0.3 mm to 0.1 mm. (©2016 IEEE [29].)

low DC resistance of 1.9 Ω/m, leading to higher conductivity textile surfaces. In addition, thinner e-threads imply lower embroidery tension and higher flexibility. Figure 2.5(c) shows an improvement in precision over the previous embroidery process using an Archimedean spiral antenna having a slot width of 2.4 mm and strip width of 8.5 mm. As such, the advantages of using Elektrisola e-threads entail higher geometrical precision and lower fabrication cost compared to the abovementioned e-threads. The drawback is that at higher frequencies (typically beyond 4 GHz), the performance of the antenna deteriorates due to roughness and imperfect metallization of the textile surface [29]. However, this frequency limit remains relatively the same for all e-threads and types of embroidery.

2.3.2 Grading the Embroidery Density for Foldable Prototypes

Surface conductivity of the embroidered e-thread surface can be controlled by adjusting the density of e-threads/mm. Typically, high surface conductivity is desired—comparable to that of copper—such that high e-thread density is pursued (e.g., a Brother 4500D embroidery machine supports a maximum of seven e-threads/mm). However, e-thread prototypes created with high density may not be flexible enough and may not lend themselves to easy folding. To facilitate folding for applications that may benefit from this attribute (e.g., foldable origami antennas), embroidery density may readily be reduced. However, reduced density will not just aid folding, but will also lower the surface conductivity and degrade the RF performance.

To address this challenge, graded embroidery has recently been reported that integrates regions of high and low embroidery density on the antenna and along the creases, respectively. This approach facilitates bending along the creases while maintaining excellent RF performance. Figure 2.6(a) shows an e-textile accordion-based dipole placed on a Styrofoam fixture; the creases created at reduced embroidery density are clearly visible. Another application that may benefit from reduced embroidery density entails the printing of ground planes. The purpose of a ground plane is to merely reflect radiation. As such, it can be implemented using a grid with a size that is not electrically seen by the wavelength of interest. Thus, low embroidery density can be incorporated into the realization of ground planes to save time and cost. Figure 2.6(b) shows a textile ground plane incorporated into an Archimedean spiral antenna.

2.3.3 Colorful Prototypes

There are various ways that enable printing of colorful e-textile prototypes. In the past, a logo-type based RFID antenna was fabricated by adhering copper tape on colorful fabric [30]. Another approach was to adopt screen printing and print conductive inks onto colorful fabrics [31]. However, these approaches lead to delamination and ink surface ruptures during flexing or folding of the fabric. Conventional conductive e-thread embroidery creates mechanically robust prototypes but is limited to a single color as attributed to the corresponding conductive material. To overcome this limitation, the authors in [32] created a colorful logo-shaped antenna using unicolor e-threads to embroider the antenna onto the fabric and filling the remaining logo with nonconductive colorful threads. However, this approach limits the area where colorful threads can be incorporated.

By contrast, Kiourti and others [9] developed a new approach for high-precision, flexible, lightweight, and colorful textile-based antennas using automated

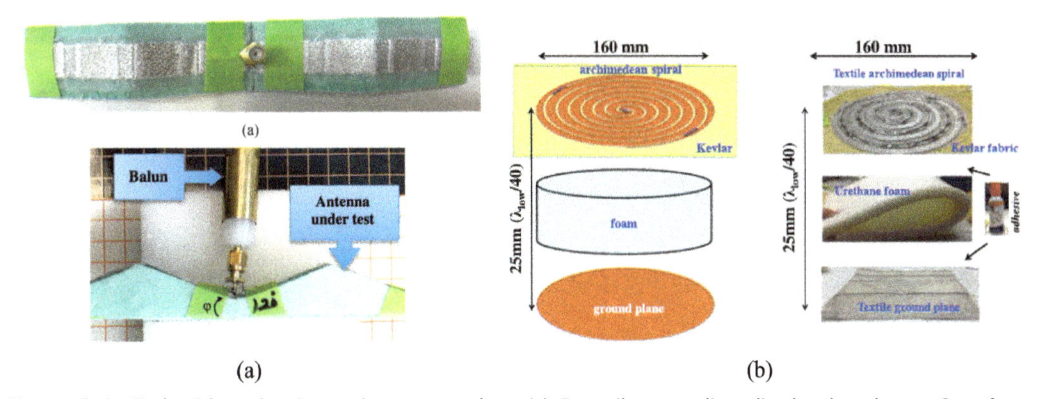

Figure 2.6 Embroidery density tuning approaches. (a) E-textile accordion dipole placed on a Styrofoam fixture. (©2018 IEEE [10].) (b) Designed and fabricated prototype of Archimedean spiral with ground plane. (©2017 IEEE [17].)

embroidery. The process uses e-threads in the bobbin case of the embroidery machine to print the desired shape of the antenna on the back of the fabric. Concurrently, a colorful assistant yarn is threaded through the needle to secure the e-threads onto the fabric. This results in unobstructed colorful shapes in the front of the fabric. Figure 2.7(a) shows a 2.4-GHz colorful textile-based dipole antenna integrated as part of a logo. Figure 2.7(b, c) shows the placement of the antenna and its RF performance validation, respectively.

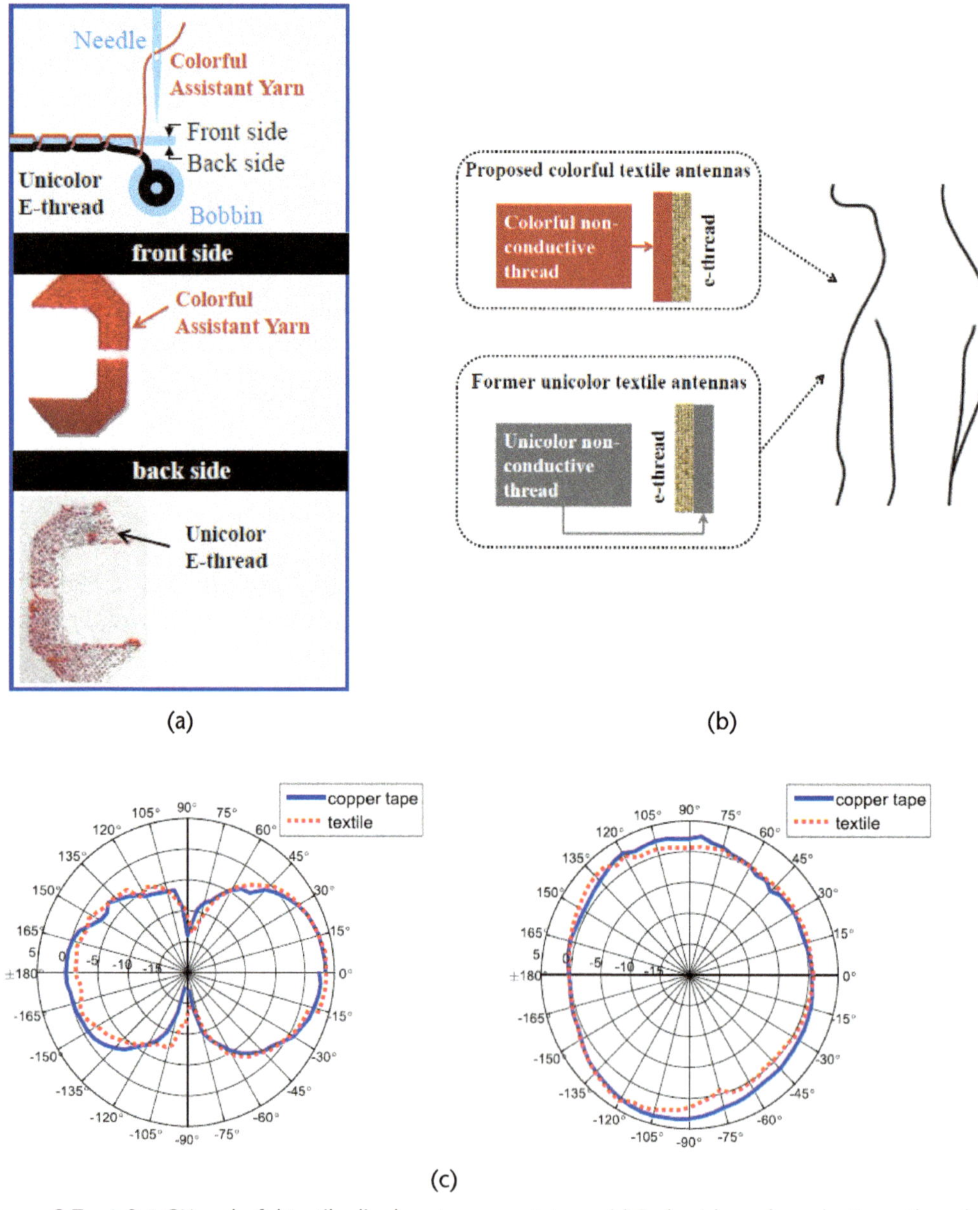

Figure 2.7 A 2.4-GHz colorful textile dipole antenna prototype: (a) Embroidery of conductive e-threads, (b) conventional unicolor and colorful prototype placement, and (c) RF performance validation. (©2015 IEEE [9].)

2.4 Polymer Integration

2.4.1 Polymer Substrates

Integration of polymers expands the application scope of e-textile antennas and sensors by enabling the realization of flexible substrates and superstrates. An advantage in this case is that polymers can either be used in pure form or they can be mixed with various materials (such as ceramics) to achieve any target performance characteristics. In other cases, polymers have been integrated into optoelectronics to produce mechanically flexible and highly efficient light-emitting diode (LED) displays called electronic paper [35]. Similarly, integration of polymers with e-threads brings forward electronics and antenna prototypes that are stretchable, flexible, and highly durable against deformations [39]. Example polymer substrates that can be integrated with e-threads are summarized next.

2.4.1.1 Liquid Crystal Polymers

LCPs are well-ordered polymer networks that are obtained using photo-initiated polymerization of liquid crystal monomers [36]. They are used in optical displays and system-on-package (SoP) applications. For SoPs, three-dimensional fabrication capability is needed and this is done via stereolithography. LCPs exhibit nearly constant relative permittivity up to 110 GHz, low loss, low water absorption (<0.04%), and low cost [38, 51]. Usually, inkjet printers are used to print patterns on LCP laminate. An example prototype is reported in [38] and entails a conformal antenna array for 5G applications.

2.4.1.2 Ceramic-Reinforced Polymers

Ceramic-reinforced elastic polymers can provide conformality in microwave applications suitable for a wide range of operating frequencies from 100 MHz to 20 GHz. Although several combinations are possible, the most common path includes PDMS polymer that is combined with high-permittivity ceramic powders, such as barium titanate (BT-BaTiO3), Mg-Ca-Ti (MCT), and Bi-Ba-Nd-titanate (BBNT). The process of preparing the PDMS, followed by cross linking the agent, and adding the ceramic powders to create ceramic-reinforced elastic polymers is shown in Figure 2.8. To introduce ceramic into the polymer, a particle dispersion process is used. By tuning the percentage concentration of ceramic powder compared to PDMS, the dielectric properties can be adjusted. However, ceramic concentration is usually kept below 30% to 40% to maintain the elastic behavior of the polymers [37].

2.4.2 Stretchable Prototypes Embedded in Polymer

A wide range of applications that entail shape deformations may necessitate prototypes that are not only flexible, but also stretchable. Embroidered electronics/antennas cannot be stretched as the e-threads incorporated are marginally stretchable and the fabric support is not stretchable either. To achieve stretchability, one solution is to design a prototype that lends itself to stretching (e.g., a wire-based

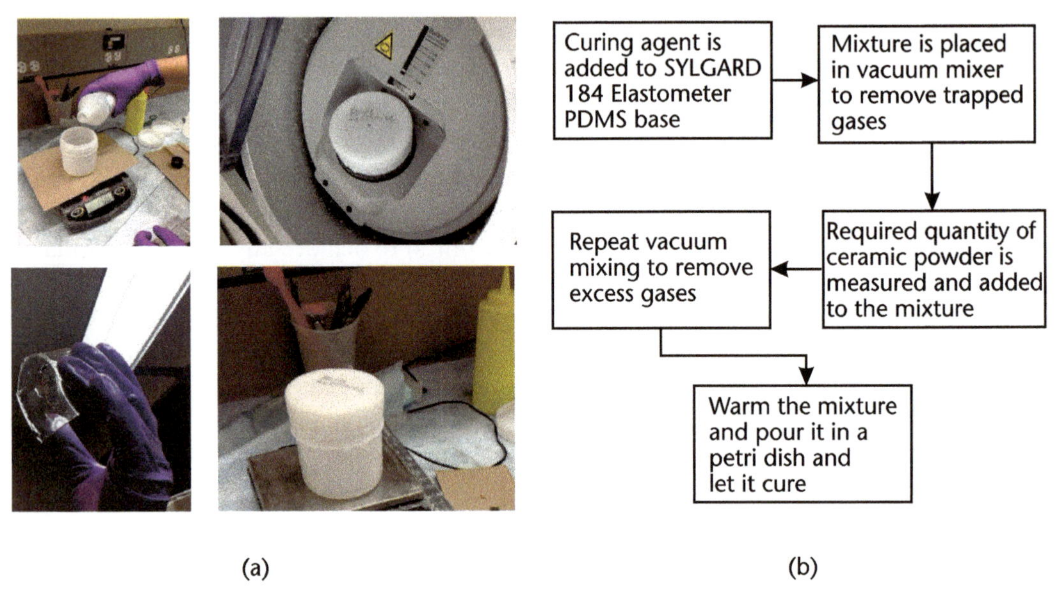

Figure 2.8 Fabrication procedure for the ceramic-reinforced polymers: (a) preparation of PDMS reinforced ceramic using PDMS base and ceramic powders, and (b) schematic representation of the procedure.

meandered shape) and then embed it into a stretchy polymer while eliminating the fabric support. For example, PDMS can be used a polymer material, although other polymer options with much higher stretchability can also be pursued.

Figure 2.9 shows the procedure to embed wire antennas into a stretchable PDMS polymer [39]. The base is prepared by mixing the polymer base and curing agent in a vacuum mixer at room temperature. Embroidered patterns (entailing both the e-threads and the polyester fabric on which they are affixed) are adhered to fixed structures, such as the sticky side of copper tapes, to maintain their shape. Later, the support fabric is melted using a fine-tip soldering iron (or other equivalent mechanism) that heats the fabric to around 250°C and creates a stand-alone

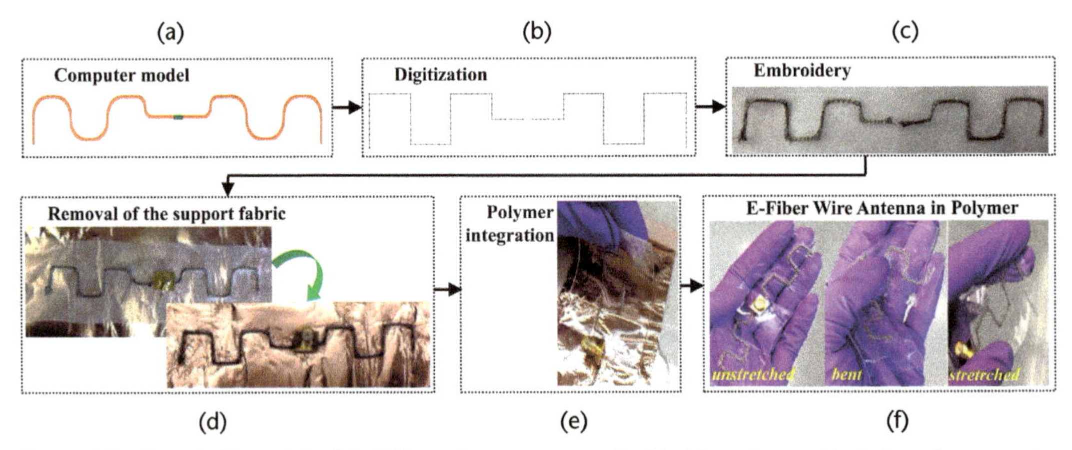

Figure 2.9 Stretchable and flexible E-fiber wire antennas embedded in polymer: (a) design of a computer model, (b) digitization, (c) embroidery using twisted 664-strand Amberstrand e-threads, (d) removal of support fabric using meting technique, (e) polymer integration, and (f) stretchable polymer embedded wire antenna. (©2014 IEEE [39].)

structure that is kept in place using the underlying adhesive. Bubble-free PDMS mixture is poured onto the structure and cured under elevated temperatures of around 120°C. After the curing process, the supporting copper tape is removed to obtain stretchable polymer prototypes.

2.4.3 Magneto-Actuated Prototypes

Reconfigurable e-textile structures can be produced using hard-magnetic soft substrates. The idea in this case is to embed the embroidered surface into the substrate and enforce a static magnetic field to get the shape altered. The approach is described in [40]. High magnetization and high coercivity neodymium-iron-boron (NdFeB) microparticles are embedded into uncured silicon rubber, also known as PDMS. The mixture is cured at 70°C for 1 hour. After curing, the composite is magnetized under impulse magnetic field (~1.5 T). E-threads are integrated with the magnetic materials during the curing process. By controlling the volume percentage of NdFeB particles, magnetic and dielectric properties as well as flexibility of the material can be tuned according to the desired operation.

Using this approach, the researchers in [40] designed a quarter-wavelength accordion-based monopole antenna that sandwiched e-threads between two layers of magnetic material. This sandwiching of e-threads provides accurate control over the bending process and protects the e-thread surface from wear and corrosion. Silicon release agent is used to facilitate extraction after the integration and curing process. Figure 2.10 shows the fabrication of a magneto-actuated monopole antenna using a 3-D printed mold to pour the prepared magnetic material.

Figure 2.10 Integrating magnetic materials with embroidered antennas: (a) graded e-thread monopole antenna, and (b) magneto-actuated antenna after integration of magnetic material. (©2020 IEEE [40].)

2.5 Performance

2.5.1 Radio-Frequency Performance

To analyze the effectiveness of e-threads over traditional copper wires, it is essential to know the contrast in terms of resulting electromagnetic performance. Since most e-threads are used for flexible electronics that operate in the RF range, this section focuses on comparing the RF performance for various e-thread prototypes and their copper counterparts.

2.5.1.1 E-Thread versus Copper Wire at Low and High Frequencies

At low frequencies (i.e., 1 kHz to 100 kHz), the researchers in [41] compared coils of 10 turns by loosely and tightly wrapping the e-threads and copper wires around a 4-cm radius Styrofoam cylinder. They used 30 AWG copper wire and 40 filament Liberator e-threads twisted at 4.5 TPI, with a diameter of 0.255 mm each. Tightly wound e-thread coils exhibited approximately nine-fold higher resistance than copper. Self-inductance values exhibited great agreement between copper wire and e-threads across all frequencies. Table 2.3 compares these cases at low frequencies. At higher frequencies (beyond 10 MHz), 4-cm long transmission lines (TL) were fabricated and tested. Specifically, a 125-mil-thick Rogers RO3003 ($\varepsilon_r = 3$, $\tan\delta = 0.001$) substrate was used, and copper backing acted as the ground plane. E-thread and copper wire transmission lines were then produced using the same wire gauge mentioned above. Performance of the copper wire and e-thread transmission line showed excellent agreement from 10 MHz to 4 GHz. Similar trend was observed beyond 4 GHz with some deviation in values. Figure 2.11 shows the fabricated model and the RF performance comparison. Such discrepancies are associated with surface roughness and material loss of the e-thread [41] (i.e., when the wavelength is comparable to the surface roughness, the performance degrades).

2.5.1.2 Embroidered E-Thread versus Copper Surfaces

To expand from a single wire study to a conductive surface area study, the authors in [13] designed transmission line prototypes of a finite width. For comparison purposes, they created prototypes with transmission lines and ground plane made of copper tape, Amberstrand-664 with double-layer stitching of 2 threads/mm, Liberator-40, and Liberator-20 with double-layer stitching of seven threads/mm.

Table 2.3 Coil Resistance and Self-Inductance Results

	Copper-Wire E-Thread				*Copper-Wire E-Thread*			
	Loosely Wound		*Tightly Wound*		*Loosely Wound*		*Tightly Wound*	
Frequency (KHz)	*R (Ω)*	*L (μH)*	*R (Ω)*	*L (μH)*	*R (Ω)*	*L (μH)*	*R (Ω)*	*L (μH)*
1	0.95	7.7	0.07	0.35	8.74	7	0.12	0.25
10	0.95	7.6	0.078	0.29	8.73	6.81	0.12	0.24
100	0.88	7.56	0.09	0.24	8.64	6.84	0.12	0.2

R = right limb, L = left limb.

From: [41].

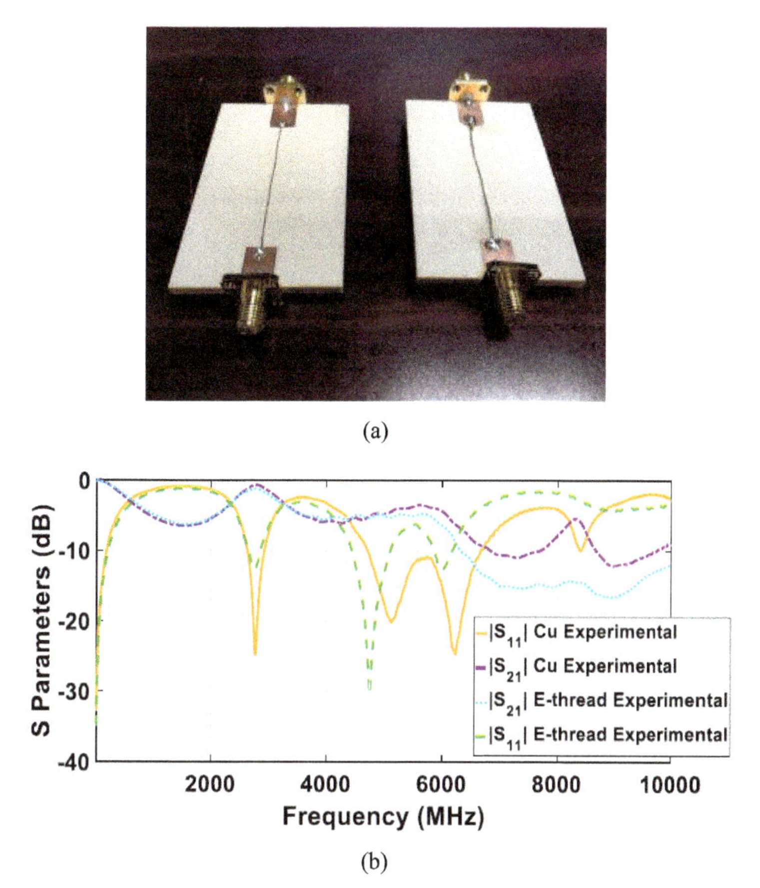

Figure 2.11 E-thread vs copper wire performance comparison at low frequencies: (a) fabricated model of copper wire and e-thread, and (b) comparison of copper and e-thread transmission lines. (©2019 IEEE [41].)

Figure 2.12(a) shows the fabricated transmission line and Figure 2.12(b) shows the S-parameter performance. All e-threads achieve good RF performance up to 3.5 GHz. However, it is essential to take into consideration that Liberator-20 provides better geometrical accuracy compared to Liberator-40 and Amberstrand e-threads.

2.5.1.3 Effect of E-Thread Ground Plane

When it comes to realizing ground planes with e-textiles rather than copper, performance degradation is almost insignificant. This performance is expected as ground planes only serve to reflect the radiation. For example, the authors of [11] fabricated three different 50 Ω transmission line structures with (a) e-thread transmission line and e-thread ground plane, (b) e-thread transmission line and copper ground plane, and (c) copper transmission line and copper ground plane. The prototypes are shown in Figure 2.13. Table 2.4 compares the insertion loss per unit length for frequencies ranging from 10 MHz to 4 GHz. As seen, good agreement is achieved. Beyond 4 GHz, the RF performance is deteriorated due to higher resistance of the e-threads [11]. As would be expected, density of the embroidery is not critical for the realization of ground planes. In fact, gridlike structures can also be used, as long as

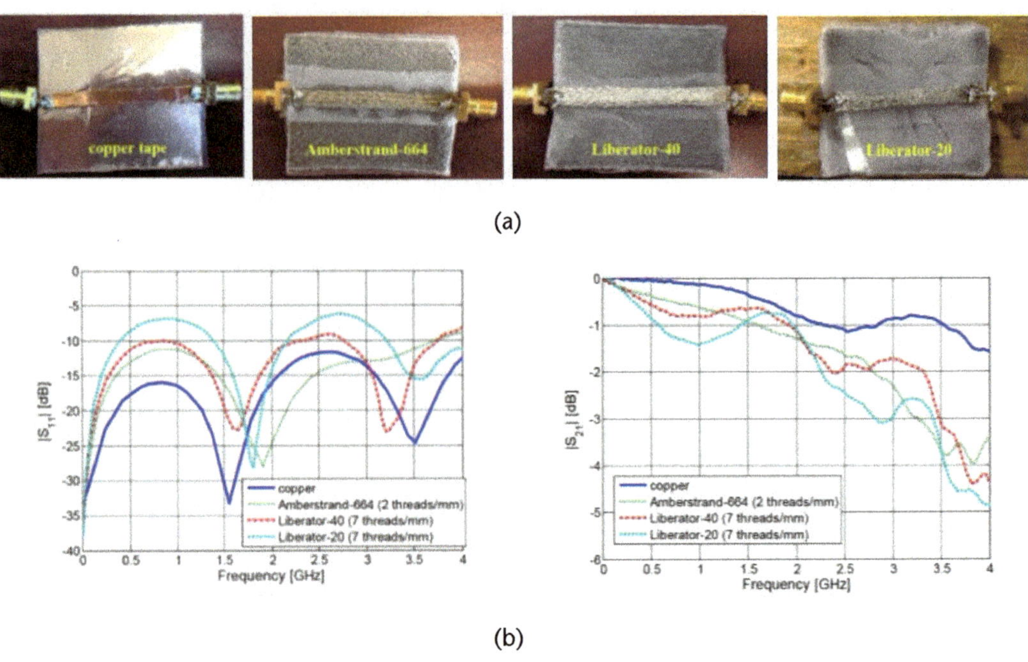

Figure 2.12 Transmission lines using e-threads: (a) fabricated model of copper wire TL and different embroidered TL, and (b) $|S_{11}|$ and $|S_{21}|$ performance transmission line prototypes. (©2015 IEEE [13].)

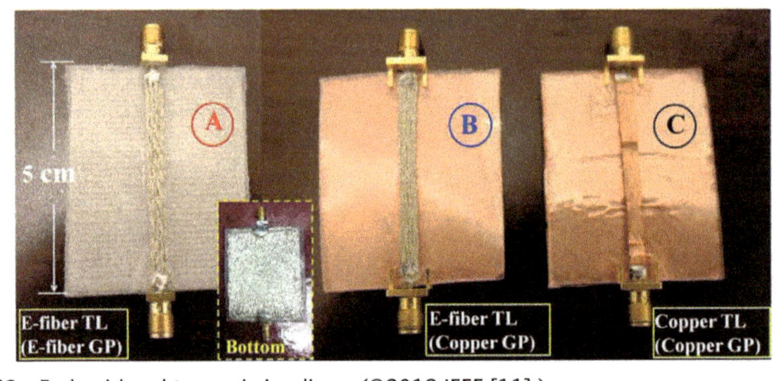

Figure 2.13 Embroidered transmission lines. (©2012 IEEE [11].)

Table 2.4 $|S_{21}|$ Comparison of Transmission Lines with E-Thread vs Copper Ground Plane

Conductor	*Ground Plane*	*Max* $|S_{21}|$ *(dB)*	*Max* $|S_{21}|$ *per Unit Length (dB/cm)*
E-thread	E-thread	1.05	0.21
E-thread	Copper	0.86	0.17
Copper	Copper	0.71	0.14

From: [11].

the grid size is not visible by the wavelength of the associated electromagnetic wave. This idea allows for flexible structures and helps save resources and cost.

2.5.1.4 Effect of E-Thread Density

E-thread density may significantly alter the RF performance of devices as surface conductivity changes with e-thread density. To confirm this, the authors in [10] fabricated 5-cm long 50 Ω transmission lines embroidered using different densities and compared their performance versus copper transmission lines. All designs employed copper ground plane and felt substrate. Figure 2.14 shows the transmission coefficient $|S_{21}|$ performance for these prototypes in the frequency range of 0 to 1.2 GHz. As hypothesized, the RF performance degrades as the density of e-threads is reduced. However, reduced density makes the prototype less stiff and easier to fold as compared to more dense structures [10]. In turn, density of the embroidery may ultimately be optimized per application scenario.

2.5.1.5 Effect of Deformation

Textile electronics, sensors, and antennas embedded within fabrics can potentially deform depending on the application scenario. For example, textile antennas placed on the arm may deform depending on the arm size and flexion of the joint. Similarly, textile antennas may crumple as the fabric deforms during motion. Such deformations may significantly alter the associated performance and must be taken into account. For example, the coplanar wearable waveguide antenna in [56] showed an excessive detuning of ~2 GHz under a typical crumpling pattern caused by bending the elbow. Similarly, the patch antenna in [62] showed deviation in the center frequency by 37 and 34 MHz for combined bending and crumpling configurations.

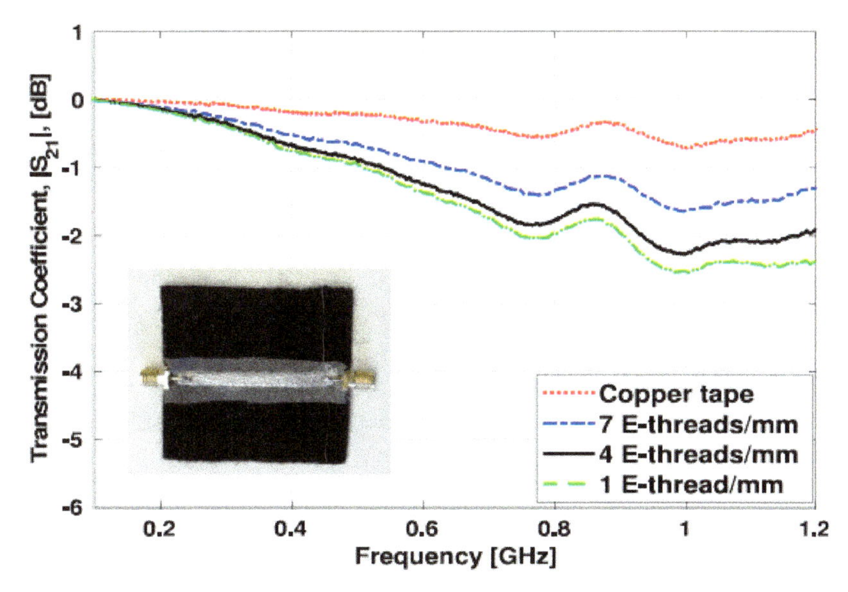

Figure 2.14 Transmission coefficient $|S_{21}|$ performance for 50 Ω transmission line prototypes. (©2018 IEEE [10].)

The authors of [64] concluded that "crumpling can have a serious effect on the resonant frequency, bandwidth and radiation from textile antennas." The authors of [65] recommended placing the antennas on flat areas of the human body to reduce the effects of crumpling. Expectedly, the latter may not always be possible.

Figure 2.15 shows change in RF performance for different crumpling cases of a planar inverted F antenna (PIFA) antenna in the *Y-Z* plane while Figure 2.16 shows similar crumpling conditions in the *X-Z* plane [64]. As the crumpling increases in the *Y-Z* plane, the reflection coefficient deteriorates significantly. The radiation pattern for different crumpling cases show rotation up to 30 degrees (case 2) and power reductions up to 8 dB in the *Y-Z* measurement plane and up to 15 degrees at some angles in the *X-Z* measurement plane compared the flat case. One of the major reasons for this is change in the separation between the top radiating element and the ground plane during crumpling.

Crumpling of a PIFA antenna in the *X-Z* plane shows a much more significant shift in the resonant frequency, as shown in Figure 2.16(e) and pattern rotation up to 40 degrees in the *Y-Z* measurement plane. Redistribution of the current during crumpling causing this shift in the resonant frequency.

It can be observed from Table 2.5 that gain and efficiency of the antenna reduces for different crumpling cases in free space. For a flat case, the antenna radiation efficiency drops from 83% to 16% when placed on the body, indicating significant absorption by the body and corresponding forward gain loss of 1.4 dB. For different crumpling conditions when the antenna is placed on the body, although there is significant absorption of the back radiation, the radiation efficiency and forward gain improves as the crumpling causes different interactions with the body. This indicates crumpling can help in the overall radiation efficiency [64].

2.5.1.6 Other factors

Embroidery requires meticulous attention to detail such as direction of stitching, stitch spacing, e-thread density, and more. These details may significantly alter the realization of the conductive portion, in turn impacting RF performance in terms of resonant frequency, gain, efficiency, and radiation pattern. Such effects have been thoroughly explored for embroidered patch antennas in [59, 60]. Embroidered/wearable antennas may also be subject to harsh environments such as abrasion, varying climate conditions, and moisture. Extensive analyses to this end have been performed in [61].

2.5.2 Mechanical Performance

Embroidered e-threads are intended primarily for wearable and conformal applications. Thus, it is essential to confirm the robustness of such embroidered prototypes and their resulting performance under mechanical stress. To analyze this, various bending tests can be performed on the conformal and load-bearing RF devices. One such test is performed by Zhong and others using three-point flexing [17]. In this work, a spiral antenna was placed flat on two supporting pins placed 160 mm apart. The third pin was lowered from above at a rate of 50 mm/min to bend the antenna, as shown in Figure 2.17(a). A maximum bending angle of 100° was considered. Overall, 50, 100, 200, and 300 flexing cycles were employed, and

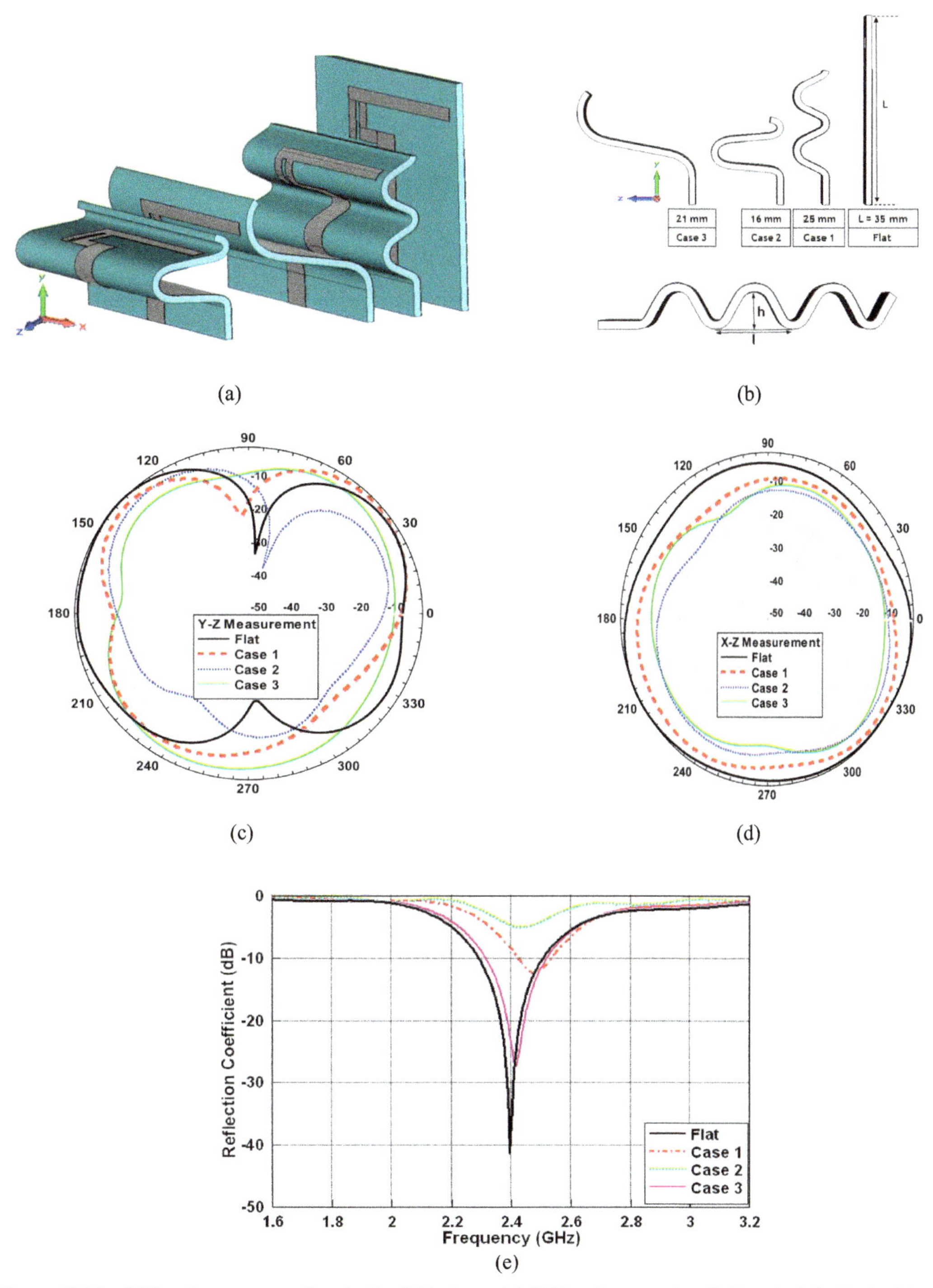

Figure 2.15 PIFA antenna crumpling in the Y-Z plane: (a) PIFA antenna using Zelt material attached to a thin felt material, (b) different crumpling cases along the Y-Z plane, (c) measured radiation pattern in the Y-Z measurement plane, (d) measured radiation pattern in the X-Z measurement plane, (e) reflection coefficient for crumpling in the Y-Z plane. (©2012 IEEE Transactions on Antennas and Propagation [64].)

the boresight realized gain was measured from 0 to 3 GHz. Figure 2.17(b) shows breaks in the surface of copper tape based spiral antenna after 50 flexing cycles

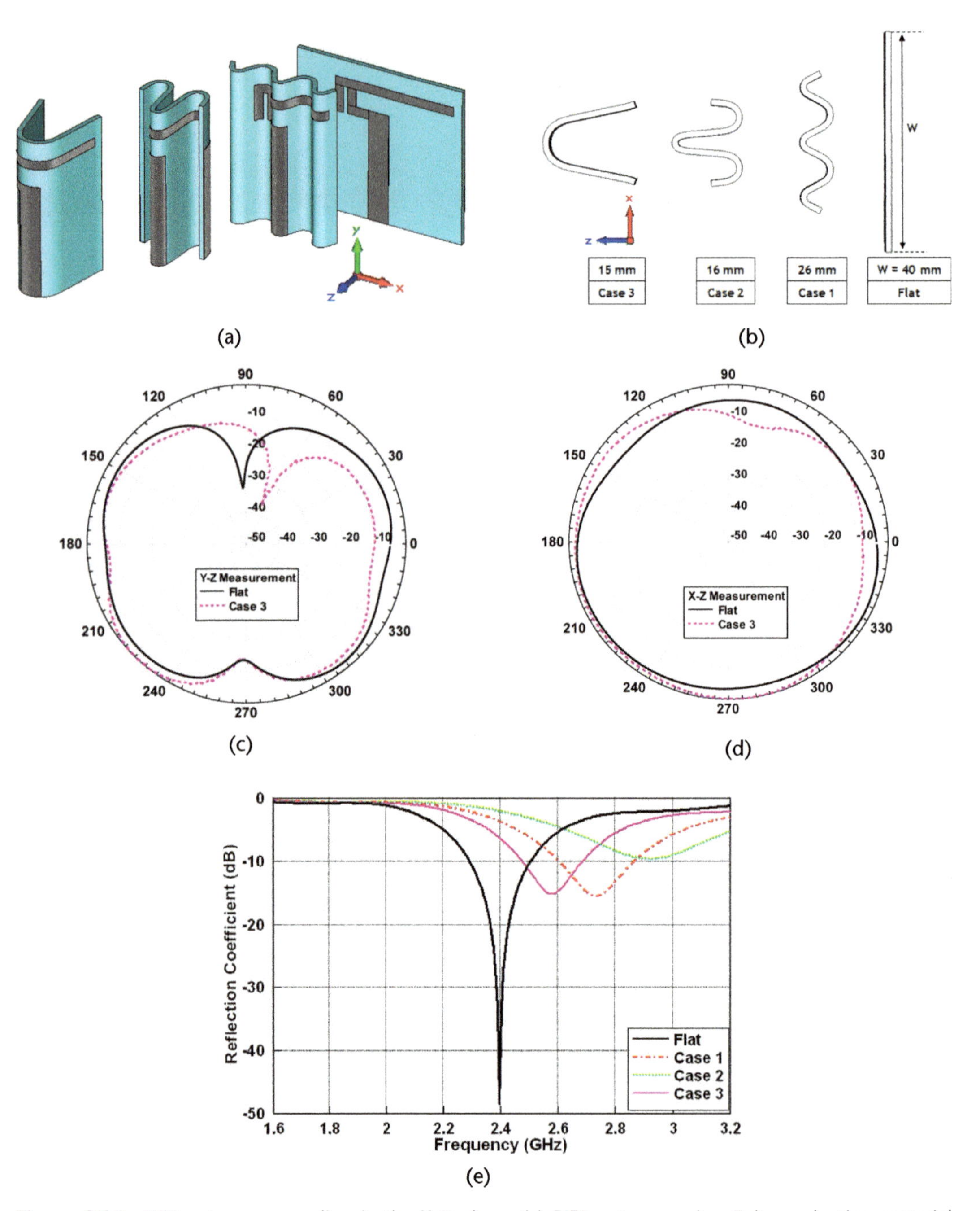

Figure 2.16 PIFA antenna crumpling in the X-Z plane: (a) PIFA antenna using Zelt conducting material attached to a thin felt material, (b) different crumpling cases along the *X-Z* plane, (c) measured radiation pattern in the *Y-Z* measurement plane, (d) measured radiation pattern in the *X-Z* measurement plane, (e) reflection coefficient for crumpling in the *X-Z* plane. (©2012 IEEE Transactions on Antennas and Propagation [64].)

with the realized gain dropping by 2 dB. By contrast, the embroidered Elektrisola e-thread based spiral antenna remained intact after 300 flexing cycles, demonstrating the robustness of the embroidered approach. Figure 2.17(c) shows boresight realized gain before and after flexing cycles.

Table 2.5 Antenna Efficiency for Different Crumpling Cases in Free Space and on Body

PIFA	*Gain (dBi)*		*Radiation Efficiency %*		*Total Efficiency %*	
	Free Space	*On Body*	*Free Space*	*On Body*	*Free Space*	*On Body*
Flat	1.184	–0.2	83	16	83	16
Y-Z crumpling						
Case 1	0.34	2	78	34	67	28
Case 2	–1.7	1.54	51	30	34	21
Case 3	1.04	2.1	82	68	81	68
X-Z crumpling						
Case 1	1.14	1.56	79	26	45	16
Case 2	0.66	2	70	41	26	19
Case 3	1.26	4.76	81	56	63	52

From: [56].

2.5.3 Launderability

Embroidered RF devices should possess the ability to tolerate repetitive soaking, washing, and drying processes. To understand the impact of washing, the authors in [15] fabricated RFID tags and repetitively washed them using detergent and three different washing programs. The read range of these RFID tags dropped from around 6 m before washing to around 3 m after 16 consecutive washing and drying processes. The loss in read range was attributed to reduced antenna radiation efficiency as a result of dissolution of conductive material from the embroidered antennas. To avoid performance degradation, RF devices and embroidered antennas can be coated with hydrophobic, flexible, and durable materials [15]. A relevant study was conducted by authors of [25] on screen-printed textile-based ultrahigh-frequency (UHF) RFID tags. Here, RFID tags were produced on 100% cotton fabric with acrylic and silicon coating materials. Different stages such as dry, wet, and damp (i.e., 10% water content), laundered, and dried after laundering were analyzed for RF performance, as shown in Figure 2.18(a) and (b). The realized gain was reduced for laundered and dried stages as compared to dry stages [25]. Figure 2.18(c) shows the microscope observations of silver-plated nylon textile electrodes after 30 washes. M1, M2, M3, and M4 in the figure stand for selected lots using 70g of powdered detergent, 60 mL of liquid detergent, no detergent under tap water, and 7g of sodium percarbonate, respectively [42].

2.6 Example Applications

E-textile prototypes may range from antennas to electromagnetic and circuit components, as well as sensors and actuators, and can be employed in applications as diverse as wearables, automotive, and airframes. This section presents a few examples.

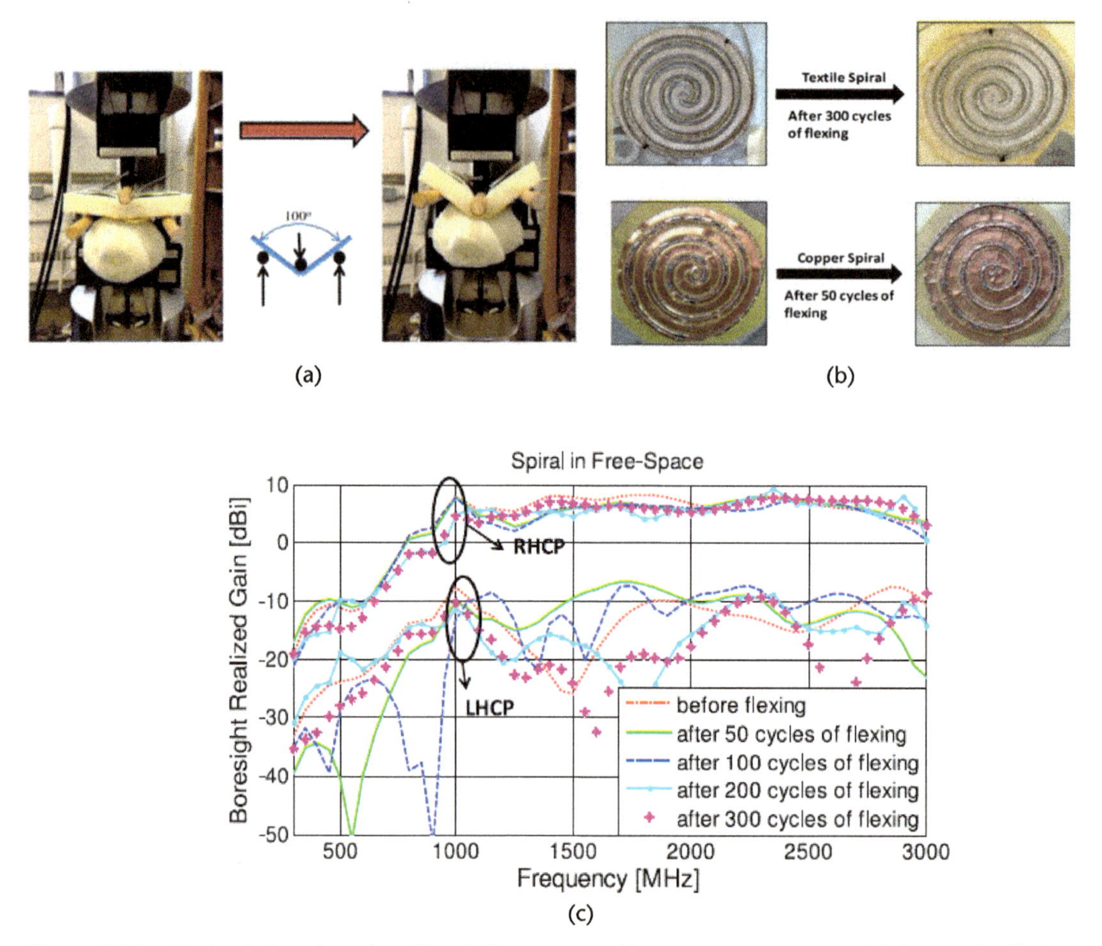

Figure 2.17 Mechanical testing of textile spiral antenna and its cooper counterpart: (a) three-point flexing test using wooden rods, (b) spiral antenna before and after flexing cycles, and (c) boresight realized gain of the textile spiral antenna before and after 50, 100, 200, and 300 cycles of flexing. (©2017 IEEE [17].)

2.6.1 Textile-Based Antennas

A conformal, load-bearing spiral antenna has been fabricated using conductive e-threads for integration on airframes [17]. Figure 2.19(a) shows the resulting prototype: it is a 160-mm diameter Archimedean spiral antenna with a textile ground plane placed 25 mm below the antenna surface. Figure 2.19(b) shows the spiral antenna conformally placed on a metallic cylinder, while Figure 2.19(c)and (d) shows its voltage standing wave ratio (VSWR) performance from 0–3 GHz. Results confirm reproducibility of the flexible prototype and indicate comparable performance to its copper counterpart.

A broadband e-thread based UHF RFID tag antenna is reported in [16] for operation in stress-prone environments, such as integration into automotive tires. The tag antenna is shown in Figure 2.20(a) and was created using conductive metal-polymer e-threads embedded in elastic PDMS polymer. This enables the antenna to withstand high pressure and stretching conditions. Tags were tested on automotive tires and compared to commercially available RFID tags. Figure 2.20(b) shows the experimental setup and defines the read range as the distance between

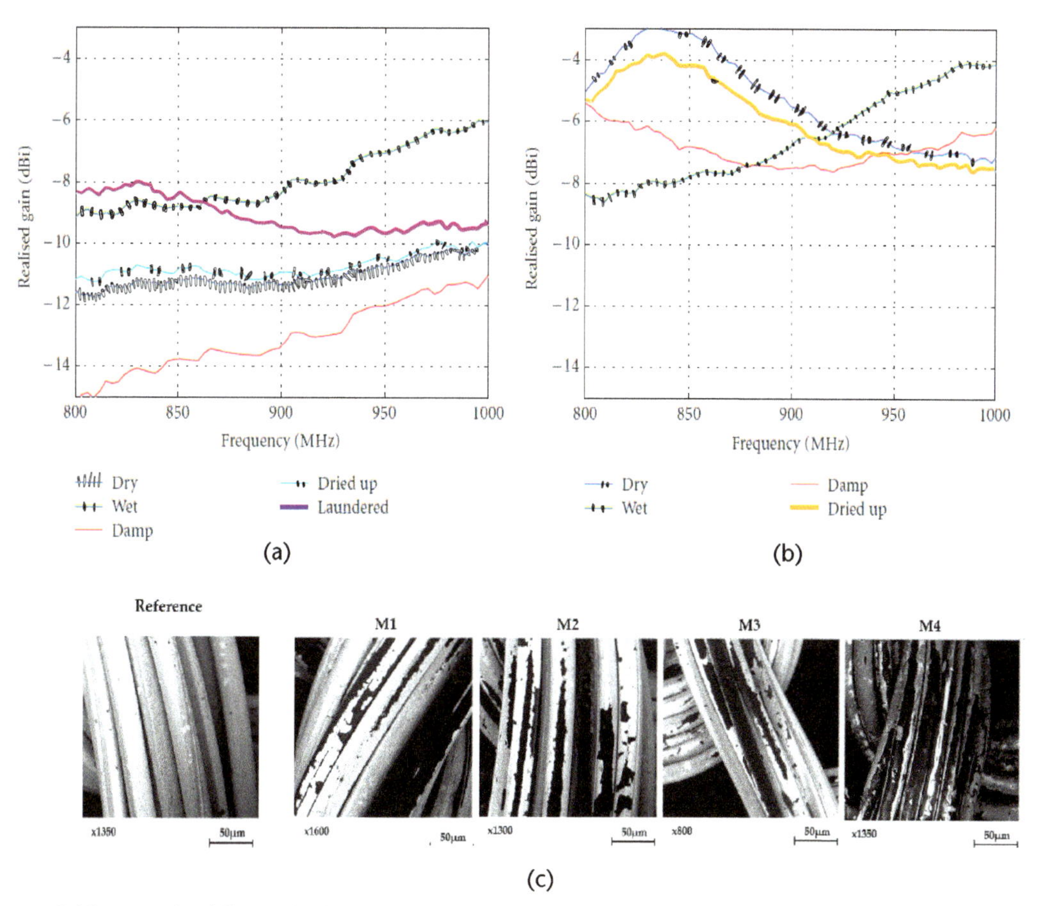

Figure 2.18 Launderability and effects on performance. (a) Realized gain of acrylic coated tags. (©2012 Int J Antennas Propag. [25].) (b) Realized gain of silicon coated tags. (©2012 Int J Antennas Propag. [25].) (c) SEM observations of electrodes fabricated using silver-coated nylon e-threads. (©2020 Sensors [42].)

the embedded tag and a remote reader. The tag was mounted on the outer surfaces S1, S2, S3, and S4. The distance between the two adjacent tires was 15 cm and the output power of the reader was 30 dBm. The RFID reader antenna was moved away from the tires until the tag could not be detected [16]. The performance is compared in Table 2.6, showing significantly improved read range for the e-textile tag.

2.6.2 Electromagnetic and Circuit Components

Embroidery can be extended to electrical components that can be sewn upon fabrics. One such application includes flexible and wearable embroidered supercapacitors, as shown in Figure 2.21(a) [43]. Embroidered coils used for near-field radiative wireless power transfer are shown in Figure 2.21(b) [44]. Embroidered transmission lines loaded with split-ring resonators (SRRs) can also be fabricated (e.g., for signal propagation control), as shown in Figure 2.21(c) [45]. Finally, Figure 2.21(d) shows embroidered prototypes of shorting vias on microstrip transmission lines [46].

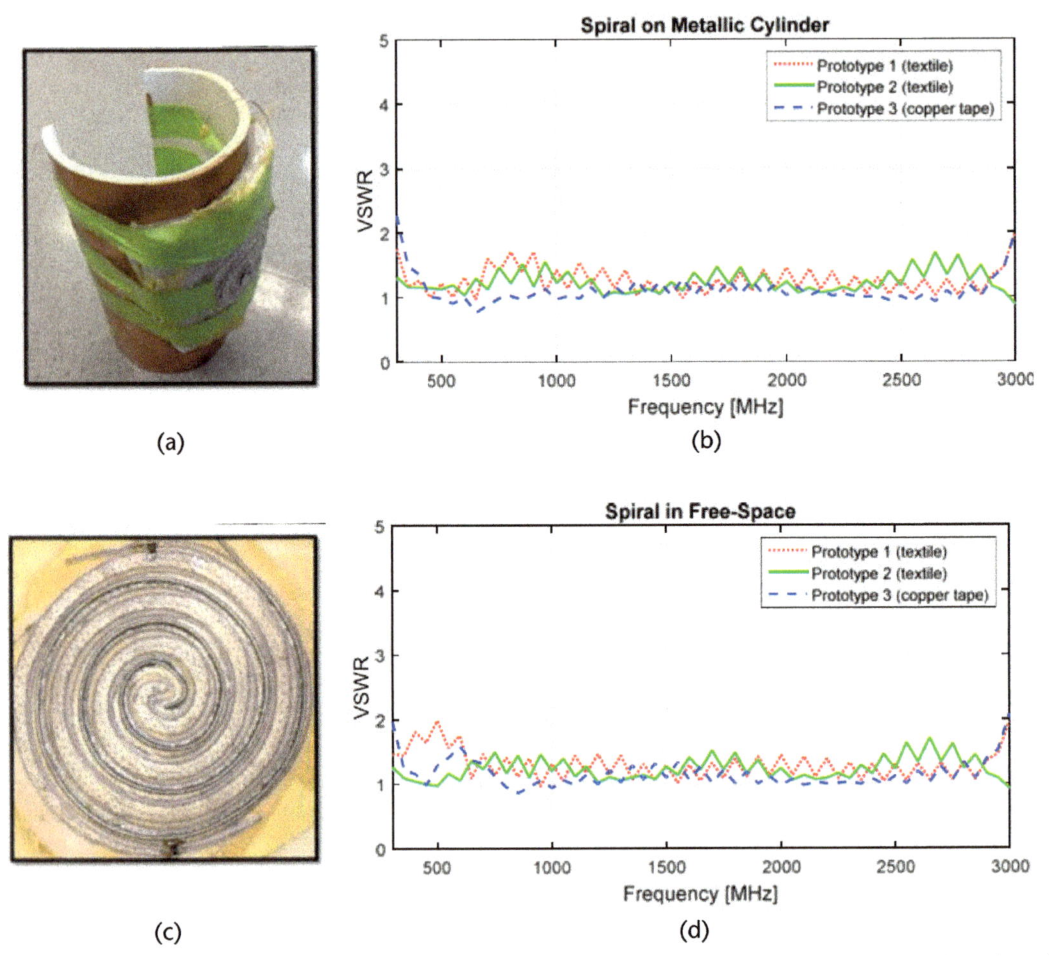

Figure 2.19 Conformal Archimedean spiral antenna: (a) planar textile spiral antenna, (b) VSWR of textile and copper-tape antenna in free space, (c) spiral antenna curved on a metallic cylinder, and (d) VSWR of textile and copper-tape antenna on metallic cylinder. (©2017 IEEE [17].)

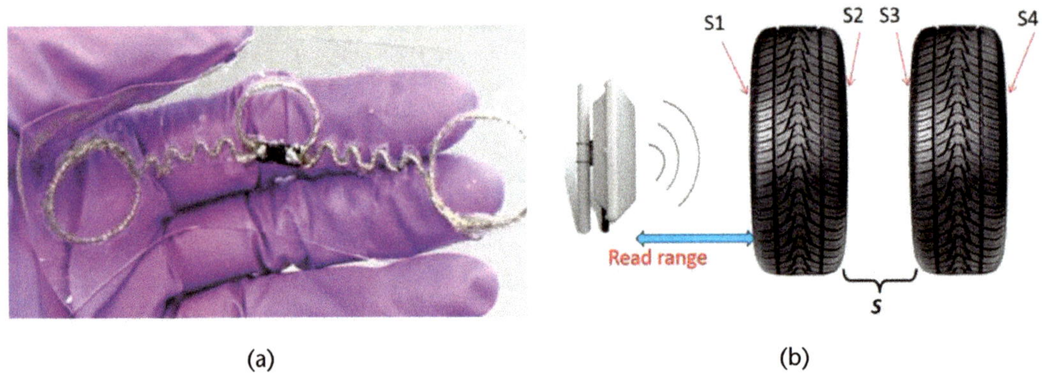

Figure 2.20 Stretchable and flexible RFID tag antenna: (a) flexible RFID tag embedded in a elastic PDMS polymer, and (b) RFID read range setup. (©2017 IEEE [16].)

Table 2.6 Read Range Test Comparison between E-Thread Based and Commercial RFID Tags

Antenna/Surface	*S1*	*S2*	*S3*	*S4*
Commercial Tags for Tires (Speedy Core)	4 ft	0.5 ft	Cannot be read	Cannot be read
Copper Wire Antenna Embedded in Polymer	13+ ft	8 ft	4 ft	4t

From: [15].

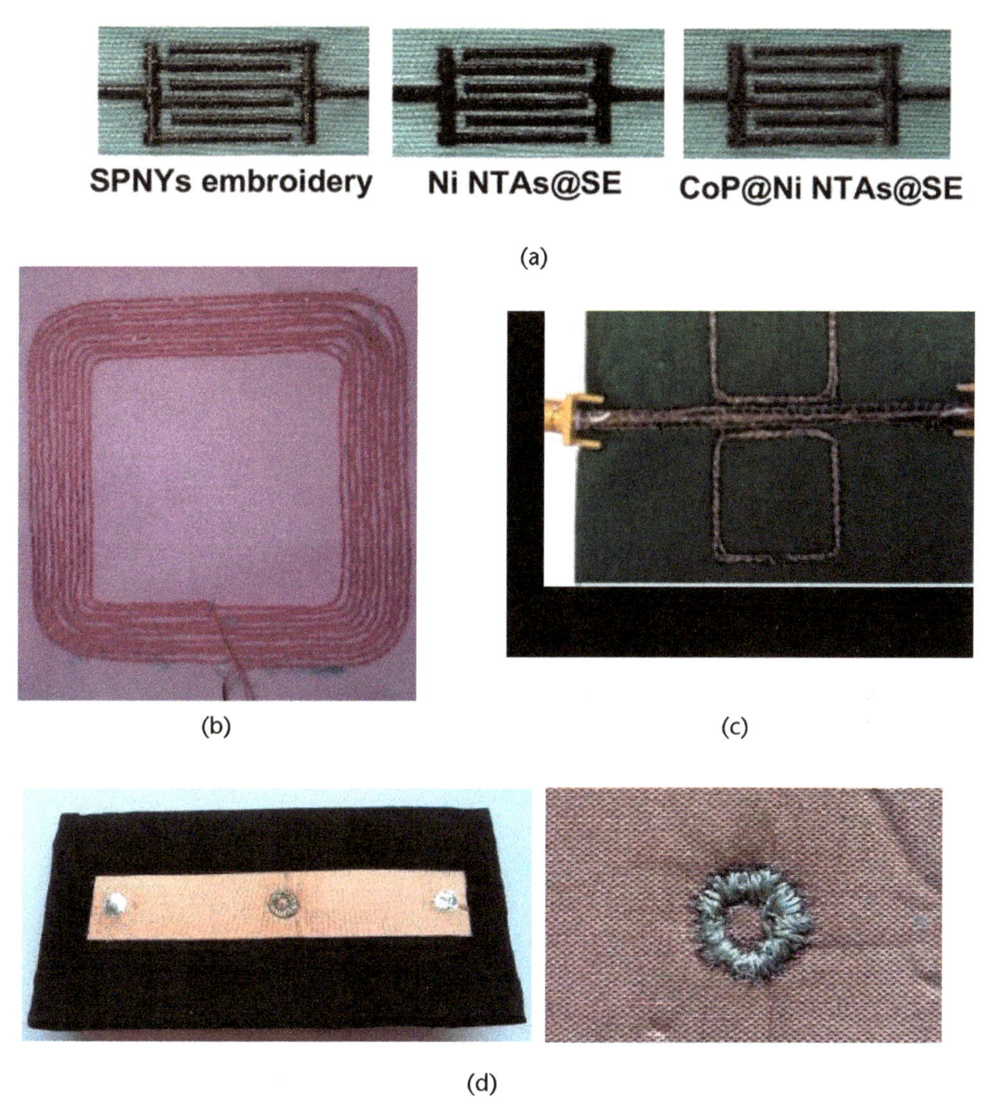

Figure 2.21 Applications of embroidered electromagnetic and circuit components. (a) Embroidered supercapacitors using silver plated nylon yarn (SPNY) and Ni and Co plated nano-thorn arrays. (©2019 Nano-Micro Lett. [43].) (b) Textile coil for near-field wireless power transfer. (©2020 Energies [44].) (c) Transmission line loaded with symmetrical SRR. (©2018 Materials [45].) (d) Microstrip transmission line with embroidered shorting via. (©2015 Sensors and Applications [46].)

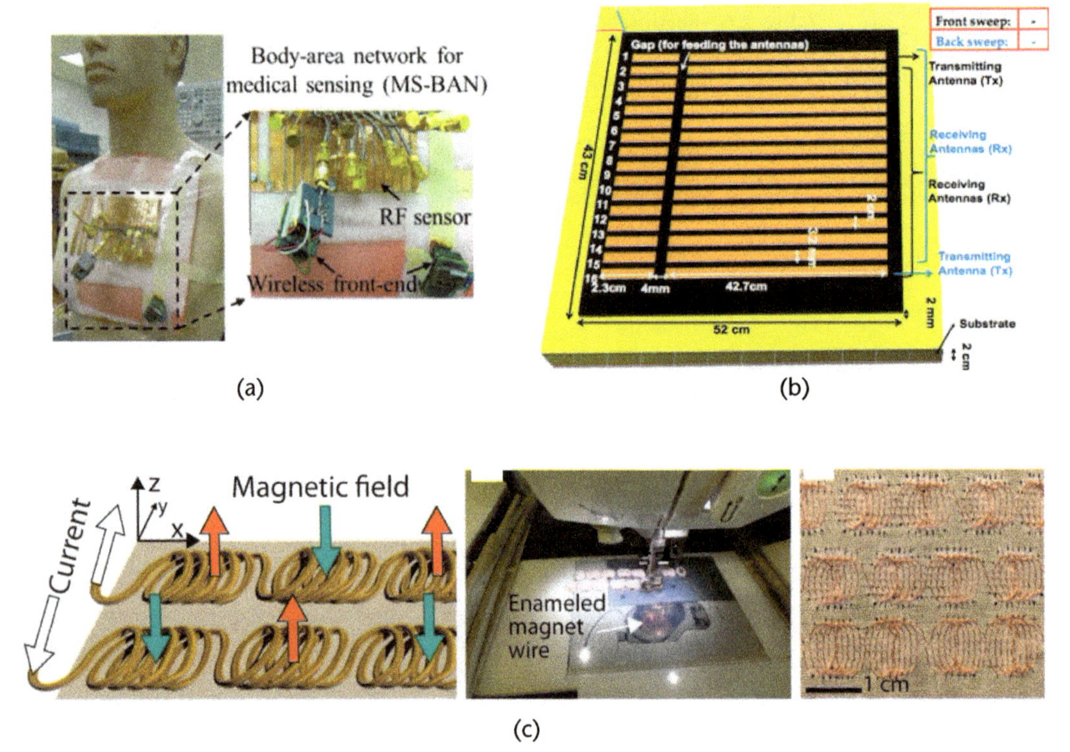

Figure 2.22 Application in sensors and actuators. (a) MS-BAN wireless link for lung monitoring sensor. (©2014 IEEE [47].) (b) Antenna-impregnated fabric prototype. (©2018 IEEE [48].) (c) Force-amplified soft electromagnetic actuators. (©2018 Actuators [49].)

2.6.3 Sensors and Actuators

A textile-based sensor for pulmonary edema monitoring is shown in Figure 2.22(a) [47]. Another wireless telemetry sensing application entails antenna impregnated fabrics for recumbent height measurement. When applied to infants and toddlers, this technology can be used to monitor and detect various health conditions, such as Turner syndrome, Crohn's disease, short stature, and growth hormone deficiency [48]. In other cases, linear actuators have been fabricated using flat embroidered motor windings in flexible fabrics, as shown in Figure 2.22(c) [49].

References

[1] Kiourti A., "Flexible, Thin Film and Wearable Antennas," in *Antenna Engineering Handbook,* 5th Edition, J. L. Volakis (ed.), McGraw–Hill Education, 2018.

[2] Nathan, A., et al., "Flexible Electronics: The Next Ubiquitous Platform," *Proc. IEEE,* Vol. 100, May 13, 2012, pp. 1486–1517.

[3] Tsolis, A., et al., "Embroidery and Related Manufacturing Techniques for Wearable Antennas: Challenges and Opportunities," *MDPI Electronics,* Vol. 3, No. 2, 2014, pp. 314–338.

[4] Liu, S., et al. "Textile Electronics for VR/AR Applications," *Advanced Functional Materials,* November 26, 2020, pp. 1–4.

[5] Stoppa, M., and A. Chiolerio, "Wearable Electronics and Smart Textiles: A Critical Review," *Sensors,* Vol. 14, No. 7, July 7, 2014, pp. 11957–11992.

[6] Meredov, A., K. Klionovski, and A. Shamim, "Screen-Printed, Flexible, Parasitic Beam-Switching Millimeter-Wave Antenna Array for Wearable Applications," *IEEE Open Journal of Antennas and Propagation,* Vol. 1, 2020, pp. 2–10.

[7] Tran, N. A., et al., "Copper Thin Film for RFID UHF Antenna on Flexible Substrate," *Advan. Nut. Scien.: Nanoscience Nanotechnol.,* Vol. 1, No. 2, 2010, pp.1–6.

[8] Cheng, S., et al., "Foldable and Stretchable Liquid Metal Planar Inverted Cone Antenna," *IEEE Trans. Antennas Propag.,* Vol. 57, No. 12, December 2009, pp. 3765–3771.

[9] Kiourti, A., and J. Volakis, "Colorful Textile Antennas Integrated into Embroidered Logos," *Journal of Sensor and Actuator Networks*, Vol. 4, No. 4, 2015, pp. 371–377.

[10] Alharbi, S., et al., "E-Textile Origami Dipole Antennas with Graded Embroidery for Adaptive RF Performance," *IEEE Antennas and Wireless Propagation Letters*, Vol. 17, No. 12, December 2018, pp. 2218–2222.

[11] Wang, Z., et al., "Embroidered Conductive Fibers on Polymer Composite for Conformal Antennas," *IEEE Transactions on Antennas and Propagation*, Vol. 60, No. 9, September 2012, pp. 4141–4147.

[12] Syscom Advanced Materials, Inc., Columbus, OH, "Syscom Advanced Materials," http://www. metalcladfibers.com/.

[13] Kiourti, A., and J. L.Volakis, "High-Geometrical-Accuracy Embroidery Process for Textile Antennas with Fine Details," *IEEE Antennas and Wireless Propagation Letters,* Vol. 14, 2015, pp. 1474–1477.

[14] Electrisola Feindraht AG Textile Wire., "Technical Brochure," http://www.textile-wire.ch/.

[15] Toivonen, M., et al. "Impact of Moisture and Washing on the Performance of Embroidered UHF RFID Tags," *IEEE Antennas and Wireless Propagation Letters,* Vol. 12, 2013, pp. 1590-1593.

[16] Shao, S., et al., "Broadband Textile-Based Passive UHF RFID Tag Antenna for Elastic Material," *IEEE Antennas and Wireless Propagation Letters,* Vol. 14, 2015, pp. 1385–1388.

[17] Zhong, J., et al., "Conformal Load-Bearing Spiral Antenna on Conductive Textile Threads," *IEEE Antennas and Wireless Propagation Letters,* Vol. 16, 2017, pp. 230–233.

[18] Zhong J., et al., "Body-Worn 30:1 Bandwidth Tightly Coupled Dipole Array on Conductive Textiles," *IEEE Antennas and Wireless Propagation Letters,* Vol. 17, No. 5, May 2018, pp. 723–726.

[19] Mantash, M., et al., "Investigation of Flexible Textile Antennas and AMC Reflectors," *International Journal of Antennas and Propagation,* Vol. 2012, Article ID 236505, May 8, 2012.

[20] Mandal, S., and N. Abraham, "An Overview of Sewing Threads Mechanical Properties on Seam Quality," *Pakistan Textile Journal,* Vol. 59, 2010, pp. 40–43.

[21] Hassan, A., et al., "All Printed Antenna Based on Silver Nanoparticles for 1.8 GHz Applications," *Appl. Phys. A*, Vol. 122, 2016, pp. 1–7.

[22] Ortego, I., et al., "Inkjet Printed Planar Coil Antenna Analysis for NFC Technology Applications," *International Journal of Antennas and Propagation,* March 22, 2012, pp 1–6.

[23] Huang, X., et al., "Highly Flexible and Conductive Printed Graphene for Wireless Wearable Communications Applications," *Scientif. Rep.,* Vol. 5, 2016, pp. 1–7.

[24] Osman, M.A.R., et al., "Textile UWB Antenna Bending and Wet Performances," *International Journal of Antennas and Propagation,* Vol. 2012, May 14, 2012, pp. 1–10.

[25] Kellomäki, T., et al. "Towards Washable Wearable Antennas: A Comparison of Coating Materials for Screen-Printed Textile-Based UHF RFID Tags," *International Journal of Antennas and Propagation,* October 30, 2012, pp. 1–11.

[26] Ali, S., et al. "Recent Advances of Wearable Antennas in Materials, Fabrication Methods, Designs, and Their Applications: State-of-the-Art," *Micromachines,* Vol. 11, No. 10, 2020, 888, pp. 1–41.

[27] Sang, M., et al. "Electronic and Thermal Properties of Graphene and Recent Advances in Graphene Based Electronics Applications," *Nanomaterials,* Vol. 9, No. 3, 2019, p. 374, pp. 1–33.

[28] Seesaard, T., P. Lorwongtragool, and T. Kerdcharoen, "Development of Fabric-Based Chemical Gas Sensors for Use as Wearable Electronic Noses," *Sensors,* Vol. 15, No. 1, January 16, 2015, pp. 1885–1902.

[29] Kiourti, A., C. Lee, and J. L. Volakis, "Fabrication of Textile Antennas and Circuits with 0.1 mm Precision," *IEEE Antennas and Wireless Propagation Letters,* Vol. 15, 2016, pp. 151–153.

[30] ElMahgoub, K., et al., "Logo-Antenna Based RFID Tags for Advertising Application," *ACES J,* Vol. 25, No. 3, March 2010, pp. 174–181.

[31] Tribe, J., et al. "Tattoo Antenna Temporary Transfers Operating on-Skin (TATTOOS)," in *Design, User Experience, and Usability: Users and Interactions* (DUXU 2015), Lecture Notes in Computer Science, Springer, Vol. 9187, 2015, pp. 685–695.

[32] Choi, J. H., et al. "Various Wearable Embroidery RFID Tag Antenna Using Electro-Thread," *Proceedings of IEEE International Symposium on Antennas and Propagation,* San Diego, CA, 2008, pp. 1–4.

[33] Liu, C., "First-Pass Introduction to Microfabrication," in *Foundation of MEMS,* Second Edition, Prentice Hall, 2012, pp. 37–38.

[34] Horrocks, R. A., and S. C. Anand, "Textiles in Filtration," in *Handbook of Technical Textiles,* Second Edition, Woodhead Publishing, 2016, pp. 89–91.

[35] Rogers, J. A., et al., "Paper Like Electronic Displays: Large Area, Rubber Stamped Plastic Sheets of Electronics and Electrophoretic Inks," *Proc. Natl. Acad. Sci.,* Vol. 98, No. 9, April 24, 2001, pp. 4835–4840.

[36] Liu, D., and D.J. Broer, "Liquid Crystal Polymer Networks: Preparation, Properties, and Applications of Films with Patterned Molecular Alignment," *Langmuir,* Vol. 30, No. 45, April 7, 2014, pp. 13499–13509.

[37] Koulouridis, S., et al., "Polymer–Ceramic Composites for Microwave Applications: Fabrication and Performance Assessment," *IEEE Transactions on Microwave Theory and Techniques,* Vol. 54, No. 12, December 2006, pp. 4202–4208.

[38] Vyas, R. et al., "Liquid Crystal Polymer (LCP): The Ultimate Solution for Low-Cost RF Flexible Electronics and Antennas," *2007 IEEE Antennas and Propagation Society International Symposium,* Honolulu, HI, 2007, pp. 1729–1732.

[39] Kiourti, A., and J. L. Volakis, "Stretchable and Flexible E-Fiber Wire Antennas Embedded in Polymer," *IEEE Antennas and Wireless Propagation Letters,* Vol. 13, 2014, pp. 1381–1384.

[40] Alharbi, S., et al., "Magnetoactuated Reconfigurable Antennas on Hard-Magnetic Soft Substrates and E-Threads," *IEEE Transactions on Antennas and Propagation,* Vol. 68, No. 8, August 2020, pp. 5882–5892.

[41] Mishra, V., and A. Kiourti, "Electromagnetic Components Realized on Conductive Wires: A Copper vs. E-Thread Comparison," *2019 IEEE International Symposium on Antennas and Propagation and USNC-URSI Radio Science Meeting,* Atlanta, GA, 2019, pp. 359–360.

[42] Gaubert, V., et al., "Investigating the Impact of Washing Cycles on Silver-Plated Textile Electrodes: A Complete Study," *Sensors,* Vol. 20, No. 6, 2020, p. 1739.

[43] Wen, J., B. Xu, and J. Zhou, "Toward Flexible and Wearable Embroidered Supercapacitors from Cobalt Phosphides-Decorated Conductive Fibers," *Nano-Micro Lett.,* Vol. 11, No. 89, 2019, pp. 1–14.

[44] Wagih, M., A. Komolafe, and B. Zaghari, "Separation-Independent Wearable 6.78 MHz Near-Field Radiative Wireless Power Transfer using Electrically Small Embroidered Textile Coils," *Energies,* Vol. 13, No. 3, 2020, p. 528.

[45] Moradi, B., R. Fernández-García, and I. Gil, "E-Textile Embroidered Metamaterial Transmission Line for Signal Propagation Control," *Materials* (Basel), Vol. 11, No. 6, 955, June 5, 2018, pp. 1–8.

[46] Lopes, C., et al., "Development of Substrate Integrated Waveguides with Textile Materials by Manual Manufacturing Techniques," *Second International Conference on Sensors and Applications,* November 2015, pp. 1–6.

[47] Salman, S., et al., "Pulmonary Edema Monitoring Sensor with Integrated Body-Area Network for Remote Medical Sensing," *IEEE Transactions on Antennas and Propagation,* Vol. 62, No. 5, May 2014, pp. 2787–2794.

[48] Zhu, K., L. Militello, and A. Kiourti, "Antenna-Impregnated Fabrics for Recumbent Height Measurement on the Go," *IEEE Journal of Electromagnetics, RF and Microwaves in Medicine and Biology,* Vol. 2, No. 1, March 2018, pp. 33–39.

[49] Doerger, S. R., and C. K. Harnett, "Force-Amplified Soft Electromagnetic Actuators," *Actuators,* Vol. 7, No. 4, 76, 2018, pp. 1–11.

[50] Wang, Z., "Embroidered Multiband Body-Worn Antenna for GSM/PCS/WLAN Communications," *IEEE Transactions on Antennas and Propagation,* Vol. 62, No. 6, June 2014, pp. 3321–3329.

[51] Thompson, D., et al., "Characterization of Liquid Crystal Polymer (LCP) Material and Transmission Lines on LCP Substrates From 30 to 110 GHz," *IEEE Transactions on Microwave Theory and Techniques,* Vol. 52, No. 4, April 2004, pp. 1343–1352.

[52] Salonen, P., J. Kim, and Y. Rahmat-Samii, "Dual-Band E-Shaped Patch Wearable Textile Antenna," *2005 IEEE Antennas and Propagation Society International Symposium,* Vol. 1A, 2005, pp. 466–469.

[53] Post, E. R., et al., "E-Broidery: Design and Fabrication of Textile-Based Computing," *IBM Systems Journal,* Vol. 39, No. 3.4, 2000, pp. 840–860, doi: 10.1147/sj.393.0840.

[54] Anagnostou, D. E., et al., "A Direct-Write Printed Antenna on Paper-Based Organic Substrate for Flexible Displays and WLAN Applications," *Journal of Display Technology,* Vol. 6, No. 11, November 2010, pp. 558–564.

[55] Kim, Y., H. Kim, and H. Yoo, "Electrical Characterization of Screen-Printed Circuits on the Fabric," *IEEE Transactions on Advanced Packaging,* Vol. 33, No. 1, February 2010, pp. 196–205.

[56] Bai, Q., and R. Langley, "Wearable EBG Antenna Bending and Crumpling," *Proc. Loughborough Antennas and Propagation Conference,* November 16–17, 2009, pp. 201–204.

[57] Szczech, J. B., et al., "Fine-Line Conductor Manufacturing Using Drop-On-Demand PZT Printing Technology," *IEEE Trans. Electron. Packag. Manuf.,* Vol. 25, No. 1, January 2002, pp. 26–33.

[58] Kim, D., and J. Moon, "Highly Conductive Ink Jet Printed Films of Nanosilver Particles for Printable Electronics," *Electrochem. Solid-State Lett.,* Vol. 8, No. 11, September 22, 2005, pp. J30–J33.

[59] Kamyshny, A., and S. Magdassi, "Conductive Nanomaterials for Printed Electronics," *Small,* Vol. 10, No. 17, 2014, pp. 3515–3535.

[60] Zhang, S., et al., "Embroidered Wearable Antennas Using Conductive Threads with Different Stitch Spacings," *2012 Loughborough Antennas & Propagation Conference (LAPC),* 2012, pp. 1–4.

[61] Lilja., J. et al., "Design and Manufacturing of Robust Textile Antennas for Harsh Environments," *IEEE Transactions on Antennas and Propagation,* Vol. 60, No. 9, September 2012, pp. 4130-4140.

[62] Elias, N. A., et al., "Bending and Crumpling Deformation Study of the Resonant Characteristic and SAR for a 2.4 GHZ Textile Antenna," *UTM Jurnal Teknologi,* November 16, 2015, pp. 17–23.

[63] Paracha, K. N., et al., "Low-Cost Printed Flexible Antenna by Using an Office Printer for Conformal Applications," *International Journal of Antennas and Propagation,* Vol. 2018, Article ID 3241581, February 14, 2018, pp 1–7.

[64] Bai, Q., and R. Langley,"Crumpling of PIFA Textile Antenna," *IEEE Transactions on Antennas and Propagation,* Vol. 60, No. 1, January 2012, pp. 63–70.

[65] Hu, B., et al. "Bending and on-Arm Effects on a Wearable Antenna for 2.45 Ghz Body Area Network," *IEEE Antennas and Wireless Propagation Letters,* Vol. 15, 2016, pp. 378–381.

CHAPTER 3

Wearable Electronics with Flexible, Transferable, and Remateable Components

Abdulhameed Abdal, Kelly Nair Rojas, Akeeb Hassan, and
Pulugurtha Markondeya Raj

3.1 Technology Drivers

Growing market demands for wearable technology in health-monitoring applications has been driving key innovations in package integration of flexible electronics. Wearable health-monitoring technologies can be classified into four categories: (1) continuous biopotential recording for electrocardiogram (ECG) or electromyogram (EMG) tracking, (2) biophotonic monitoring for blood flow and oxygenation, (3) electrochemical monitoring for glucose, uric acid, and other biomarkers such as sweat analytes, and (4) wearable imaging systems. These are illustrated in the first row of Figure 3.1. However, as the functional demands grow for wearable electronics, so does the need for component integration for power telemetry, signal processing, and wireless communication. Advanced component integration technologies with flex and fabric integration is the key enabler for this technology evolution. Packaging of wearable electronic systems is broadly classified into different categories: rigid packaged components on thick circuit boards, rigid packaged devices on flexible carriers, embedding thin bare chips in flexible substrates, fabric-integrated electronics, hybrid flex-fabric integration, transferable electronics as on-skin e-tattoos, and so on. These are shown in the second row of Figure 3.1. The evolution from packaged rigid components to embedded thin chiplets in flexible and fabric substrates is primarily driven by the need for seamless power and data connectivity, new sensor integration for superior health monitoring, and user comfort in terms of unobtrusive monitoring and high conformability. Package downscaling with traditional approaches such as wafer-level packaging, high-density coreless substrates, package-on-package, or 3-D chip stacking with interposers and packages for wearable health monitoring have played a major role in the proliferation of smart watches with continuous ECG recording, monitoring of glucose, and other biomarkers. With the emergence of stretchable and flexible or fabric electronics,

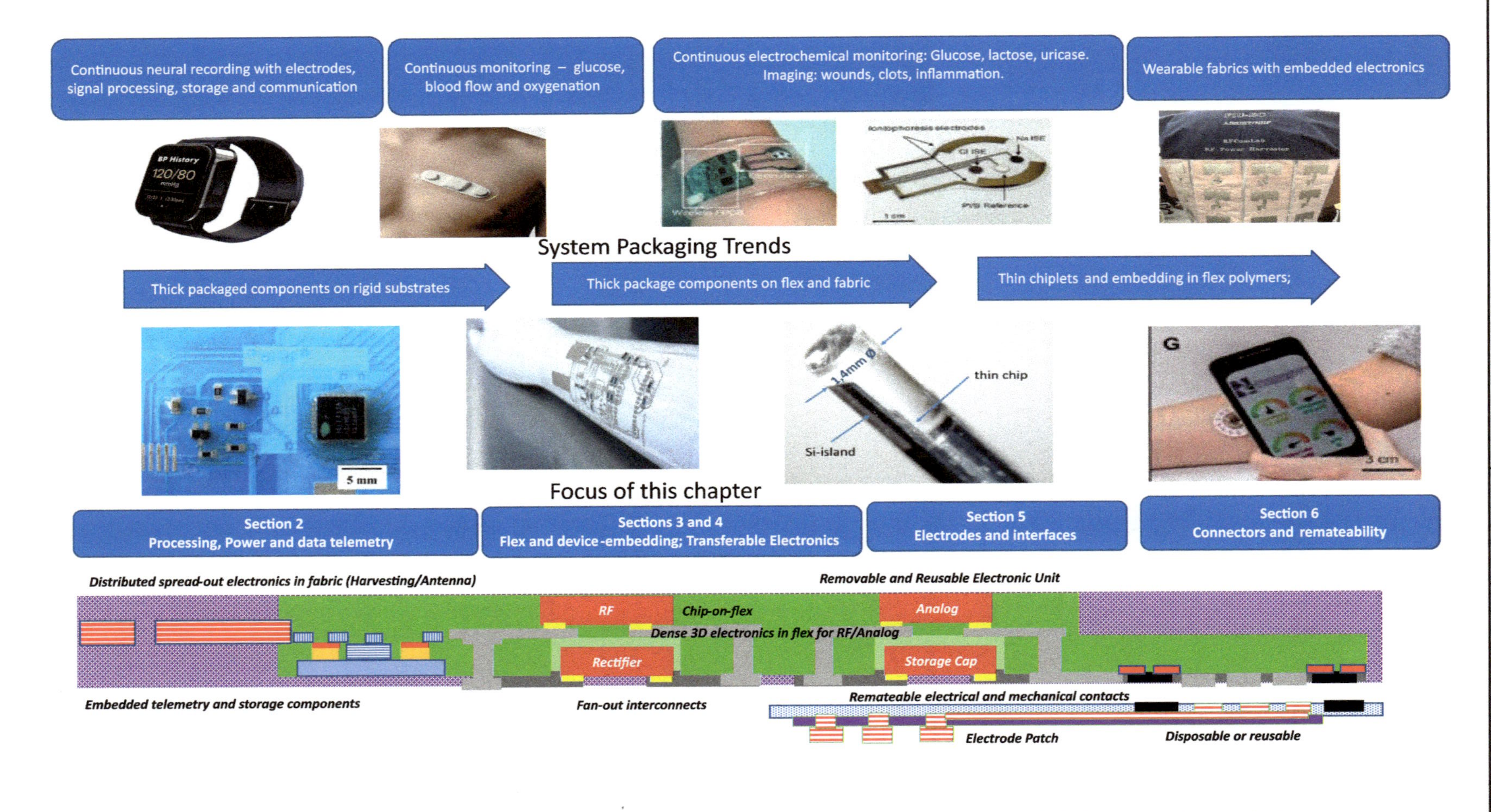

Figure 3.1 Biomedical wearable health-monitoring products, system packaging architectures, and basic packaging building blocks that are the focus of this chapter. Top row: N Squared technologies, John Volakis, and Shubhendu Bhardwaj (FIU) [2]. Middle row: G. Wable (Jabil Circuits), Subu Iyer (UCLA) [2, 4].

these electronics are now proliferating towards higher conformability to the human skin for efficient and unobtrusive health monitoring. Advanced biosensor devices are expected to deliver signals with high sensitivity and specificity to avoid artifacts and external artifacts that typically occur during motion and to promote a high-quality signal. As well, these wearable electronics utilize thin-film circuit transfer with low-modulus adhesives that are highly conformal to skin, eventually appearing as tattoos. Furthermore, to provide adequate modularity and reusability, the electronic subsystems are interfaced to each other with remateable connectors for easy assembly and disassembly. This chapter will introduce the key system components of flexible electronics, process integration in flexible substrates, key electrode technologies, remateability, and subsequent evolution in each segment. The key building blocks of the image are shown in Figure 3.1. Readers are referred to [1–4] for more examples of wearable sensor technology elements and building blocks.

3.2 Functional Building Blocks

Wearable electronics are based on three essential system blocks: signal processing, power and data telemetry, and power storage. Figure 3.2 shows a hardware package cross section with key features such as prepackaged or bare die components on a flex substrate and embedded dies inside the flex substrate for 3-D high density packaging solutions. This section will briefly review the key system components for health-monitoring systems.

3.2.1 System Architecture and Components

In a typical wearable electronic system, the biosensor quantifies a physical measurement from neural action potential or biomarker concentration and converts it to an electrical signal. The analog-digital converter (ADC) processes the amplified analog signal sent by the sensor to a digital signal. The digital signal is either stored in the memory or transmitted to an external reader through a wireless transceiver. In certain cases, the output digital data is also converted back to analog data through a D-A converter and sent to an actuator as a response. Figure 3.3 shows a block diagram for biomedical signal acquisition, processing, and communication to a central data receiving unit. The key active components are amplifiers, ADCs, and microcontrollers. The main purpose of a microcontroller is to acquire data, process and filter it by removing any unnecessary artifacts, and also generate the output signal for external communication or feedback toward a therapeutic response. The choice of microcontrollers is guided by power consumption and performance. Since microcontrollers contain many components in an integrated embedded system, it is critical to design them with efficient signal processing, power consumption, and form-factor. By using an integrated microcontroller with built-in signal processing and RF transceivers through system-on-chip (SoC) devices, the package footprints can be substantially reduced. The main advantages of SoC microcontroller devices are the integration of peripheral components, low cost, and reduced power consumption. The key innovation in the integration of such active system components is the evolution of embedded thinned bare chips as opposed to the assembly of prepackaged devices. Thin bare die functional blocks that are interconnected with

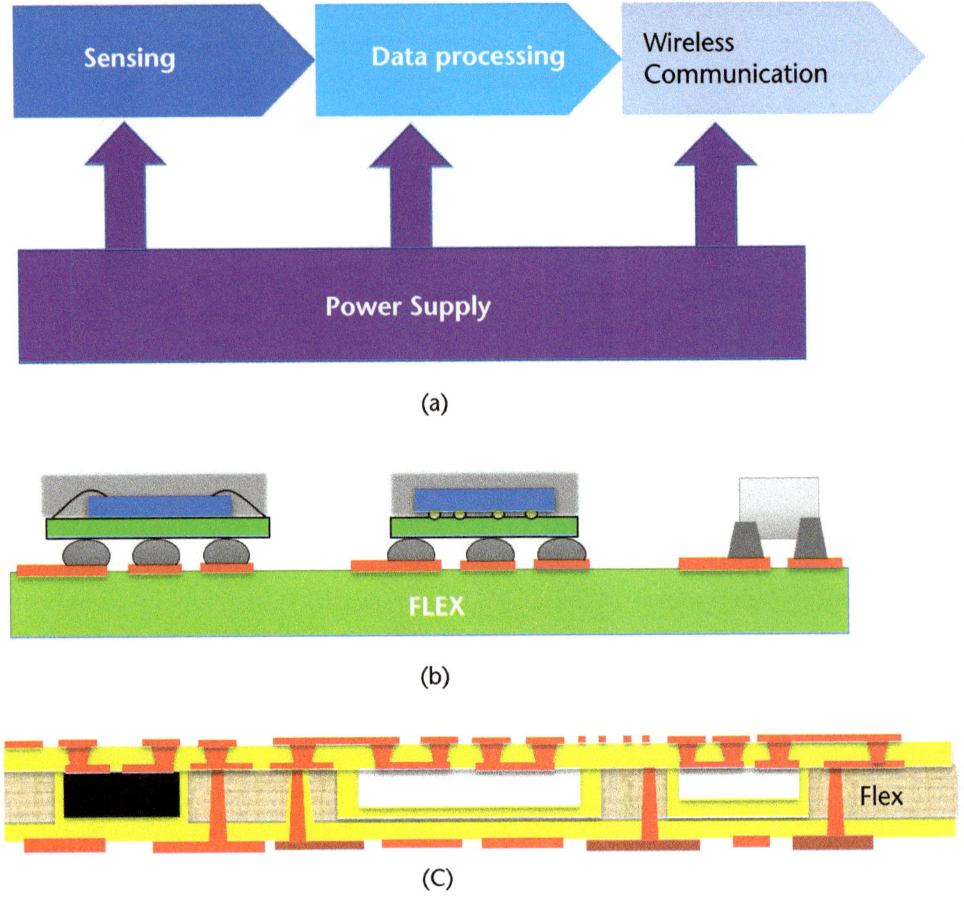

Figure 3.2 (a) Key system blocks, (b) components on flex, and (c) embedded components inside flex.

high-density interconnects will bring major paradigm change to emerging wearable electronic systems. Simplified system topologies can reduce the component count and also lower the power. In the most simplified form, a passive sensing-communication system is most effective as it only requires a wireless communication interface and direct signal modulation from the sensor with no additional signal processing. These approaches are discussed in other chapters of this book.

While semiconductor device assembly is the key approach to integrate power-signal processing-communication functions in a package, thin-film active devices that are directly grown on flex are emerging as an alternative approach [5]. The thin-film transistor (TFT), like any other field-effect transistor (FET), controls current flow between the source and drain contacts by the voltage applied to the gate electrode. The resultant electric field would create a channel for charges to move freely and easily between source and drain contacts. In other words, these are three-terminal (source, drain, and gate) voltage-controlled devices where the output current is the function of input voltage. TFTs can be manufactured using organic and inorganic materials such as amorphous silicon and polycrystalline silicon and are used for amplification and switching. Organic materials are preferred because of characteristics such as lightweight, flexibility, compatibility to plastic,

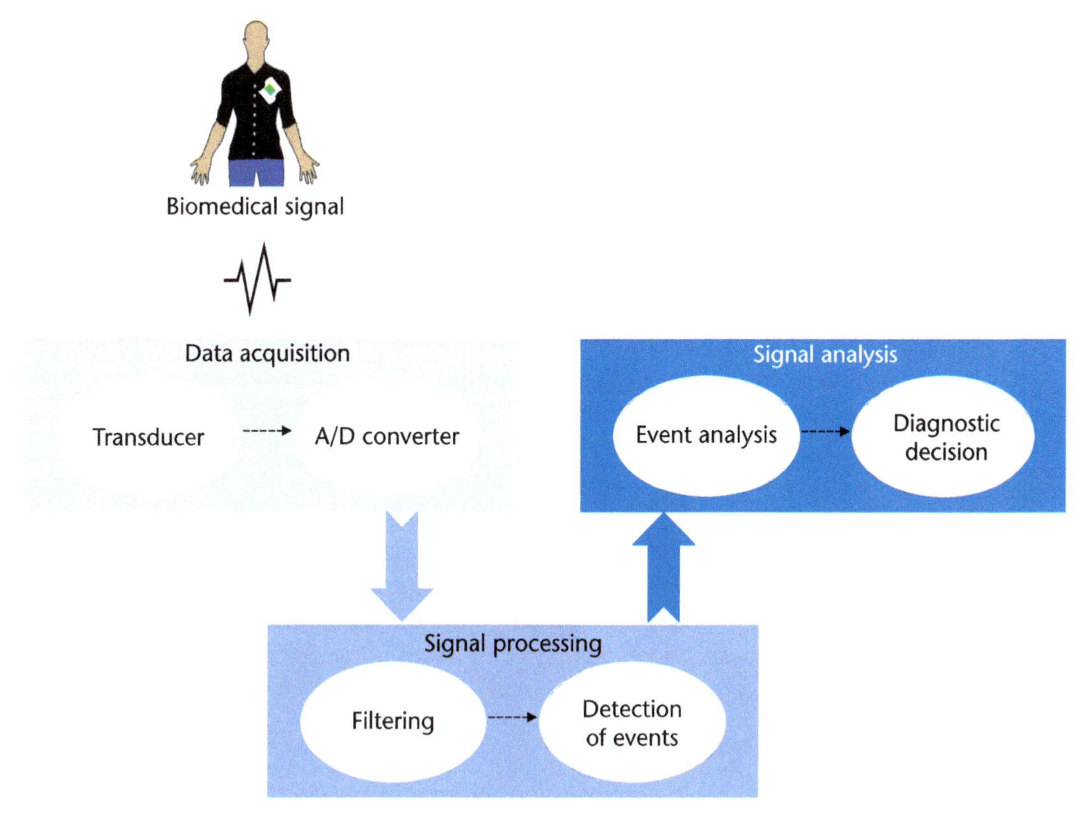

Figure 3.3 Block diagram from biomedical sensing to wireless communication.

and cost efficiency. In addition, TFTs can be manufactured as a top or bottom gate structure with a gate insulator separating the gate from the source. In the top gate structure, the active layer is deposited first on the substrate, followed by the gate contact and gate insulator. However, in a bottom gate structure, the gate contact and insulator are deposited on the substrate first, followed by the active layer. Controlling the width and channel length of the transistor is important to achieve the target capacitance. The key parameter is the electron mobility, which relates the drain current, drain and gate voltages, threshold voltage, and channel width and length. High mobility effect permits higher cut-off frequencies and higher current to be passed through. Materials such as amorphous silicon and polycrystalline silicon have attractive mobility effects of 1 cm^2/V_s and 500 cm^2/V_s, respectively, charge carrier mobility, ratio of currents in the on/off states (I_{on}/I_{off}), threshold voltage (V_{th}), and the subthreshold slope that describes how readily the device switches between on/off states in response to small variations in gate voltage. Organic semiconductors that have 10 times smaller mobility compared to polycrystalline silicon are continuously being enhanced to beyond 60 cm^2/V_s.

Wearable systems require advanced passive components for several reasons: Power supply noise filters to smoothen the output voltage;

- Matching network components for better signal or power transfer;
- Filter the RF signals to select the target frequency band;

- Modulate the impedance to pull up or pull down the voltage levels.

Unlike active components, passive components do not amplify the signal but instead modify the impedance to filter or efficiently transmit or process the signal. Since they do not provide amplification, power supply is not needed for their basic function. Passive components typically outnumber the actives by more than five times and determine the package size and manufacturing constraints. Noticeable passive component building blocks to form these filter or matching functions in flexible wearable electronics are capacitors, inductors, resistors, couplers, transformers, and other combinations of these. Passive components are integrated as either surface-mounted prepackaged components or embedded thin-film components. Incorporation of surface-mount capacitors is not effective because of the component thickness, routing and layout issues, and layout-dependent resistance and inductance. Flexible packages with 3-D thin-film passives will thus have a pervasive role in all future electronic systems. Embedded passives eliminate package parasitics and enhance system performance, reduce package footprints, eliminate supply-chain issues, and minimize the bill-of-materials. They are also preferred to enable miniaturization and offer high reliability due to elimination of solder joints. A detailed review of embedded passives can be found elsewhere [6].

3.2.2 Power and Data Telemetry

A key function of wearable systems is the communication of the collected data to an external reader or a local hardware through near-field links. These links are also effective in transmitting power to the active devices for signal amplification and modulation, thus providing both power and data telemetry. There are many means of near-field telemetry such as using inductive links, piezoelectric, and recently piezomagnetostrictive or multiferroic telemetry. In the power delivery chain, the power from the telemetry link is then fed to rectifiers and then to storage capacitors. In inductive links, transfer of power is possible when a source generates transient magnetic fields that would induce voltage at the receiver. The key factor in such coupling is the magnetic flux linkage between the transmitter and receiver coils. In an inductive link system, power is fed to the transmitting loop to facilitate maximum energy transfer at the resonant frequency. Coils work as resonators and store the energy around the transmitter at resonant frequency, which subsequently induces a voltage in the receiving coil or the loop antenna. Optimizing the coil design and tolerance to misalignments would increase the wireless power transfer (WPT) efficiency. A major limitation with inductive links arises because the power transfer efficiency dramatically reduces with the size of the receiving coil. This is because both the mutual inductance and the quality factor degrade with smaller receiving coils. One way to address this limitation is through the incorporation of magnetic lenses for directing the magnetic fields. Recent work at Florida International University (FIU) has shown the benefit of incorporating flexible magnetic lenses to improve the efficiency of power transfer through improved mutual coupling. However, the transmitter powder densities are low because of the constraints associated with the maximum allowed magnetic fields at MHz frequencies.

To tackle the constraints with inductive links, ultrasonic power telemetry using piezoelectric transducers has been considered as an alternative. This is because of the lower wave velocities that naturally lead to smaller wavelengths, resulting in smaller resonator dimensions. Ferroelectrics such as $BaTiO_3$ and lead zirconate titanate (PZT) are highly favorable because of their high piezoelectric coefficients. Fabricating telemetry links that use ultrasonic waves to induce voltage is one of the predominant methods to enable miniaturization. This is because voltage is directly proportional to the thickness of the ferroelectric and its dielectric constant. Low frequency ranges are advantageous compared to current technologies that utilize 13.56-MHz inductive links and Bluetooth (2.4–5 GHz) bands as a means for wireless communication. However, maximum power densities that are allowed in tissue are as high as 100 mW/cc with ultrasonic piezoelectric structures. To reach the next level of wireless power telemetry, multiferroic coupling with piezo-magnetostrictive mode is considered to be a promising approach. Magnetostriction is observed when magnetic fields induce volumetric change or strains in ferromagnetic layers. When coupled with an adjacent piezoelectric film, this strain induces a voltage. The induced voltage is a function of the applied magnetic field (H), magnetoelectric coefficient (α), and its thickness ($V_0 = \alpha \times H \times T$). Because of the human body's tolerance to higher magnetic field densities at lower frequencies and the correspondingly smaller resonator dimensions at 200 kHz, higher power densities can be transmitted with this telemetry mode. Figure 3.4 shows the achieved power densities with piezo-magnetostrictive power transfer mode. Such telemetry films have recently been integrated into flexible packages at the FIU Packaging Lab for applications such as neural stimulation, neural recording, and biophotonics. Piezo-magnetostrictive interfaces offer higher power because the PVDF/Metglas® stack can support as the flex carrier for the whole package, as shown in Figure 3.5. Exploring piezomagnetic coupling can meet the demand for wireless power data telemetry with energy densities reaching ~20× of ultrasonic piezoelectric energy harvesters.

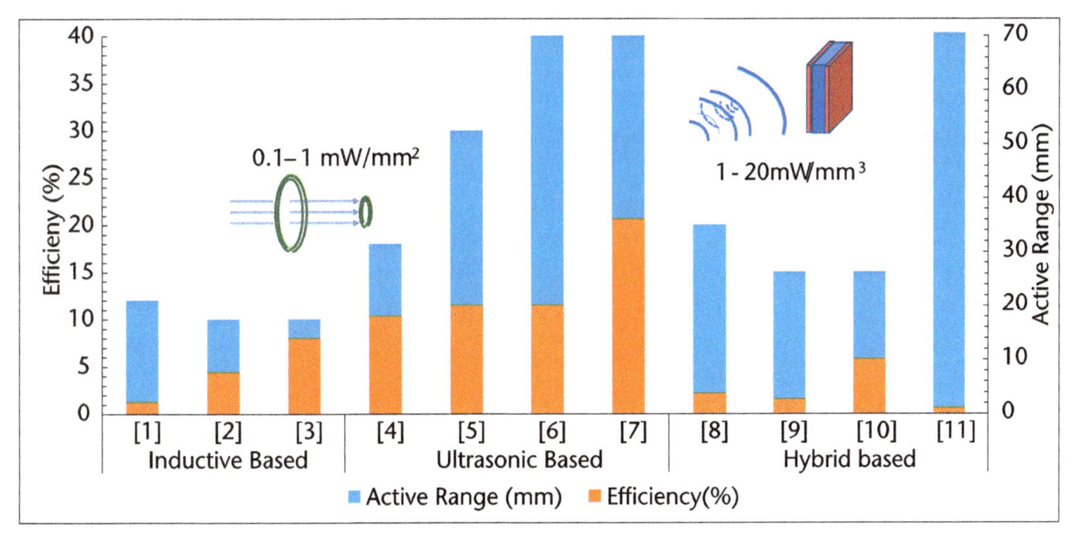

Figure 3.4 Efficiency and active range for inductive, ultrasonic, and hybrid technologies and representative power densities.

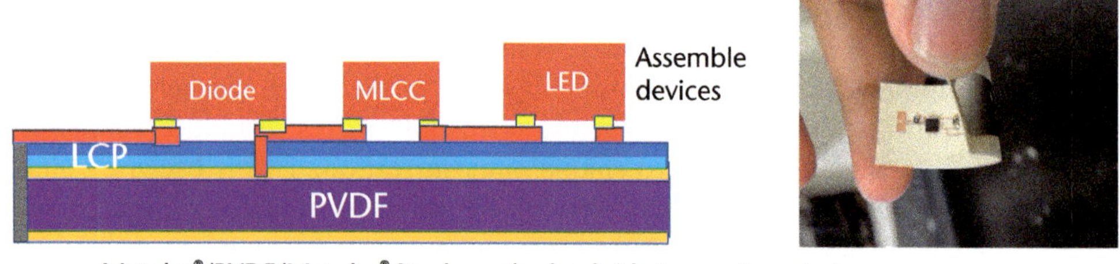

Figure 3.5 3-D heterogenous integration of piezo-magnetostrictive interfaces for power data telemetry developed by the Packaging Lab at FIU.

3.2.3 Energy Storage: Batteries and Supercapacitors

Power is delivered through various charge storage devices such as batteries, supercapacitors, and capacitors. To meet market demands, energy storage and power supply must be of high-power density, long reliability life, and low cost. Conventional batteries are limited to sealed rigid cases in order to meet reliability and safety constraints. In traditional approaches, portable devices incorporate stiff and brittle materials and are favorable because of their high energy density, high power, and rechargeability. These traditional batteries are made of one or more galvanic cells containing a cathode, anode, separator, and current collectors. Flexible batteries tend to have better adaptability for wearable devices since they are designed to be conformal and sustainable under arbitrary tension or compression by using intrinsically deformable materials to meet technological advancement. Therefore, the components of a flexible battery must also be durable to maintain desired electric performance. Many approaches are studied to fabricate flexible batteries in various sizes and configurations such as using polymer binders to fabricate composite electrodes, or even incorporating nanoparticle complexes for high surface area electrodes. Graphene or carbon nanotubes (CNTs) on porous nanowires can host compounds with a higher concentration of lithium ions to achieve 1 Ahg^{-1} and 1,000 cycles of operation. The anode candidates are comprised of inorganic oxides and sulfides (CoO, NiO, FeS , MoS_2, SnO_x) and multicomponent oxides such as cobalt molybdate ($CoMoO_4$), lithium titanate ($Li_4Ti_5O_{12}$), among others, and lithium cobalt oxide ($LiCoO_2$), spinel-structured lithium manganese oxide ($LiMn_2O_4$), and lithium iron phosphate ($LiFePO_4$) as a cathode with graphene collectors [7]. Along with higher carrier mobility and improved conductivities, and higher surface area because of their 3-D porous structure, they can handle the volumetric changes during ion extraction and insertion for minimal electrode degradation, higher ionic conductivity to deliver higher power, and prevent dissolution. The fundamental packaging structure of a flexible battery source includes a flexible anode with a thin layer of printed cathode electrodes on a stainless-steel foil or conducting polymers.

In supercapacitors, charges are stored by electrical double-layer capacitance (EDLC) and pseudocapacitance at the electrode/electrolyte interface under charging voltage. The former relies on angstrom-scale separation between the electrode and ions of electrolytes while the latter operates via different class of surface reactions, such as reversible redox reaction of electrolyte ions and electrode surfaces, or intercalation/deintercalation of ions at electrodes. This process is slower but can store much larger amounts of charges than the former one. For both types, the specific surface area and conductivity of electrodes determine the electrochemical performance of the microsupercapacitor. Supercapacitors are able to replace flexible batteries because of their ability to provide even higher power densities, longer life cycles, and very fast charge-discharge rates compared to Li-ion batteries. Relatively small areal capacitance <1 to ~10 mF/cm^2 with high power density of EDLC MSC while large capacitance >3,000 mF/cm^2 with very low power due to poor conductivity have been reported. Also, supercapacitors are deemed to be environmentally friendly, which broadens the aspects of using such systems in low-power applications safely.

The primary components in supercapacitor are electrodes and electrolytes. Supercapacitors usually contain high-surface electrodes to create electric potential difference between two plates. For EDLC, various carbon-based electrodes, such as activated carbon (AC), onion-like carbon (OLC), carbide-derived carbon (CDC), CNT, and graphene, or recently Si-based Si nanowires, silicon carbide, and Si-coated carbon sheath have been also employed [7, 8]. The schematic representation of an enhanced surface area with a graphene extension is shown in Figure 3.6. For pseudocapacitance, metal oxides such as RuO_2, MnO_2, or conductive polymers (Polypyrrole (PPy), Poly(3,4-ethylenedioxythiophene) (PEDOT)) and transition metal oxides (RuO_2, MnO_2, $NiO/Ni(OH)_2$) have been studied as electrode candidates [9]. For electrolyte, solid-state materials can facilitate the packaging of the microsupercapacitor (MSC) devices, eliminating concerns of leaking and instability of liquid-based ones. Gel types such as polyvinyl alcohol/ sulfuric acid (PVA/ H_2SO_4), polyvinyl alcohol/phosphoric acid (PVA/H_3PO_4), and polyvinyl alcohol/ lithium chloride electrolytes or ionic liquids/composites such as PVA/ 1-butyl-3-methylimidazolium tetrafluoroborate ($BMIBF_4$) and 1-ethyl-3-methylimididazolium bis(trifluoromethylsulfonyl) imide/fumed silica represent the typical solid-state electrolytes [8, 10, 11]. The power and energy densities of supercapacitors and batteries are compared in Figure 3.6.

Recent innovations focus on building flexible supercapacitors using paper substrates because of their porous and fibrous characteristics. This enables superior adhesion to metallic electrodes and better charge accumulation because of the high surface area. Flexible hybrid supercapacitors also adhere well to deformable materials such as CNTs and CNFs. Materials such as graphene, paper, and textiles have

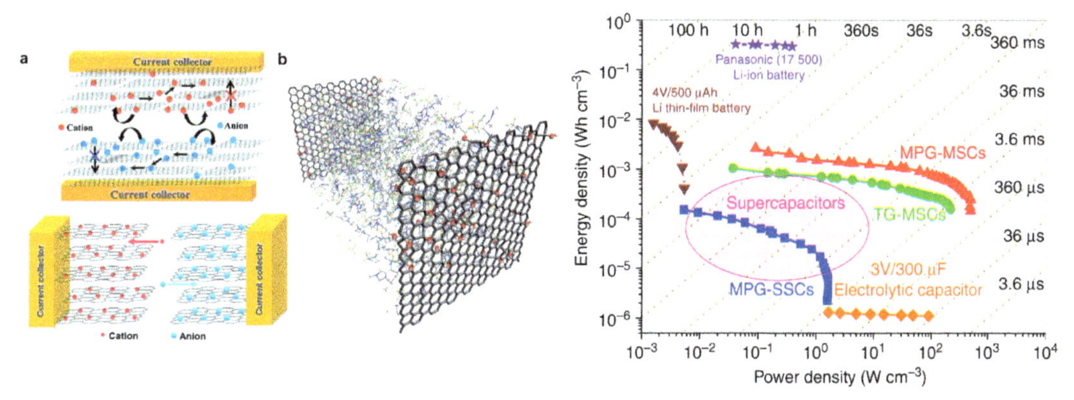

Figure 3.6 (a) High surface area graphene electrodes (left) [12, 13] to increase capacitance density, and (b) Ragone plot showing the energy and power densities of batteries and supercapacitor technologies [14, 15].

been demonstrated as electrodes for supercapacitors to achieve better performance at a lower cost. Most importantly, new era concepts focus on smart materials based on piezoelectric and piezomagnetic relationships because of their ability to self-charge and result in high power density. For example, ferroelectrics with metallic electrodes such as CuO are used as supercapacitor electrodes because of their versatility in mechanical designs and electrical outputs. Although they are emerging as a substitute for flexible batteries, higher power densities and ultra-miniaturization will be required in the near future.

3.3 Technology Building Blocks for Heterogeneous Component Integration

A wearable system consists of an electronic hub on a flexible carrier with extended biosignal interfaces such as electrodes. The electrodes are printed onto a polyurethane substrate that is laminated onto a fabric with a hot-pressed adhesive [16], as shown by a schematic cross section in Figure 3.7. The electronic hub houses the power, processing, and communication functions as described earlier. Packaging of wearable electronic systems is broadly classified into different categories, such as rigid packaged components on thick circuit boards, rigid packaged devices on flexible carriers, embedding thin bare chip in flexible substrates, fabric-integrated electronics, hybrid flex-fabric integration, and transferable electronics as on-skin e-tattoos. Heterogeneous high-density integration will result in enhanced functional component densities through component miniaturization, multilayered and finer interconnects, and advanced assembly technologies. With the eventual access to bare chips in close partnership with semiconductor device manufacturers, this strategy with reduce the overall size of the finished module with true heterogenous

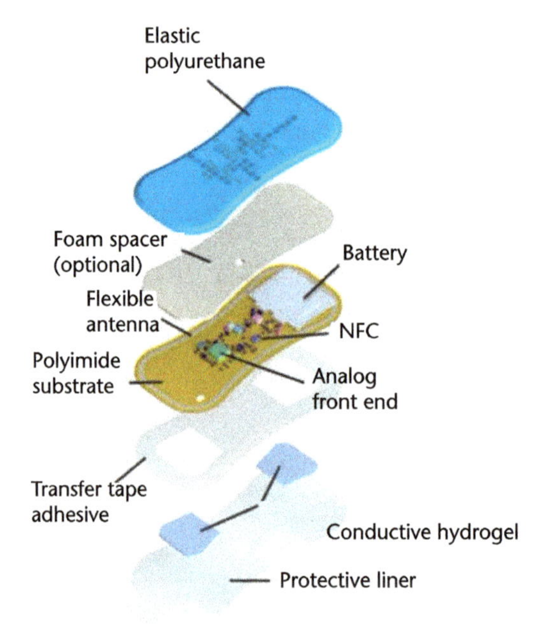

Figure 3.7 Electronic hub on a flexible electrode extension with TPU substrate laminating onto a fabric [18].

integration of power, photonic sensors, and wireless communication front-end module with antennas in a single substrate. The next key step is the replacement of surface-mount discrete passive components with embedded thin-film passives. This trend is shown in Figure 3.8. This is the key not only for health monitoring on humans, but also for emerging IoT, structural health-monitoring, as well as many other applications where remote 100% wireless sensing is required. This section describes the material and fabrication technologies of the flex carrier and device assembly. The fabrication of the high-density electronics hub with 3-D and fan-out packaging is not emphasized here but can be found in the previous works of the authors [6, 11, 17].

3.3.1 Thin Substrates

The migration from rigid electronics to flexible electronics was made possible due to a group of flexible substrates and their unique attributes that enable optimum bendability without cracking. Certain characteristics must be considered for flexible substrates in terms of electrical, chemical, optical, magnetic, mechanical, and thermal properties. Chemical properties require substrates to be inert and not release contaminants under chemical process. Thermomechanical properties are important when designing a flexible substrate such as compatibility to maximum fabrication temperature. In general, flexible materials must have the appropriate elastic modulus to adhere to stiffness requirements when being processed and handled. Also, high thermal mismatch between the thin-film and substrate is one of the common problems that would cause mechanical failures such as propagating a crack or interfacial delamination under thermal cycling. Another critical factor to note is dimensional stability at processing temperatures. Thermal stress affects plastic foil substrates when processing at upper processing temperatures of the film and could lead to undesirable changes such as expansion and shrinkage. In terms of optical properties, optical clarity may be required to view through the film and must be low or not birefringent when used as base substrates. In addition, substrates must have low electromagnetic interferences and incorporate surface quality to minimize defects and change in electrical operability. Four main categories of flexible substrates are available to meet current desired properties and they are based on polymers, papers, and metal foils.

3.3.1.1 Polymers

Polymer films are ideal for flexible electronics because of their physical properties and processability. The key properties of thermoplastic polymers, such as high

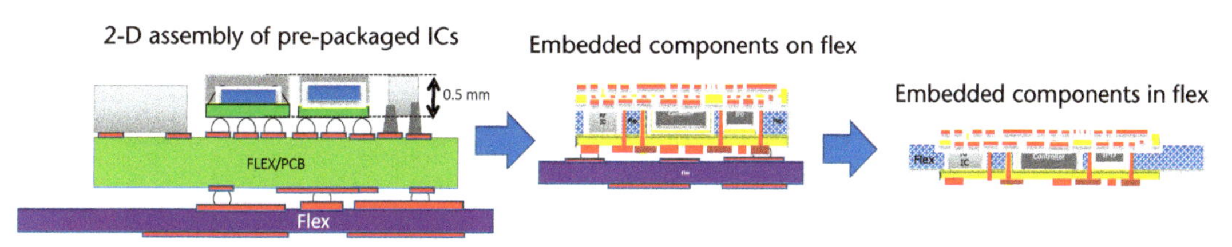

Figure 3.8 3-D heterogenous integration trend to realize future wearable electronics [2–4, 19].

elongation to failure, strength, and processing at 25 microns and below, make them foldable, rollable, conformable, and easy to handle. Many applications tend to use polymer films because they are inexpensive and exploit this flexibility and performance. In addition, their low dielectric constant and low dielectric loss provide a lot of RF design space. For example, LCPs are known to retain low loss to 110 GHz [20]. Plastic films can be categorized into three groups: (1) glass transition materials, (2) thermoplastic semicrystalline, and (3) thermoplastic nanocrystalline. Glass transition temperature (T_g) is the temperature when an amorphous polymer transforms into a melted state (T_m) when heated. When transformed from a glass state, the polymer becomes soft and flexible, exhibiting viscoelastic and rubbery characteristics. Some common examples of high-glass transition materials include polyacrylates (PAR), polyimide (PI), and polycyclic olefin (PCO). High-glass transition materials present superior properties as they have high thermal stability up to 300°C, low outgassing under vacuum conditions, and chemical resistance. It is important to note PI has a high T_g ~350°C but has other limitations from its high moisture absorption and high loss, along with its yellow color. On the other hand, thermoplastic semicrystalline materials have sharp melting points, which make them rapidly change into low viscosity liquid when heated above the T_m. Polyethylene terephthalate (PET), some forms of polyamide (nylon), and polyethylene naphthalate (PEN) are examples of thermoplastic semicrystalline and are widely studied because they offer good mechanical strength with high elastic moduli, low permeability to water, and are optically clear. Clarity is vital in emissive displays and such materials offer high transmittance >80%. In addition, major differences between amorphous and thermoplastics is that amorphous plastics are not birefringent and have poor solvent resistance. PET and PEN films have low CTE—typically <25 ppm/°C—to avoid mismatch with other inorganic devices. However, a disadvantage of thermoplastic semicrystalline materials is their dimensional stability. Because of their biaxial structure orientation, noncrystalline regions achieve the desired molecular orientation with negligible extension of molecular chains. As a consequence, PET and PEN experience undesirable changes such as shrinkage at T_g, requiring thermal relaxation. Thermoplastic nanocrystalline plastics such as polycarbonate (PC) generally have higher glass-transition temperatures when compared to PET and PEN. However, these types of plastic foils have poor resistance to process chemicals and large CTE >50 ppm/°C. Ideal plastic material should have key characteristics such as resistance to chemical properties, low CTE, good dimensional stability, and high surface quality. Barrier coatings can enhance these characteristics by offering reduced gas permeability and surface toughness.

Table 3.1 summarizes the four categories of materials with corresponding advantages and disadvantages.

New packaging technologies require higher reliability of next-generation multichip structures. The trend to high-density packaging and thinner housings causes compromised interconnect reliability because of thermomechanical failures. Epoxy was considered a gold standard for die-attachment adhesives where integrated circuits (ICs) were smaller in size. However, with regard to the demand for larger dies and thinner packages, cross-linked epoxy adhesives are not a key solution. This is due to its inflexibility and large modulus of elasticity causing critical issues in reliability. Researchers shifted to a new spectrum of creative materials to supply the demand for large-density packaging technologies by exploring thermoplastic

Table 3.1 Flexible Substrates Available and Corresponding Advantages and Disadvantages

Substrates Material	*Advantages*	*Disadvantages*
Paper	• Good adhesion • Low cost • Optimum flexibility	• Not flame retardant • Disposable and one-time use • Coating required for powered electronics
Thin glass (30–200 microns)	• Conformable • High transmittance and refractive index • Transparent • Very low permeation • Good dimensional stability • Smooth surface quality	• Requires careful handling • Very fragile • Requires reinforcement by polymer to reduce cracks
Metal foils (25–100 microns)	• Good barrier material • Resistance to chemical processes • High temperature processes • Good dimensional stability	• Capacitive effect • Further coating required for isolation • Rough surface • Opaque in nature • Sharp rolling marks
Polymer	• Bendable, flexible, and rollable • Low cost • Transparent • Low CTE	• Viscoplastic dimensional changes • Poor dimensional stability • Sensitivity to temperature and humidity • Low process temperature

elastomers under conventional additive manufacturing techniques. Thermoplastic adhesives are non-cross-linked materials with linear molecules that are flexible in nature and exhibit softening behavior above glass-transition temperatures. In addition, these types of elastomeric adhesives can be melted and remelted without any chemical changes, making them suitable for attachment of components because of their simplicity to remove and rework. Unlike thermosetting adhesives, thermoplastic area-area interconnections can become molten at increased temperatures to promote replacing and repairing components. Moreover, thermoplastic elastomers offer high strength and flexibility, low modulus, good adhesion, and rapid curing.

One of the common materials studied as a replacement for epoxy is thermoplastic polyurethane (TPU). Polyurethane is an elastic material that can be thermosetting or thermoplastic, while TPU consists of interconnected hard and soft segments. Soft segments of TPU are linear molecules of polyester chains that influence its elastic properties. Because of its biomedical compatibility, researchers found TPU can be stretchable to high strains. Fatigue behavior testing was addressed to maintain requirements of long-term stability for in vivo by understanding mechanical integrity under low and high cyclic loading. It has been demonstrated that TPU does show increased fatigue strength under cyclic loading because of undergoing reversible deformation, which intercepts development of cracks when absorbing deformation energy [14]. Other thermoplastic polymers considered in wearable electronics are silicone adhesives. Stretchable conductive silicone adhesives have demonstrated superior electrical performance under cyclic bending cycles when mixed with 70 wt% trimodal Ag mixture. However, it is important to note that this study concluded that a higher percentage of trimodal Ag mixture (~80 wt%) offered close relationship bending and releasing strains when stretched. This is not always suitable for long-term reliability since excessive metal conductive fillers can result in poor adhesion.

Paper substrates: Because of its porous and fibrous nature, recent research shifted to paper-based substrates to enable potential implementation for biomedical applications. The cost of fabrication is inherently cheap and is fabricated by a combination of electroless plating and electrodeposition. Physical properties illustrated the possibility for paper to become a good candidate for adhesion to conductors and has been used as disposable sensors recently. However, the main challenge of paper is that it is not flame-retardant as many electronic devices require. This broadens the spectrum for research to find applicable materials for coating on paper and how it may affect flexibility and thermal conductivity.

Ultra-thin glass: Thin glass substrates lower than 100 μm thickness have gained popularity for next-generation wearable displays, especially in organic-LED (OLED). Optical properties are highly desirable with luminous transmittance of 90%, high refractive index ~1.5, and high ultraviolet (UV) resistance. As well, ultra-thin glass with a thickness of 30 μm has low stress birefringence, making it the perfect candidate as a base substrate in LCD technologies. One of the distinguishable properties of glass from polymeric layers is its thermal stability. Glass has better dimensional stability at high elevated temperatures when compared to polymer-based foils. However, it is important to note that the material properties of glass are also a function of cooldown speed. If the process of annealing is slowed down, glass tends to become denser. At higher elevated temperatures, glass may experience some level of thermal shrinkage and must be controlled for thin films. Thermal shrinkage depends on the temperature, time, and the type of glass used. In addition, CTE of solid glass is generally linear in behavior from temperatures –50°C up to 450°C. For any temperature above 600°C for high strain, the CTE will depend on the glass composition. Another attractive property of glass such as borosilicate is its resistance to chemical processes. It is impermeable to water and oxygen, which promotes a longer lifetime. Moreover, glass has excellent surface properties allowing a smooth surface quality created by tension during the solidification process. Careful handling of glass is vital to avoid breakage probability. The only disadvantage of using thin glass is that it is fragile, which can make it difficult to handle without suitable equipment. Reinforcement of glass by polymer surface coating can elevate glass strength by reducing microcracks and existing propagated cracks.

Metal foils: Stainless steel and titanium are the most common materials used for flexible light and display applications and they are considered good barrier materials. Metal foils are demonstrated in electronic devices such as OLED, TFT, and photovoltaic devices. Important characteristics of flexible metal foils are high resistance to chemical processes, suitability for high-temperature processes near 1000°C, and low permeation to oxygen and water. As well, thermomechanical properties offer excellent dimensional stability at elevated temperatures and could also serve as a heat sink in electronic devices. They are lightweight, flexible, and unbreakable substrates (~125 μm thickness). Although metal foils seem very attractive, common disadvantages with metal foils are that they are opaque in nature, do experience rough surfaces, and most important, they are electrically conductive. For circuit isolation applications, metal foils must be coated with an insulating layer such as SiN_x and SiO_2, which affects the adhesion layer and barrier properties. In addition, stainless steel foils exhibit sharp rolling marks that would require further polishing or planarizing.

3.3.2 Circuit Formation: Metallization, Photopatterning, or Additive Deposition

Circuit formation involves metal and dielectric deposition and patterning to form the circuit traces. Dielectric deposition is achieved through physical vapor deposition (PVD), liquid coatings, or polymer dry-film lamination. Metallization is achieved with the laminated, sputtered, or plated metallic films or printed metal particle paste. These technologies are used to form high-density electronic modules in laminates, silicon and glass interposers, or wafer-embedded fan-out packages and are widely reviewed by the authors elsewhere and not discussed here [6, 17]. Due to the enormous growth in flexible electronics in the past, these technologies are continuously enhanced to overcome technical and manufacturing challenges for lower cost at high volumes. Blanket metal or dielectric films are patterned through etching processes with a suitable mask. Metal patterning is shifting from subtractive or semiadditive electroplating to complete additive manufacturing processes for lower cost and compatibility with different polymer carriers. The most prevalent techniques are screen printing, inkjet printing, and thin-film fan-out patterning. These are illustrated in Figure 3.9 and the key process parameters and technology parameters are compared in Table 3.2. These will be reviewed in this section. The extension of these processes to transferable electronics are described in the next section.

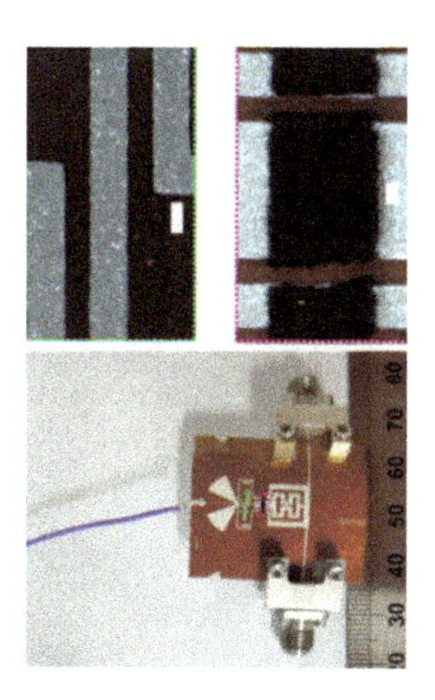

Screen-printed fine traces

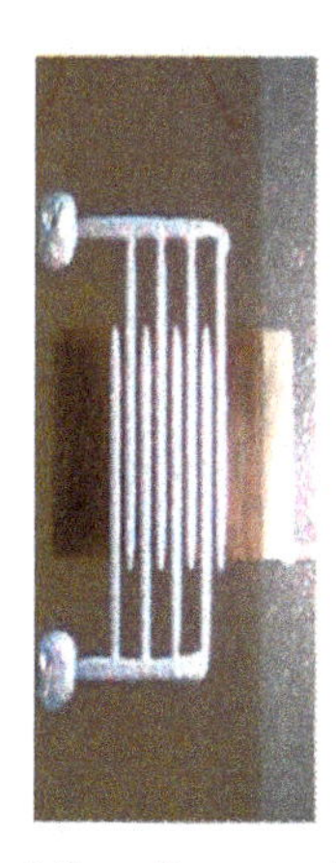

Microdispensed traces

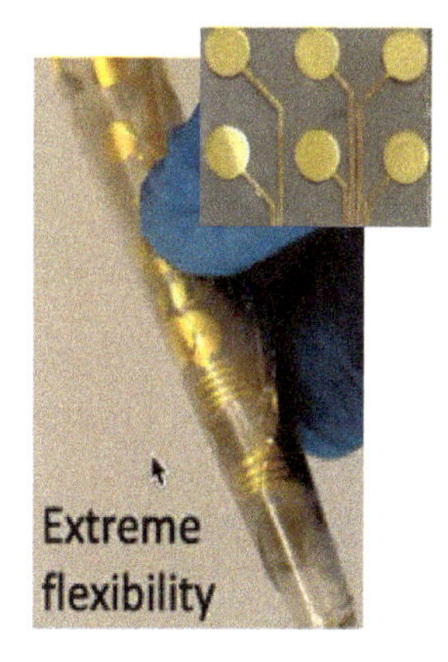

Lithographically patterned copper traces

Multilayered and multimaterial inkjet traces

Figure 3.9 Circuit patterning advances with screen printing [21], nozzle microdispensing (Mike Newton, nScrypt), lithographic copper patterning (Subu Iyer, UCLA), and inkjet printing (Jaim Nulman, Nano Dimension).

Table 3.2 Additive Manufacturing Options and the Technology Capabilities

Additive Manufacturing Processes		*Resolution (Linewidth)*	*Viscosity (Pa·s)*	*Write Speed*	*Complex Curvature Printing*
Large-area paste printing	Screen-printing	30–50 microns Thickness: 12 microns or thicker (depending on emulsion thickness)	1–50	0.15–0.25 m/sec	No
Droplet	Inkjet	25–200 μm	<0.01	0.3mm^3/s (single nozzle) 0.258–m/sec	Moderate
	Aerosol jet	10–150 μm Thickness 10 nm–5 μm	<2.5	0.25mm^3/s (single nozzle) 0.2 m/sec	Excellent
Flow	Micropen	100 μm	<1,000	50 mm^3/s	Excellent
Tip	AFM	12-nm linewidth 5-nm spatial resolution	Macromolecules, (ex. thiol) or nanoparticles	0.2–5 μm/s	Limited

From: [17].

3.3.2.1 Screen-Printing

Screen and stencil printing are low-cost scalable manufacturing techniques and are ideal for forming interconnections and passivation. This technique is based on pushing a viscous paste through a mesh or openings onto the underneath substrate. The pattern is controlled through the open areas in the screen. The screen is composed of a frame with a mesh stretched in between. The openings are defined by a pattern of emulsion, which is a negative of the required printed pattern. Stainless-steel wires are used to form the mesh as they are more stable, unlike polyester or nylon wires that are prone to dimensional instabilities from permanent absorption and moisture. Thread diameter and opening along with number of threads per unit area are the key metrics that determine the fraction of open area and influence the quality of the printing. Up to 400 threads per inch are typically used. The screen mesh is at a certain snap-off distance from the substrate but not directly in contact with the substrate. The snap-off distance is also a critical process parameter as it can affect the quality of the line definition. For example, larger distance may not transfer the material from the screen to the substrate, while smaller distance may disturb edge definition because there is no adequate snap-off action. The distance between the screen and substrate is about 0.4% of the screen width for stainless steel mesh. Printing with a squeegee, which is a rubber or steel blade between a movable holder, creates tension in the screen. This tension is critical to enable the screen to snap back and create a well-controlled edge definition as the squeegee moves over the screen. Screen-printing is inherently considered to have limitations in achieving resolution that aerosol jet printing or nozzle microdispensing can achieve. Recent advances have enabled screen printing to reach the design capabilities of other advanced printing technologies. This technology is applied to print silver-polymer (epoxy, polyurethane) composites for flexible electronics but also extended to silver nanowire–poly dimethoxy siloxane composites for transparent conductors [22].

3.3.2.2 Inkjet Printing

Unlike laser ablation, inkjet printing is considered as an additive technique used to manufacture 3-D structures. The basic premise of inkjet printing is the breakdown of liquid streams into single droplets, where the channel is actuated by a piezoelectric, allowing control of the shape and size of the ink droplets. This allows the ink droplet to be deposited on the solid substrates in a precise location, and some factors to consider in this process are the physicochemical properties of the liquid droplets used, such as viscosity of the precursor material, impact velocity, and surface properties. Notable advantages of utilizing inkjet printing for patterning are cost-efficiency, environmental friendliness since it is waste-free, accurate deposition of scaled-down volume up to picoliters, high-quality films compared to other techniques, and most important, no requirement for a clean room environment. Details of inkjet printing and other additive manufacturing technologies such as aerosol jet printing and microsyringe dispensing can be found in [23].

3.3.2.3 Flexographic and Gravure Printing

Flexographic printing is a continuous technique for patterning and is attractive because of its high-throughput process of mass volumes. It is commonly used for label printing and packaging. It features four cylinders—a fountain roller, an anilox roller, a plate cylinder, and the impression cylinder. The technique encompasses a fountain roller with flexible rotary plates where ink is fed and transferred to a metal anilox roller, carrying the thin layer to the plate cylinder. The plate cylinder is made of a flexible resin or rubber and acts like an image carrier. Then the image carrier transfers the ink to the substrate, allowing precise distribution of ink via the impression cylinder's pressure. The key step is the preparation of the flexible plate cylinder that transfers the image to a cylinder. The image layer is typically prepared as a flat substrate with UV irradiation of a masked photopolymer layer, followed by subtractive removal of the uncured polymer to create the relief layer (raised surface of the print pattern). This is followed by wrapping around the plate cylinder. This technique is widely used for nonporous materials and flexible thin layers such as metal foils, paper, and polymers. In contrast, gravure printing utilizes an engraved cylinder where the image is acid-etched to different depths of recesses. This cylinder is partially dipped in an ink fountain. The ink from the recess of the engraved cylinder is directly applied to the substrate. This technique produces prints of higher quality and is commonly used for printing magazines. In terms of similarities, both rotary printing technologies require equipment such as cylinders, plates, and sleeves. In addition, they both can perform continuous long-run fabrication for high-quality mass production that is cost effective. Depending on the application intended, generally flexographic printing is more favorable because of the ability to print on porous and nonporous substrates, the ability to use versatility of inks, and higher resolution output of prints.

3.3.2.4 Lithographic Photopatterning, Nanoimprint Lithography, and Self-Aligned Imprint Lithography

Conventional photolithography for circuit patterning has dominated the bulk of the material cost during production because of photoresist applications and the multiple etching steps involved during the process. In the most common practice for low-cost flexible laminate fabrication, dry-film or liquid photoresists are applied to the metal surface to be patterned. This is followed by UV exposure through a PET photomask, followed by patterning of the photoresist that acts as an etch mask. The metal is then etched with acids and the photoresist is stripped. The UV exposure dose is typically of 80–100 mW/cm^2. A weak alkaline solution is used for photopatterning. Typical acid etch-rates can be controlled to 1 micron/minute. The next step toward this is the emergence of nanoimprint lithography (NIL). In this process, a silicon template is used to emboss the patterns in the resist material. This eliminates the need for photosensitive material as patterning is done with hot embossing. Another step is the evolution UV-NIL, where the resist is UV cured through a transparent PDMS mold to solidify the patterned resist. PDMS is again the most common mold favored because of its elasticity, simple fabrication, and the ability to create compact contacts. In this process, when a material is deposited, the photopolymer is heated to above the glass-transition temperature to enable a flow into the stamp and then cured with UV light to harden the polymer. However, these techniques are sequential and create complications in alignment tolerances and process steps when multiple layers of films need to be patterned to form a 3-D active device stack such as a TFT on a substrate. By reducing the process steps, advanced nanoimprint techniques based on soft lithography has revolutionized the conventional manufacturing technique to achieve reduce costs for mass production and potentially enhanced efficiency. In one such approach, known as self-aligned imprint lithography (SAIL), multiple mask levels and the associated geometric information are incorporated and imprinted as a single or monolithic three-dimensional structure (3-D). An elastomeric stamp, typically PDMS, is utilized to deposit the monolithic 3-D mask on a substrate. The PDMS stamp is then peeled off cleanly and wet and dry etching processing complete the cycle by producing high-resolution structural patterning of the 3-D TFT multilayered film stack on the substrate. Many other applications utilize SAIL rapid-prototyping technique because of its benefits such as cost-efficiency, flexibility, and simplicity. However, a critical drawback of SAIL is the possibility of contamination that may cause defects due to physical contact between the elastomeric mask and substrate. This problem is alleviated through self-cleaning processes.

3.3.2.5 Laser Ablation Patterning

In this subtractive technique, structural patterning of the substrate is based on a high-intensity laser, where it directly writes into the polymer layer, eliminating the standard etching processes and photoresist applications. The laser ablation process results in removing material on the surface of the substrate. This is possible because the focus of the laser functions is breaking the chemical bonds of the polymer and causing generation of heat when absorbing the photonic energy. It is important to note that ablation only occurs when a specific material absorbs enough energy to vaporize or melt. The generation of heat causes melting or vaporization of the layer resulting in removing macroscopic materials through the resultant plasma plume. A plasma plume is formed when the polymer transitions from a solid state to a gaseous state. When irradiating the polymer with pulsed xenon-fluoride (XeF), a melt pool is formed, where the heat generated causes a liquid vapor phase transition at increased temperature. Basically, the polymer is fractured into shorter units and ejected during vaporization because of high pressure. Key parameters involved in laser ablation of polymers are laser wavelength, pulse duration and repetition, and laser fluence such as monochromaticity and directionality. Polymer materials that can undergo laser ablation are PI, PET, PDMS, PTFE, and are highly desirable because of characteristics such as minimal wear and frictional properties, light weight, and resistance to corrosion. This approach is now adapted to also structure metals benefiting from the high absorption of infrared wavelengths, relatively lower cost to make IR fiber lasers at high power (1–15 W, 27-μm beam diameter, 1.06 μm fiber laser), and high-precision 2-μm positioning through galvanoscanners [13, 24]. To achieve further improvements in precisions, UV lasers were developed by LPKF (ProtoLaser U4 (355 nm) with 6W power of 20-μm laser diameter and 2-μm scan resolution) to remove 1.18 sq in/min of 1 oz (36 μm) of copper.

3.3.2.6 Roll-to-Roll Manufacturing

The processes discussed above can be scaled to high-volume and high-through roll-to-roll (R2R) manufacturing to transform conventional batch processing to continuous processing in order to manufacture flexible substrates. R2R employs two moving rolls along a processing line to transfer flexible substrates through additive and subtractive processes and the output ensues a finished product. For example, a raw material is perceived on one end and the other end results in a finished product. Other similar manufacturing techniques could also be found today such as sheet-to-sheet or roll-to-sheet. However, the pivotal benefits of using the R2R process is to manufacture continuously in a cost-efficient manner, which would increase production yields and rates. The impact of this technique allows mass production and reduces the total cost of manufacturing when compared to the traditional manufacturing techniques by performing several sequential steps. Many applications take advantage of the R2R process to manufacture flexible electronics, flexible photovoltaic cells, batteries, metal foils, and textile substrates to meet market demands. A variety of substrates can be used in R2R depending on the application intended and the processing steps required. Polymeric films can be used for optical applications as they are required for transparency, stainless-steel foils can be used for higher temperature tolerance, and other material substrates such as textile and paper can be used for optimum flexibility. The processing technology is able to coat, embed, laminate, and print substrates at high volumes when transported from one roll to another. The three sequential steps involved in a typical R2R manufacturing process flow broadly are deposition, patterning, and packaging as shown in Figure 3.10. These steps were described earlier. The adoption of this novel technology enables manufacturers to revolutionize manufacturing efficient and flexible electronics.

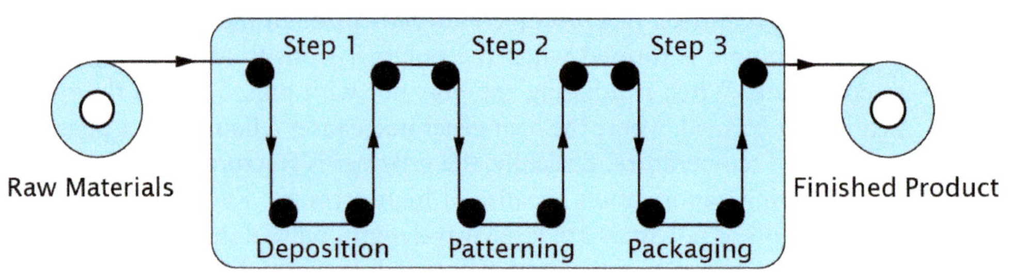

Figure 3.10 Process flow of R2R manufacturing.

3.3.3 Device and Component Assembly

Assembly on substrates is the key step toward hardware integration. The key requirement for assembly is to have compatible metal pads on the substrates, metal terminations on the chip side, and a bonding material to provide the interconnection. The bonding materials are either solders, conductive adhesives, or direct metal fan-out interconnections. Solder reflow is a well-established and the most prevalent technology to assemble dies and passive or active components on rigid substrates with copper pads. The copper pads need to be modified with Ni-Au finish or organic surface protection (OSP) layer. However, solder reflow is not directly compatible with printed conductors. Since solder reflow demands high-temperature processing, it is not suitable for various low-temperature polymer substrates that are mainly utilized in flexible electronics. Wire-bonding, isotropic conductive adhesives (ICA) or pastes, anisotropic conductive films (ACFs), and fully additive techniques (sintering and dielectric ramp interconnection) are the options for surface-mount device (SMD) assembly onto flexible substrates. Table 3.3 summarizes the different types of assemblies studied today [17].

3.3.3.1 Low-Temperature Soldering

Since soldering is the workhorse technology for device assembly, innovations in solder assembly are widely sought for advanced flex packaging. Solder assembly innovations in traditional high-performance electronics are directed toward reducing the interconnection pitch or enhancing the current-handling ability with advanced solder barriers. For wearable electronics that only require coarse pitch assembly and low power, the innovations are oriented toward reducing the solder assembly temperature. This is with alloys of lower melting point (e.g., Sn-Bi alloys or liquid gallium alloys), or low-temperature solder heating. Recently a novel photonic soldering has been introduced that is compatible with flex packaging [25]. Photonic soldering utilizes a high-power xenon flash lamp to solder ultrathin bare chips on copper pads on flexible substrates within a few seconds. A PulseForge 1300 xenon lamp system (Novacentrix, Austin, Texas) with energies covering 1 J/cm^2 to 7 J/cm^2 is used to vary components and substrates. Two sets of energy pulses are utilized for preheating components above the decomposition temperature of the solder flux and soldering process. Low-intensity pulses are chosen for decomposition while high-intensity pulses with 5-ms duration are implemented for soldering components. During the photonic soldering process with short temporal pulses, solder

Table 3.3 Assembly Types Available with Corresponding Performance and Process

Assembly Type	*Cross Section*	*Performance*	*Assembly Process*
Soliders: High-temp: Ex. Sn-Ag-Cu Low-temp: Ex. Sn-Bi		High conductivity High current-handling Self-alignment	High reflow temp (>260 C) with lead-free solders Gold surface finish with nickle barriers (or organic solderability perservatives)
Isotropic conductive adhesives Sliver-epoxy Silver-elastomer Silver-nanoinks		Higher bump resistance Pitch limited by the printing	Printed bumps ex stencil-printing Low-temperature assembly
Anisotropic conductive films Epoxy film with Au-shell particles		No alignment needed Acts as in-build underfill	Higher resistance Cross-talk between bumps cause high RF losses
Ramp interconnections		High inductance from long interconnects Design flexibility	Epoxy ramps Printed silver adhesives
Fan-out connections		Lowest parasitics Thinner packages Ideal for mm wave and THz interconnects	Via on the IC pads Printed vias and conductor traces

joints reach to around a liquidus temperature of lead-free alloy (210°C), without raising the substrate temperature. In conventional reflow, soldering long holding times around the liquidus temperature damages polyester foils. Since substrate and components have different absorption coefficients, filter masks are incorporated into this novel technique to selectively heating these components. Thus, it enables mass production. The basic principle of photonic soldering is shown in Figure 3.11.

3.3.3.2 Anisotropic Conductive Adhesives

Anisotropic conductive adhesives (ACAs) are films composed of insulating resin binder and sparse volume loading of conductive fillers (5%–20%). Electrical conduction of these composite materials occurs solely at z-direction and conductive particles are confined between bonding pads and bumps. These adhesives or films are dispensed or laminated between surfaces of interest. An electrical path is established between two conductive surfaces by applying heat and pressure on conductive particles. One of the advantages of ACA material is that it nullifies the requirement of patterning since it does not need to be selectively applied. Furthermore, it does not need any underfilling as ACA itself acts as underfill. Bumps are essential for ACA bonding. Cu, Au, or Ni-Au bumps are usually implemented on pads. Harder bumps are more beneficial than conventional solder to confine conductive particles. With fine fillers, ACAs are suitable for fine-pitch applications when the particle movement during assembly is prevented with an anchoring polymer layer. Paik and others have shown an ultrashort pitch of <20 μm with nanofiber-based ACF interconnection by utilizing conducting filler particles. Modified ACF sheets are formed

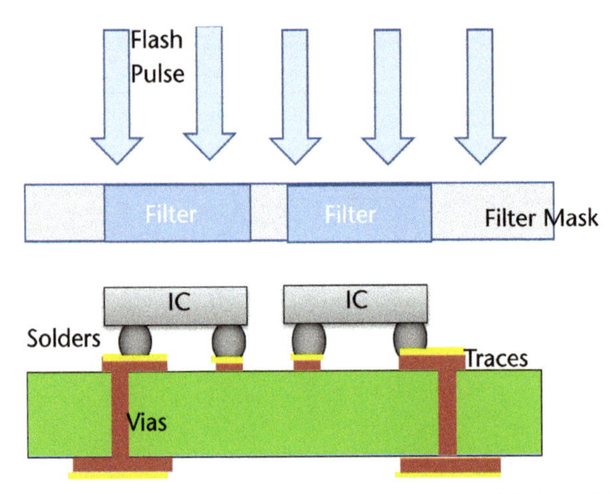

Figure 3.11 (a) Unmasked photonic soldering, and (b) photonic soldering with filter masks.

with Sn-Bi solder balls and electrospun polyvinylidene fluoride (PVDF) nanofibers that have an average diameter of 400 nm, including insulated conductive particles [26]. Since nanofibers or other similar anchor polymer layers of polyacrylonitrile (PAN) restrict the movement of conductive fillers, it prevents the bridging at fine pitch. Furthermore, it ensures significant capture rate (>80%) of conductive particles between the interconnection pads to form a reliable and robust electrical junction to chip-substrate. Thermocompression and plasma etching were implemented to fabricate nanofiber sheets with conductive particles. Following that, high-viscosity-NCF layers were laminated on the top and bottom of NS (nanofiber sheet) to fabricate NS-ACF, as illustrated in Figure 3.12.

3.3.3.3 Isotropic Conductive Adhesives or Films

ICAs or isotropic adhesive films (ICFs) consists of high filler content, and thus are conducive in all directions. Significant enhancements were made in the electrical

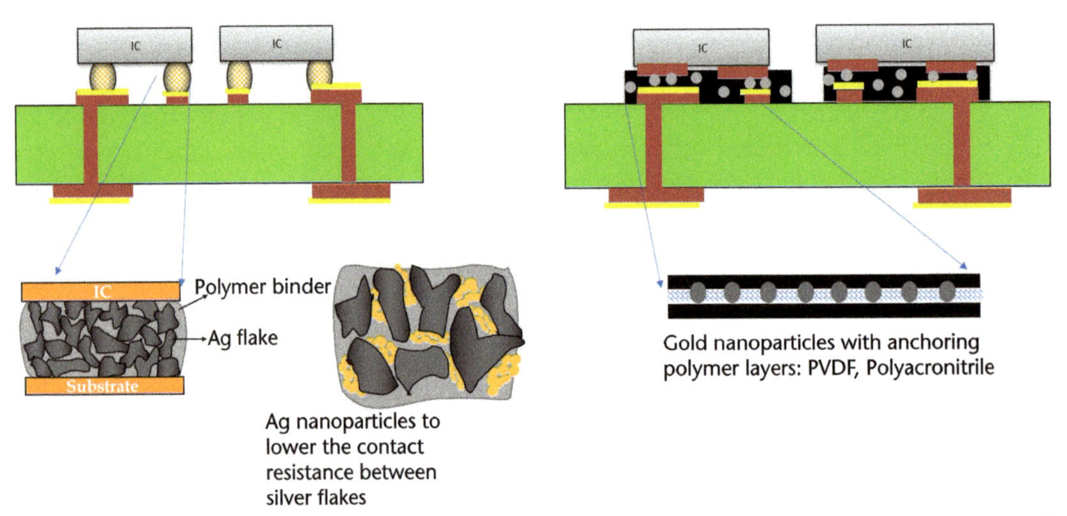

Figure 3.12 (a, b) ACFs with nanofibers to retain particles under the bumps, (c) schematic of assembly with ICAs, and (d) nanoparticle-enhanced adhesives.

conductivity of ICAs through (1) removing the surface-coating layers on the conductive particles for more close metallic contacts, (2) enhancing the cure shrinkage of the resin system, (3) removal of oxides, (4) implementation of nano-sized Ag flakes, and (5) metal-to-metal bonding between the conductive particles with a low-melting-point alloy coating. ICA bonding using Ag flakes is shown in Figure 3.12. Flexible polymer molecules are blended into epoxy resin for stress relief to enhance thermomechanical reliability and avoid interfacial delamination of the ICA joints. Han et al. [27] demonstrated a significant decrease in electrical resistivity from 1.14 mΩ.cm to 0.137 mΩ.cm by mixing Ag nanoparticles to 80 wt% Ag micro-flake- (Ag-MF) filled ICAs. The Ag nanoparticles are sintered at 250°C for yielding strong and reliable bonds for bridging the Ag-MF to themselves and conductive (metal) pads. Embedded electronics into textiles are analyzed and tested by utilizing ICAs from Creative Materials, Inc. Kelvin probe structures were developed to investigate interconnect flexing and moisture permeation of 8 mm × 8 mm chips [9]. The contact resistance test vehicle was based on a four-point probing for a single 200 μm in diameter and 17.5 μm thick bump (Figure 3.13) and found to be stable with bending tests.

3.3.3.4 Dielectric Ramp Interconnections

Dielectric ramp connections can be perceived as an extension of wire bonding. In this process, a die with its active side up is affixed on the substrate prior to printing the dielectric ramp [28]. Chip-to-package interconnects are formed by printing interconnects from the chip pads to the traces. However, the key is to form a ramp between the chip edge and the package in order to support the printed interconnects. The dielectric ramp surrounding the edges of a chip to make interconnections from die pads to the substrates, as shown in Figure 3.14. The interconnections and wiring are created in one single step. However, this technique is only applicable to

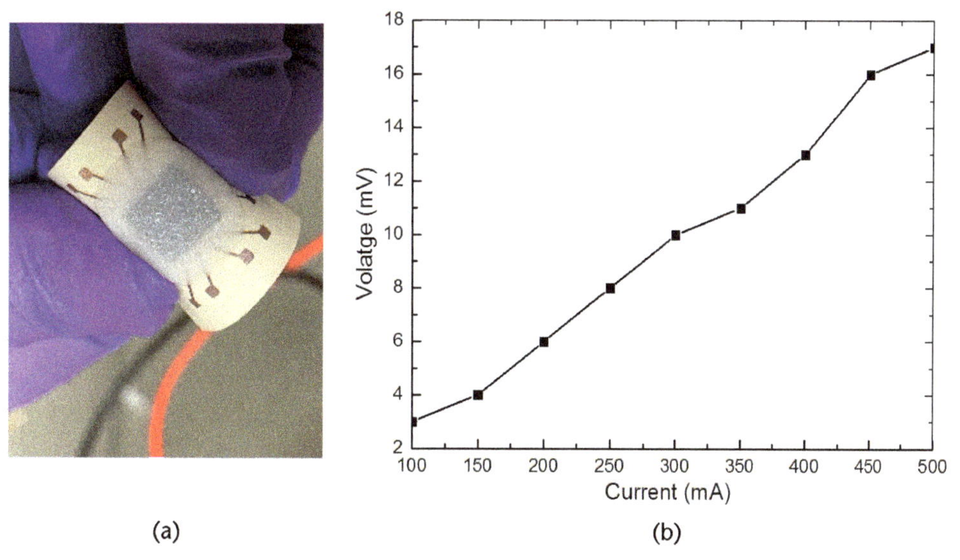

Figure 3.13 (a) Flexing of flip-chip structure with fluoroelastomer encapsulation, and (b) current-voltage relationship of Kelvin probe structure adapted from [9].

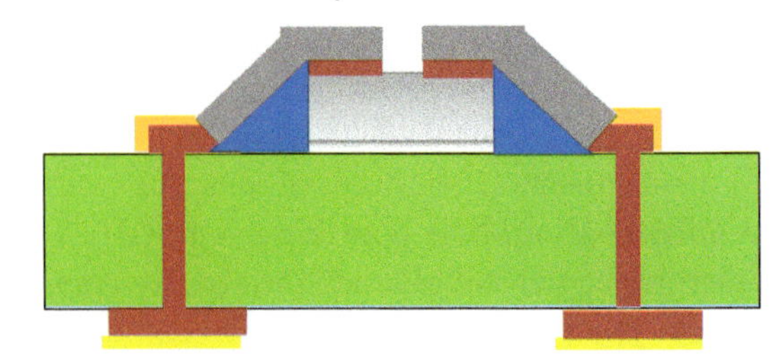

Figure 3.14 Dielectric ramp interconnection.

the peripheral layout, and it has large footprints as well as parasitic due to longer trace length.

3.3.3.5 Chip-Embedding in Flex

In a chip-embedding process, interconnects are formed from the chip terminations to metal traces with direct metallization from printed interconnections. This technology has several advantages: (1) high electrical speed due to shortest interconnect path, (2) low cost, (3) ultrathin package due to embedded ultrathin dies and solder joints assembly, and (4) smaller footprint than existing technology, namely flip and wire-bonded packages. The process-flow starts drilling cavities in flexible organic carriers. Hot-pressing is used to realize adhesive bonding between two ultrathin substrates. After bonding, dies are assembled with a high-speed pick and place tool. Redistribution layer (RDL) lamination and curing are done on both sides of the substrates to reduce warpage of the ultrathin package. In order to reveal the copper microbumps on the die, plasma etching or laser via drilling is implemented; finally, a semiadditive process or completely additive screen-printing process is utilized to form the chip-package and RDL, as shown in Figure 3.15.

The representative cross section of a flex-embedded package is shown in Figure 3.16. Microwave interconnect loss because of elastomer adhesive is low, at 0.1 dB/cm and loss performance remains the same under a bending test [29]. With such elastomer adhesive interconnects, the fabricated flexible fan-out package prototypes were bent over a 1-cm radius cylinder and resistance change is measured before and after bending cycles. Mechanical deformation does not change the via resistance significantly (<5%) and transmit and receive conductor traces, and vias with a backside-assembled fan-out package show less than 2% change in resistance [30]. This novel packaging provides a pathway for flexible wearable wireless sensing system.

An alternative form of chip-embedding in flex is through wafer-level fan-out packaging. In this process, the chips are assembled facing down on a temporary carrier (first handler) with a release tape, followed by encapsulation with PDMS. This stack is now bonded to a second wafer. The release tape can be softened through heating in order to release the layers on the top. The devices are now on the processing wafer (second handler), while facing up, which is convenient to form the redistribution layers on the top with processes that mimic back-end-of-the-line (BEOL). This involves planarization, deposition of biocompatible polymers such as Parylene C and photopatternable SU-8. After the wiring redistribution, the stack is removed from the second handler and is ready to be integrated with the flex

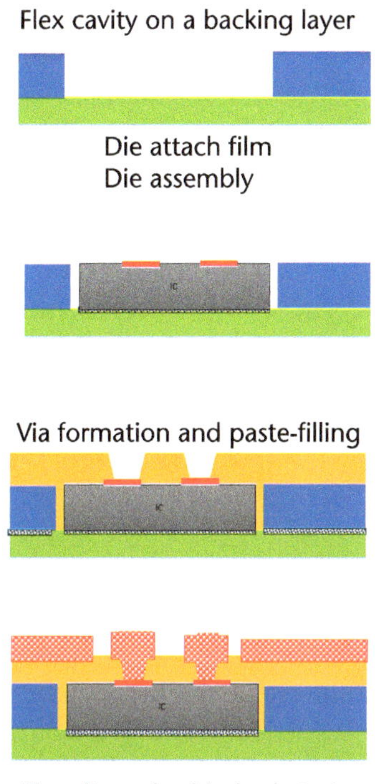

Figure 3.15 Process flow and assembly of chip-in-flex.

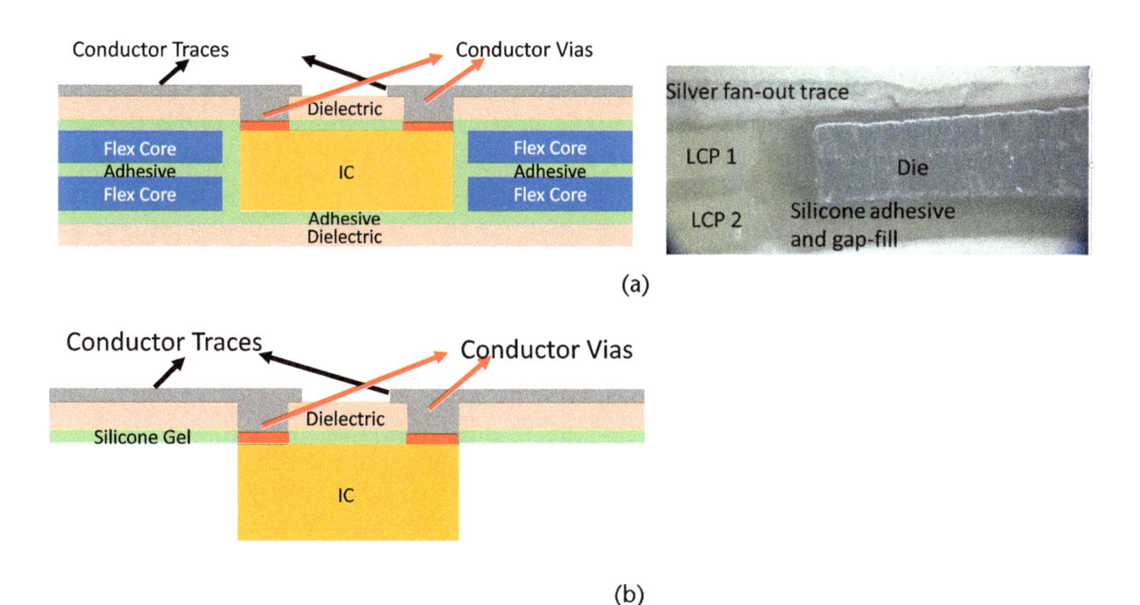

Figure 3.16 (a) Cross section of flexible packages with embedded dies and printed fan-out interconnections, and (b) cross section of backside-assembled flexible fan-out interconnection.

electrode arrays for biosignal recording. The resulting embedded-flex package can be so flexible that it can be rolled into small pipes as shown in Figure 3.9.

3.3.4 Encapsulation

Encapsulation seals out water vapor and oxygen and protects the devices from dust, sweat, and other contamination. Thin-film encapsulation (TFE) eliminates moisture and oxygen edge permeation allowing flexibility owing to its thickness [31]. Encapsulation of electronics by molding of the final system package is one of the key technologies that protects the electronics from mechanical damage and moisture penetration. In a rapid manufacturing potting technique, the board or component is normally placed in a shell or capsule. Subsequently, the encapsulant material is filled up until the component is completely covered and reaches the top of the capsule. The fundamental factors such as the wetting behavior, mechanical properties, biocompatibility, and optical clarity are taken into consideration while selecting the encapsulant material. The adhesive bonded joints should withstand high shear strength and the peel forces. Testing cured adhesives for biocompatibility as per healthcare standards UPS Class VI (The United States Pharmacopeia) and ISO-10993 is necessary for its allowed use in medical and healthcare applications. Curing the encapsulation epoxy using visible light instead of ultraviolet light may also result in improved device performance. The encapsulation must provide adequate adhesion and ensure thermal management, electrical isolation, and biocompatibility needs are met. In certain cases, the encapsulation must also provide electromagnetic interference (EMI) shielding to protect the device from signal interference and ensure proper operation of the device.

Epoxies and silicones are the most common encapsulants. Adhesives such as silicone elastomers such as DOWSIL™ 3-4207 insulate electronics and provide stress relief for interconnections under applied displacements. These are ideal for encapsulating emerging packaging structures that evolved from the current flip-chip assembly to contemporary techniques such as fan-out assembly and embedded chip structures. With higher interconnect density and thinner packaging layouts, larger stresses act on the interconnections. The use of a dielectric gel as a gap fill can enhance the flexibility by providing stress relief when bending at high curvatures or under thermal stress. The aid of additional mechanical strength enables higher component densities for next-generation electronic packaging. Figure 3.17 shows the effects of PDMS encapsulant on a chip-on-flex using mechanical modeling adapted from [18]. Hydrophobic fluoroelastomer was utilized to encapsulate the flip-chip packaging structure to resist moisture permeation of embedded electronics under washing cycles. The contact resistance of silver adhesive increased by 33.3% from from 30 mΩ to 40 mΩ due to fluroelastomer's hydrophobic nature.

3.4 Transferable On-Skin Electronics

High-density electronics with fine features need rigid inorganic substrates such as silicon and glass for fabrication. The rigidness provides dimensional stability, thermal compatibility, surface smoothness for fine-line lithography, tool compatibility, and other advantages that are key for fabrication. On the other hand, the final devices need to be flexible and conformal to skin, or as wearable fabrics for several applications, while also providing breathability and other attributes of a skin-like elastomer. One approach to address both needs is to fabricate electronic circuits on

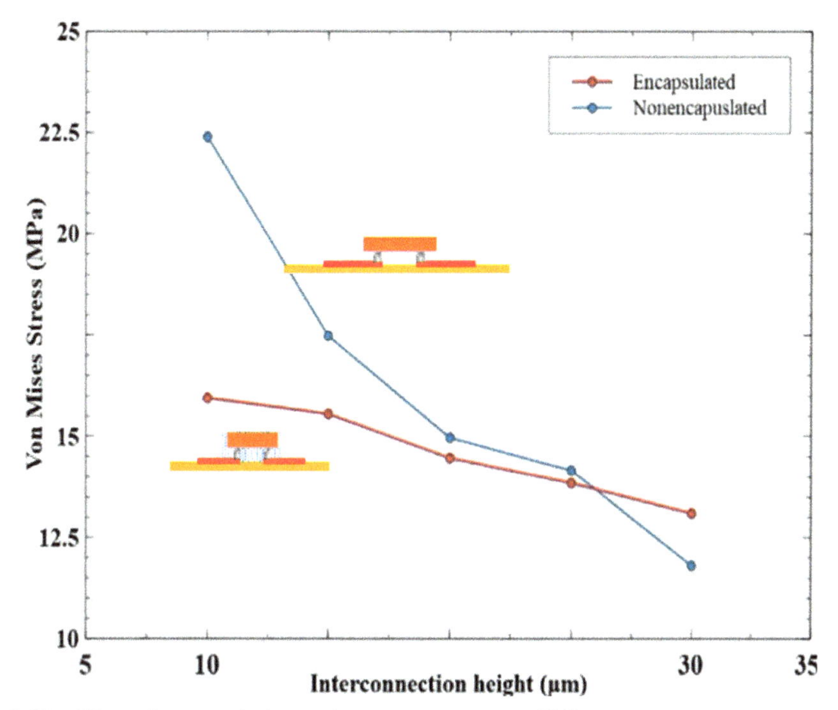

Figure 3.17 Effect of encapsulation on interconnect stresses [18].

a rigid substrate and release them onto the flexible elastomer layer through a transfer process. The circuits can then be transferred to the skin. Transferable electronics from donor substrate are pursued with multiple options: (a) laser or thermal-assisted release and transfer, (b) microtransfer print with elastomeric stamps, (c) transfer onto water-soluble tapes with an etching-based release process, (d) transfer of a flex substrate to skin with adhesives, or (e) direct transfer of flex to fabric. These approaches are classified in Figure 3.18.

3.4.1 Laser or Thermal-Assisted Release

Transfer printing is typically a two-step process that incorporates a receiver and donor substrate to mediate the transfer of traces. The device, circuit, or electrode layers are first fabricated on an inorganic donor substrate with a sacrificial layer. Circuit patterning is achieved with subtractive etching or additive deposition through screen printing, inkjet printing, or even drawing with a pencil. The circuit

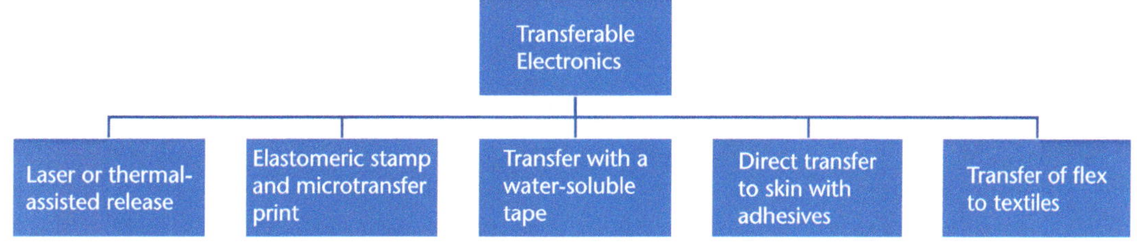

Figure 3.18 Broad classification of circuit transfer processes.

layer, with or without the carrier, can be transferred in a selective or nonselective massively parallel manner as arrays. If only the circuit layers are transferred and not the carrier, a critical key parameter in microscale transfer printing is adhesion modulation. The underlying competing interfacial delamination allows the transfer process between the donor substrate and the final or intermediate substrate. Different adhesion modulation techniques are extensively studied to enhance transfer reliability of printing such as kinetic control of stamps, thermal mismatch of interfaces based on the laser noncontact method, and water-soluble-assisted tapes. The most common approach is laser heating or using a thermally softened film to detach the film.

3.4.2 Transfer with an Elastomeric Stamp

In this approach, an elastomeric stamp is used to transfer the device from a diced wafer or a trace array from a carrier donor substrate to the receiver system substrate, The stamp facilitates the pickup and transfer. The adhesion between the devices and donor substrate need to be lower than the bonding strength to the elastomeric stamp. Similarly, if traces are to be transferred, the stamp-ink interface must exhibit stronger adhesion than the ink-substrate interface to ensure switchability in the retrieval process. However, during the removal process, the stamp-ink interface must be weaker to print the inks on the substrate. The transfer of devices from wafer to target substrate is shown in Figure 3.19.

3.4.3 Transfer with a Water-Soluble Tape

An example of a hybrid manufacturing technique to fabricate e-tattoos is by conventional photolithography followed by transfer printing. The transfer process is facilitated by a sacrificial release layer that is dissolved with water. Typical fabrication

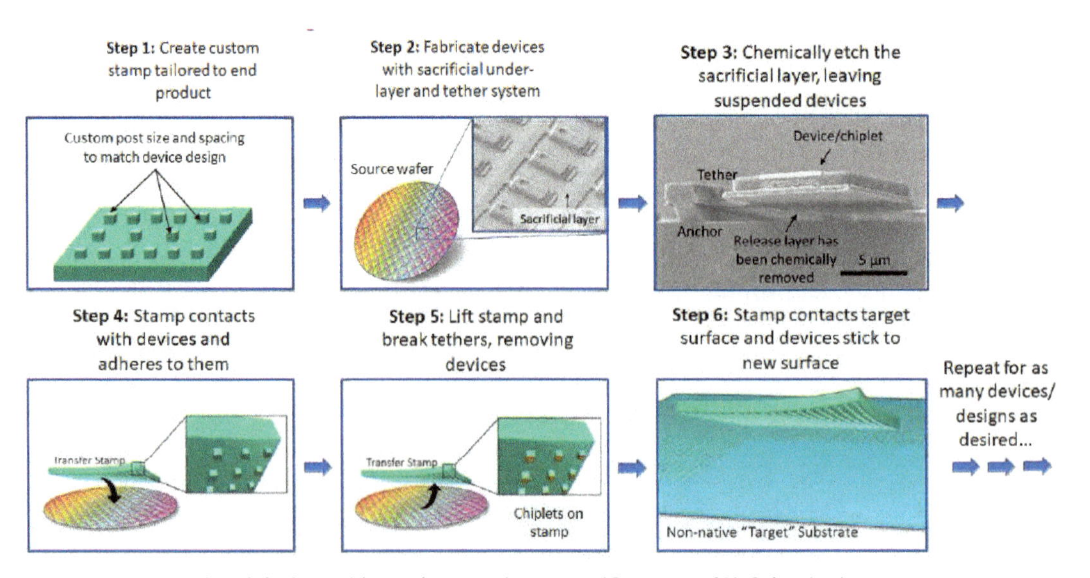

Figure 3.19 Transfer of devices with an elastomeric stamp. (Courtesy of X-Celeprint.)

process starts with cleaning the glass layer, adding a primer for adhesion, and spin-coating the release layer such as polymethyl methacrylate (PMMA), which is followed by electrode patterning. Through the process of etching and dissolving the PMMA, electrodes are transferred onto a sacrificial tape. The electrode layers are then released onto the skin and the sacrificial water tape is dissolved. The process flow is illustrated in Figure 3.20. A connector is hot pressed onto the contact pads, which allows on-skin placement and measurements.

3.4.4 Direct Flex Transfer onto Skin: Cut, Paste, Peel, and Release

Hybrid techniques of conventional additive manufacturing techniques and transfer printing are commonly used to fabricate stretchable conformal electrodes, now labeled as e-tattoos. These are highly bendable and stretchable wearable electronics that adhesively conform on the skin for batteryless health monitoring. The benefits exceed current noninvasive monitors in many ways:

Accurate real-time electrical activity;

- Versatility in size and shape of electrodes;
- High quality signal-to-noise ratio (SNR);
- Long-term connection without conductive gel;
- Pain-free, minimal adhesion, and easily removable from skin.

These substantial benefits support the rapid development of e-tattoos by offering better understanding of human physiology in a conformal and socially acceptable manner. Other hybrid additive manufacturing techniques involve rapid prototyping stretchable electrodes by inkjet printing. Casson and others [32] fabricated the first inkjet-printed e-tattoo ECG using silver nanoparticle electrodes. The versatility of inkjet printing electrodes enables personalized shapes and sizes for patient desirability and large surface area attachment on the skin. After printing the electrodes, a temporary adhesive tattoo-like film was added to the surface for attachment to the skin. Results show promising potential of low-cost e-tattoo ECG

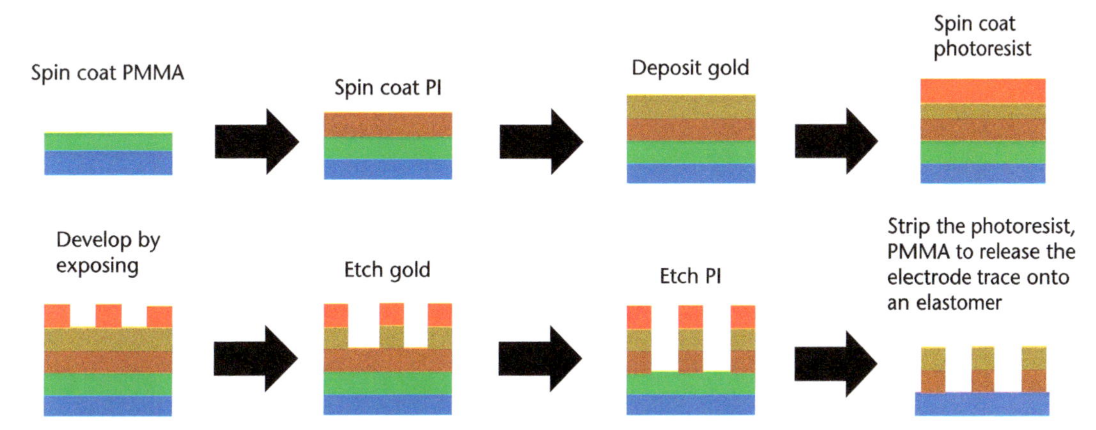

Figure 3.20 Circuit transfer from substrates with sacrificial PMMA layer.

with high fidelity of quality signals when compared to conventional Ag/AgCl ECG electrodes. In one such study, it was shown to function continuous monitoring for 5 days with acceptable SNR above 30 dBm. This is advantageous over noninvasive heart monitoring based on photoplethysmography (PPG) for users of smart watches, where the heart rate interval averages from PPG are affected by motion.

As well, flexible e-tattoos were demonstrated for seismocardiography (SCG) applications to monitor chest vibrations. Ha and others [33] fabricated an ultrathin e-tattoo SCG based on piezoelectric PVDF film with Cu-Ni electrodes. Based on the design of the vibration sensor and 3-D digital image correlation (3-D DIC) method, mapped chest deformations localized near the sternum show low elastic mismatch is necessary for high sensitivity. The PVDF was patterned into flexible serpentine patterns such that they are stretched in the longitudinal direction to have high sensitivity to vibrational amplitudes. A temporary weak tape was utilized as a support to prevent thermal stresses on the film during the serpentine patterning. This allows transferring the PVDF film onto medical tape adhesives for on-skin attachment. These top circuit transferable techniques modernized how e-tattoos are perceived for next-generation noninvasive stretchable monitoring.

3.4.5 Flex Substrate Embedding into E-Textiles

E-textiles are smart garments with embroidered conductive threads (e.g., Elektrisola® conductive thread) to allow complex circuit designs into wearable electronics. These have all the required attributes such as washability, flexibility, and breathability to offer comfortable and high-quality sensing for users. However, hybrid integration of thin embedded silicon chips, passive components, or other prepackaged devices into e-textiles requires innovative device assembly or embedding techniques. High-density and fine-pitch flexible interposers can address the challenge by allowing seamless integration from device to e-textile. In this approach, the devices are integrated into the flexible substrate first, which is then transferred into the fabric and printed connections are formed between the flex and conductive threads. Transferring the whole flexible circuit into an electronic conductive textiles or e-textiles has several fabrication advantages because there is no need to release the films from the flex carrier. This approach from FIU creates flexible interconnects between the fabric and flex, enabling deformable and remateable characteristics. ACAs are utilized as the conducting path between flex and fabric because of their adherence integrity between interfacial surfaces. 3-D hybrid integration of the fan-out chip-in-flex e-textile fabrication process is demonstrated in Figure 3.21. The process integration steps are photolithographic patterning of copper traces on LCP substrates, die assembly onto copper traces, flex interposer assembly on the textile, printing flex-to-textile silver interconnects, and encapsulating with hydrophobic fluoroelastomer. Mechanical reliability study of the redistribution layer demonstrated minimal resistance changes during bending in 1.5-mm dies. Further miniaturization in flex-to-textile technology by decreasing interconnect pitch and assembling thinner dies is presently under investigation and development at FIU Packaging Lab to enable higher flexibility under mechanical reliability requirements.

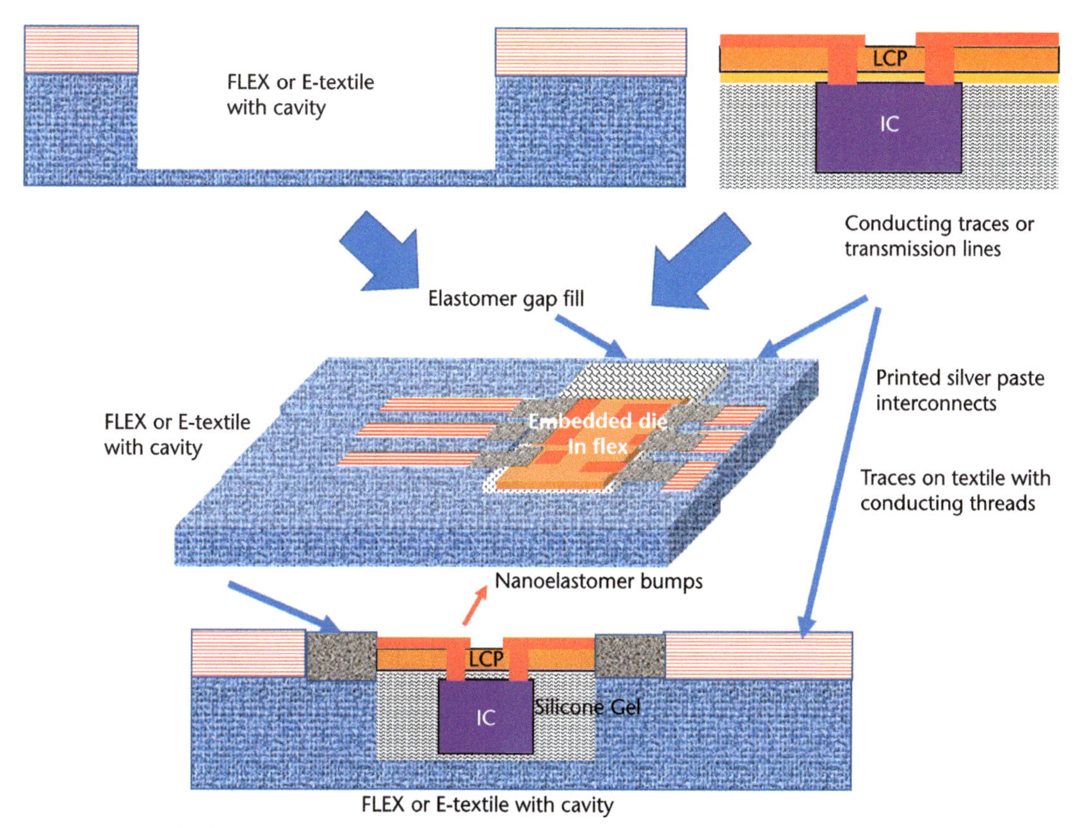

Figure 3.21 Hybrid integration of high-density and fine-pitch interposer flex into e-textile.

3.5 Biosignal Interfaces: Electrode and Photonic Interfaces

Accurate biosignal reading is crucial in fields such as medicine, nutrition, occupational health, and even sports. The goal of wearable biosignal interfaces is to reduce the need for invasive procedures and still obtain accurate biosignal readings. Biopotential signals (e.g., ECG, EEG, EMG) in wearable systems provide appropriate biofeedback or control commands to also provide appropriate therapeutic response. The accuracy of biosignal reading helps physicians to understand patients and provide helpful diagnosis and treatments. The interfaces are typically electrical, piezoelectric, or photonic in nature. Electrode technology is a key building block for wearable electronics as the electrodes need to interface with the skin, and therefore generate practical commercial products into the consumer electronics market [34]. For example, rapid development of EEG-based wearable healthcare devices and brain-computer interfaces, reliable and user-friendly EEG sensors for EEG recording, especially at forehead sites, are all dependent on electrode advances [35]. Current research focuses on investigating advanced electrode materials, skin-electrode interface, form factor design, and appropriate electrode body placement for a wearable sensing platform. This chapter focuses predominantly on electrodes, whose requirements are classified below as (a) low impedance with skin, (b) skin compatibility, and (c) long-term stability.

Low impedance with skin: Advanced electrodes with superior biosignal sensitivity and skin compatibility are the key requirements for on-skin electrodes. This is contingent on achieving low impedance. Design and fabrication of low-impedance, biocompatible electrodes have been a main challenge in the field of wearable devices. Low impedance matters because it helps to obtain solid signal-to-noise ratio and sensitivity. This impedance is determined by exposure area and electrical properties such as the ionic conductivity through the skin and at the electrode-skin interface. Skin-electrode impedance is thus influenced by various geometric factors that include the printed electrode area, skin-electrode interface material, and applied pressure. By lowering the impedance, the sensitivity of electrodes increases and so does the electric change transfer during stimulation. Conventional wet electrodes are often used for EEG measurement for their low impedance. However, wet electrodes require skin preparation and conduction gels to reduce skin-electrode contact impedance. Based on the recent findings, existing EEG sensors cannot meet the requirements because wet electrodes require tedious setup and conductive pastes or gels. Reusable, dry electrodes may provide new ways to obtain physiological measurements. These electrodes may also provide improved usability and comfort during physiological measurements with smart watches and smart fabrics. Dry electrodes may be incorporated into various wearable form factors. Nevertheless, most dry electrodes are limited in many aspects, starting from fabrication. These electrode types are often built on rigid substrates that do not allow them to be highly flexible and conform to the human skin. As a result, the current focus is on developing flexible (or sometimes even stretchable) substrates. These types of substrates could allow electrodes to be more useful in the wearable devices field. This is because these dry electrode types could provide more robust skin-electrode contact and less skin irritation. As a result, such development could have a vast set of biomedical applications. Despite having a better understanding of dry electrodes, recording bioelectrical signals such as EEG, ECG, and EMG during patients' daily activities continues to be a challenge because of their high impedance with skin and instabilities described in the subsequent sections.

Skin compatibility: Most common problems between living cells and electronic devices include corrosion, inflammation at electrode sites, reduced long-term stability, foreign body response, low-charge injection limits for certain applications, and mechanical mismatch between living tissue and electrodes. As a result, improvements to current electrodes are linked to aspects such as low-impedance over long time scales at the electrode-tissue interface, reduced tissue inflammatory responses, high-charge injection limits, high flexibility, mechanical strength, and matching with tissue. Therefore, skin-electrode compatibility is important to developing accurate readings in tools such as ECG and EEG. However, another significant aspect of skin-electrode compatibility is high breathability. High breathability of epidermal electronics is crucial because it provides long-term attachment and nonirritating contact to the skin.

The structure of the skin is made up of three important layers: hypodermis (subcutaneous layer), epidermis, and dermis. In general, the hypodermis is the deepest layer of the human skin that consists of fat. Epidermis is the outermost layer above the dermis and made up of mostly keratin-producing cells. Since the epidermis is the outermost layer, it protects the skin from possible harmful interactions with the exterior environment but would also obstruct critical collection of information. Therefore, it is important to discern the mechanical properties of the epidermis to validate proper functionality of sensors. The epidermis is stretchable and soft in nature that produces a damping effect in response to mechanical forces [36]. Also, the skin creates a nonlinear stress-strain curve because of its anisotropic composition and the Young's modulus (0.10–18 MPa) is variable on constraints such as location, age, and hydration [37, 38]. The human skin is considered as a spring-mass damper system because of its frequency response to mechanical loads. For example, the resistance of the skin increases as the mechanical load increases, making the spring and mass effect of skin elasticity even more significant. In addition, water impacts viscoelastic properties of the human skin, which can directly alter the frequency response due to a load [39].

Because of the skin's protective layer and highly variable mechanical properties, it is important to choose materials that match similar characteristics. The leap to innovative miniaturized multifunctional sensors is only possible because of skin-compatible polymer-based materials. It is important to note that polymer plays an essential role in determining the functionality of the biosensors in terms of integrating mechanical and electrical media in a complex multilayer package. Specifically, the polymer layer promotes mechanical flexibility and stretchability into any arbitrary configuration without causing degradation to performance. Thin-film materials used as skin attachment must adhere to similar properties of the skin in terms of stretchability and must withstand large strains to avoid any failures by delamination or breaking of the substrate. Most familiar types of polymer support layers used in recent advances are polyurethane and PDMS, since they exhibit compatible mechanical attributes that offer compliant bendability, stretchability, and maintain stable electric functionality. Basic research in electrode materials and their architectures continues to enhance current electrode technologies by exploiting materials such as Ag/AgCl and carbon electrodes, gold fractal structures on elastomers, and electrochemical electrodes. Current research is focused on solving these issues to create flexible and organic electrode materials, functionalized and nanostructured electrodes, hydrogels and carbon materials, and transistors as the recording electrodes.

Long-term stability: Electrode instabilities occur from several reasons, such as motion artifacts, sweating, skin irritation, and other factors. Existing electrodes cannot absorb sweat effectively and that sweat causes cross-interferences, and even short circuits, between adjacent electrodes, especially when individuals move or in a hot and humid environment. Previous studies showed nocturnal sweating is a common problem that causes low-frequency artifacts in measured EEG signals. Dry electrodes lead to relatively high electrode-skin impedances, significant impedance changes due to sweating, and increased risk of sweat artifacts. Recent electrochemical in vitro investigations revealed sweat artifact tolerance of EEG electrodes can be improved with appropriate electrode material design. This specific research investigated in vivo electrode-skin impedance and the quality of EEG signals and interference due to sweating in 11 healthy volunteers [32]. The researchers used commercial Ag and Ag/AgCl inks to test electrode sets with differently constructed ink layers. Besides the fact that the electrode set in this study was designed to be used without skin abrasion, electrode-skin impedances and EEG signals were recorded before and after exercise-induced sweating. Extensive variation was noticed in electrode-skin impedances between volunteers and electrode positions. As a result of their limitations, researchers have focused on developing flexible and comfortable dry electrode designs. Prominent examples are silver nanowire/elastomer, graphene/epoxy, and conductive textiles, all of which will be described in the subsequent sections.

3.5.1 Ag/AgCl Electrodes

The neuropotentials in the body are transduced as electronic signals through faradic or electrical double-layer interfaces between the electrode and skin. The electrical circuit models for electrodes are shown in Figure 3.22. With faradic reaction interfaces, electronic current is generated through electrochemical electron-exchange reactions between the gel and the biological tissues, thereby generating current through the gel path. Such nonpolarizable or faradic electrodes are advantageous as they do not filter the signals unlike capacitive interface electrodes. With capacitive interfaces, the electrical double layer between the skin and electrode converts the ionic flux to electronic current. In this case, no actual current flows but change in the ionic concentration at the electrode skin interface creates a capacitive (or displacement) current. The capacitive displacement current flows through the moisture or sweat at the dry electrode-skin interface. Adhesive Ag/AgCl electrodes form the workhorse nonpolarizable or faradic electrode array technology. This type of electrode is generally inexpensive to manufacture, has a stable potential, is nontoxic, and has a simple construction. The electrode works by converting ion current at the surface of human tissues to electron current. This is sent through the lead wire to the tool/instrument used for reading. A crucial aspect of the Ag/AgCl electrode is its electrolyte gel. This gel is applied between the electrode surface and tissue, and contains free Cl ions. These ions help carry charge through the electrolyte. The chloridization of silver facilitates the redox reactions by converting the chloride

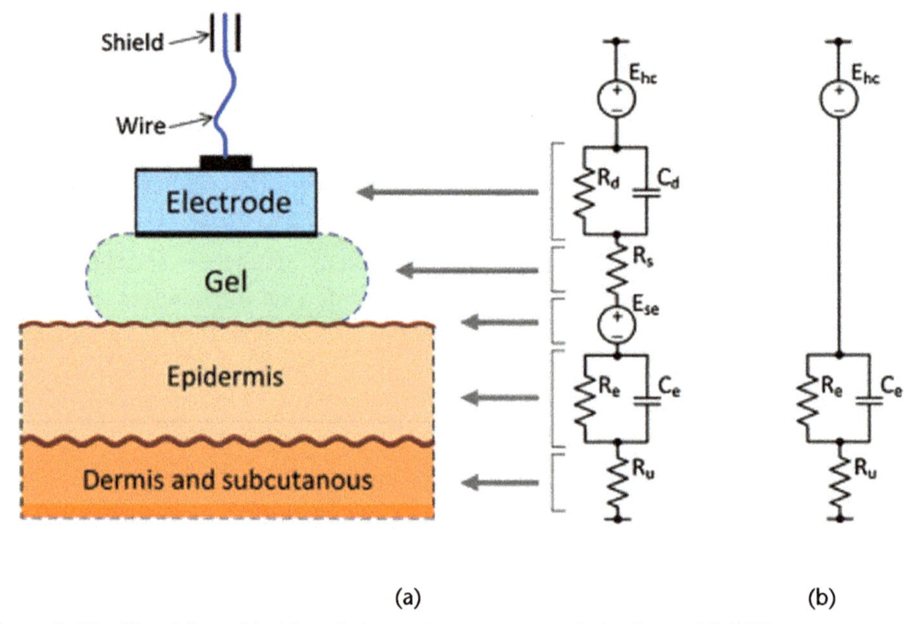

Figure 3.22 Silver/silver chloride gel electrode structures and circuit models [40].

anions to zero valence state and vice versa, as well as efficient generation of current. However, the wet electrodes reduce the electrode life and are associated with various skin reactions causing irritation. Ag/AgCl electrodes have demonstrated stable electrochemical potential throughout different measurement conditions. Some of the applications include medical tests, wearable sensors, wearable devices, EEG and EKG sensors, and sweat analysis. As a result, these electrodes so far are excellent for use in simulation, precision bioelectric recording and electrophysiology, electrochemistry, and as reference electrodes.

3.5.2 Dry Electrodes

Despite attempts to improve the effectiveness of wet Ag/AgCl electrodes, they continue to face limitations. Commercial Ag/AgCl electrode use is less preferred for continuous monitoring because they cause skin irritation from the conductive gels. Ag/AgCl electrodes require skin preparation or incorporation of an electrolytic gel and may also have a limited shelf life. Emerging real-world EEG applications require gel-free electrodes that incorporate systemically (concerning magnitude and stability) the electrode-skin impedance of bioelectrodes (wet, semidry, and dry.) Dry electrodes that do not require gel, adhesive, or skin preparation have been a research focus but not used widely in the medical field yet. Different types of dry electrodes are prevalent in the market. These are insulated metal dry electrode, flexible or soft material (made of conductive polymer and foam), and fabric electrodes [40]. Examples of dry or gel-free electrodes are silver nanowires/polydimethoxysiloxane, silver nanowire/polyurethane, or the carbon-based versions of these as described in the next section. Results show that dry electrode impedances are significantly higher and unstable when compared to impedances of wet electrodes and semidry electrodes. Also, dry electrode impedances are also lowered significantly under pressure or after skin abrasion [41]. Besides the promising future that dry electrodes could have in the wearable device realm compared to Ag/AgCl electrodes, they continue to have limitations, specifically in skin-electrode contact. For example, silver-nanowire (AgNW)-based dry electrode was fabricated for noninvasive and wearable ECG sensing. The signals from the AgNW electrode and Ag/AgCl electrode were collected simultaneously in two different conditions such as sitting and walking. According to the observations, quantitative comparisons demonstrated AgNW electrodes could collect just as acceptable ECG waveforms as the Ag/AgCl electrode during sitting and walking conditions. However, baseline drift and waveform distortions existed in the AgNW electrode and were likely due to the electrode motion. In conclusion, if the skin-electrode contact is improved, then dry electrodes can be promising substitutes for Ag/AgCl electrodes.

Experimental results show the flexible dry electrode array has reproductible electrode potential, relatively low electrode-skin impedance, and good stability. EEG signals can be effectively captured with a high quality comparable to that of wet electrodes. These results provide the feasibility of forehead EEG recording in real-world scenarios using the proposed flexible dry electrode array, with a rapid and simple operation and advantages of self-application, user-friendliness, and wearer comfort. The factors investigated include electrode type, skin locations, pressure, skin abrasion, and electrode contact area. An innovative approach in advancing dry electrodes is currently under development that incorporates a dry electrode that is fully textile. These dry electrodes could be used in long-term monitoring of ECG signals. The ECG signals obtained with dry and wet electrodes were comparatively studied as a function of body posture and movement and shows that skin-electrode impedance can be influenced by the printed electrode area, skin-electrode interfacial material, and applied pressure [42]. In conclusion, fully textile printed dry electrodes present an inexpensive health monitoring platform solution for mobile wearable electronics, providing a possible solution to limitations currently faced by Ag/AgCl electrodes.

3.5.3 Carbon- or Conducting Polymer-Based Electrodes

Carbon-based electrodes show exceptional structural, optical, thermal, electrical, and mechanical properties. As a result of these properties, they are being studied in various fields including electronics, energy, catalysis, sensing, and biomedical electronics. Carbon (including its composites) is a material of high performance. As a result, it is particularly useful in electrochemistry due to its extreme properties. Moreover, carbon is a good conductor. Therefore, it is used successfully in electrolysis or in applications where electron transfer is required. The electrons move freely within the carbon's structure. This electron movement causes the material to be highly conductive. Moreover, overall, carbon is inexpensive. It can also be stable at high temperatures and is a durable material. When compared to commercial electrodes, such as screen-printed or sputtered electrodes, graphite powder-based electrodes can also be fabricated easily with the form and desired characteristics. These allow modification to enhance selectivity and their easily renewable surface. As a result, the fabrication process would not need hazardous acids or bases for their cleaning, making these electrode types an interesting option for obtaining a large variety of electrodes to resolve different types of analytical problems.

In addition, experimentation on the performance of printed electrodes was investigated by testing MWCNT/PDMS composite conductivity and measuring electrode-skin impedance for electrode radius varying from 8 to 16 mm. Dry ECG electrodes with the largest area demonstrated better performance in terms of MWCNT/PDMS composite conductivity, ECG signal intensity, and correlation when compared to the commercial wet Ag/AgCl electrode. Also, capability of dry ECG electrodes for monitoring ECG signals in both a relaxed sitting position and while the subject is in motion was also investigated and results compared with Ag/AgCl ECG electrodes. However, while the subject is in motion, printed dry electrodes were less noisy and could better identify typical ECG characteristics in signals due to their better conformal contact at the electrode-skin interface. Such development demonstrates how reliable involving the conventional screen-printing process may be for the development of flexible dry ECG electrodes for biomedical applications. Furthermore, these studies show the usefulness of carbon electrodes in the biomedical realm.

In neural recording devices, enhancing single neuron-sensing with high-sensitivity performance requires low-impedance electrodes. Standard requirements for next-generation flexible electrodes must be biocompatible, provide enhanced surface area to achieve large capacitance, be mechanically complaint for patient safety by preventing tissue scarring, and maintain low impedance under scalability of dimension. At FIU Packaging Lab, graphene and PEDOT-PSS of nanostructured electrode arrays were developed for neural recording. Process integration involved electrode pads assembled on the backside of the neural recording hub on an LCP flexible substrate. Conductive ink traces were designed at 200 μm thickness of printed silver traces, followed by graphene–PEDOT-PSS coating with silicone binders. Initial characterization by immersing electrodes into phosphate buffer solution (PBS) show promising lower noise results when compared to planar silver electrodes, demonstrating better enhanced sensitivity of the heterogenous system.

3.5.4 Fractal Gold Electrodes

Fractal is derived from the Latin word "Fractus," which means fractured segments into complex geometries. Emerging fractal designs of electrodes to develop multiband antennas have evolved the trend from conventional microstrip antennas and Euclidean shapes due to their limitation in efficiency and miniaturization. Specifically, in neurostimulation, conventional Euclidean microelectrodes have shorter lifetime, higher impedance, low current density gradient, and are prone to mechanical failures. However, numerical simulations have shown that fractal structures can penetrate deeper to achieve better neurostimulation performance by achieving higher current density gradient due to the available surface area. Fractal designs have gained an immense popularity because of properties such as multiband characteristics and spatial resolution, high surface area in limited space with engineering control, and optimum efficiency and stretchability. Geometric designs are based on recursive infinite networks and categorized in three complex shapes such as lines (Hilbert, Koch, Peano), loops (Moore and Viscek), and mesh structures known as the Greek cross. Corresponding gold fractal shapes bonded to elastomeric substrates under elastic tensile strain is shown in Figure 3.23. These fractal designs enable hard-soft materials integration and vary in terms of mechanical strain percentages and stretchability.

Peano curves and Greek cross fractal layouts are recently used in skin electrodes for applications in ECG, EEG, and EMG due to their robustness and conformal connectivity. One of the key advancements in recent developments is integrating gold fractals on elastomeric substrates to reach peak stretching strain. For example, in electrophysical applications, metal-based sensors are demonstrated by covalently bonding gold-polyimide Greek cross fragments to an elastomer by chromium-silica

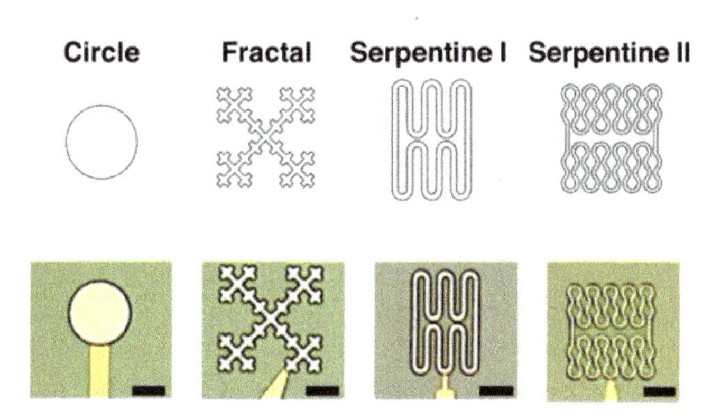

Figure 3.23 Different types of fractal electrode designs [43].

bonding layer through UV treatment. This fractal design integration is favorable for temperature sensing since it allows high-quality signals and precision [44, 45]. Another approach is fabricating thin (20 nm thickness) building blocks of gold chloride meshed fractals frameworks by wet-chemical templateless and adhering to PDMS substrates without any further treatment at room temperature. Such 2-D block synthesis enables stretchability of 110% strain without degrading electrical conductivity and maintaining 70% optical transmittance [46].

3.5.5 Electrochemical Electrodes

Electrochemical electrodes enable biomarker sensing through electron-transfer-based chemical reactions, typically in the presence of an enzyme or catalyst that is surface-immobilized in a ionic medium. Electrochemical sensing typically utilizes three electrodes, with the working electrode acting as the main site for the analyte reaction and current generation in the presence of an enzyme. The enzyme acts as a biochemical recognition element and is immobilized with an inorganic or organic gel. It catalyzes the redox reaction to initiate an electron transfer that can be read through amperometry. Reaction mediators are often utilized to further increase the sensitivity. Higher response current with enhanced electrocatalytic activity is the key desired characteristic. The most clinically relevant electrochemical electrodes are utilized for glucose detection while other biomarkers, such as pH, lactose, uricase, cortisol, testosterone, and ionic concentration of sweat, are gaining increasing attention. The advances in these electrodes are again classified as higher surface area, enzyme immobilization, and carbon-based electrodes. MoS_2 microflower modified by Au nanoparticles could improve the glucose sensor electrochemical performance. Such enhancement occurs as a result of better electronic conductivity and electrocatalytic activity of the metal. By detecting different glucose concentrations in 0.1 M PBS solution using Au-MoS_2, results show solid electrocatalytic activity with high sensitivity of 932 μA mM^{-1} [47]. Furthermore, its linear range was 0–15 mM for the glucose detection, showcasing Au-MoS_2 as a useful candidate in electrochemical sensing applications. Carbon powder-based electrodes also have electrochemical performance of quasi-noble metal electrodes and hence are also used for electrochemical biosensors. Commercial screen-printed carbon electrode arrays are developed by Zimmer Peacock. Figure 3.24 illustrates the state-of-the art product with performance in measuring pH levels.

3.6 Remateable Connectors

Remateability is the ability to disconnect the system and reconnect by the user without any tools/processes. It is the ability to assemble and disassemble electronics without sacrificing parts or discharging them. This is very critical for wearable electronics as the subsystems are spread into the fabric and on-skin electronic/photonic/ultrasonic interfaces. The subsystems need to be disassembled when the user removes the fabric or disposes the on-skin electronic units. This is often associated with washing cycles where the electronic unit in the fabric is subjected to severe mechanical loads in a high-temperature water environment. The subsystems, thus, need to be electrically and mechanically connected and disconnected to achieve

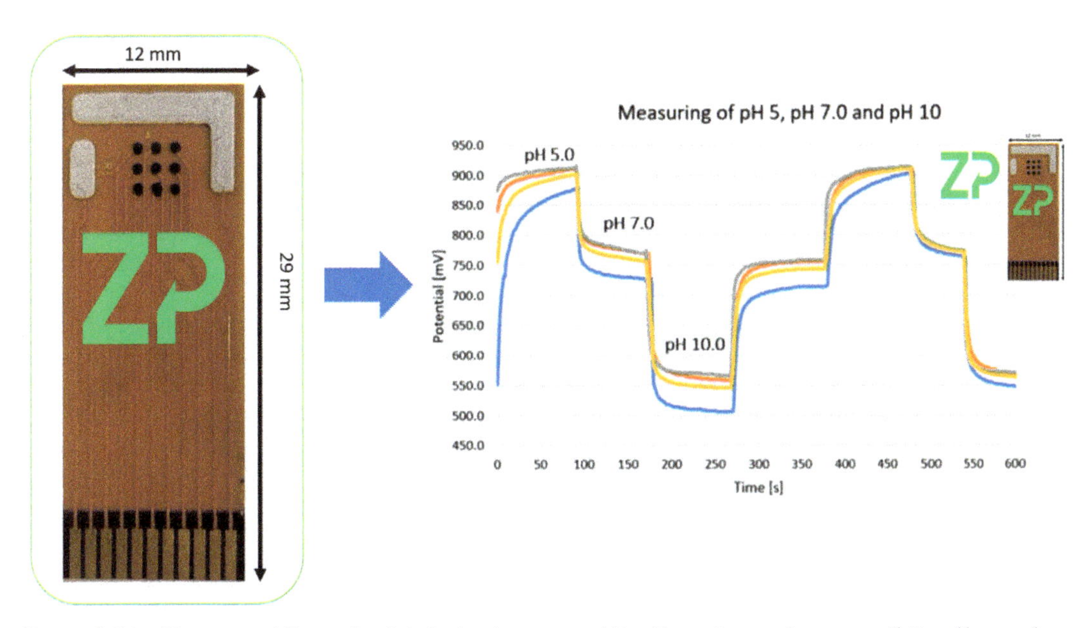

Figure 3.24 Zimmer and Peacock mini electrode array and its pH sensing performance. (https://www.zimmerpeacocktech.com/products/electrochemical-sensors/ph-sensor/.)

reversible assembly. Introducing remateable connectors between the subsystems will aid the reliability and lifetime in multiple scenarios:

Remateability between the electronic processing unit and the sensor interface: The electronic hub that does the signal processing, storage, and communication is detached from the on-skin sensors. The on-skin sensors are typically disposed.

- *Remateability between fabric and on-skin flex:* A wearable electronic system may compromise a reusable fabric component and a disposable component on the skin. The reusable component, for example, incorporates the wireless power harvesting and broadband communication system while the disposable component incorporates the flex circuit that is disposable.
- *Remateability within the fabric:* The different subsystems within the fabric, such as the high-density communication unit that is more susceptible to failures during washing, is disassembled during washing. The fabric with robust conductive threads in fabric is subjected to washing after disassembly.

Several options have been explored for remateability. These are classified into (1) a pin-socket snap-in mechanism where the disposable component integrates a mechanical snap-in connect mechanism to detach from the reusable component, and (2) flat snap-on connectors with hold-and-release categories where magnetic links or microscrews can be used as remateable connectors to clamp between mechanical interconnection and electrical contact. The third category that focuses on detachable adhesives such as ACAs when assembling flex-to-flex are traditionally targeted to be reworkable but not remateable. With advances in polymer chemistry, these adhesives can be optimized to achieve remateable functionality. These approaches are discussed in this section. A flowchart that classified these mechanisms is shown in Figure 3.25.

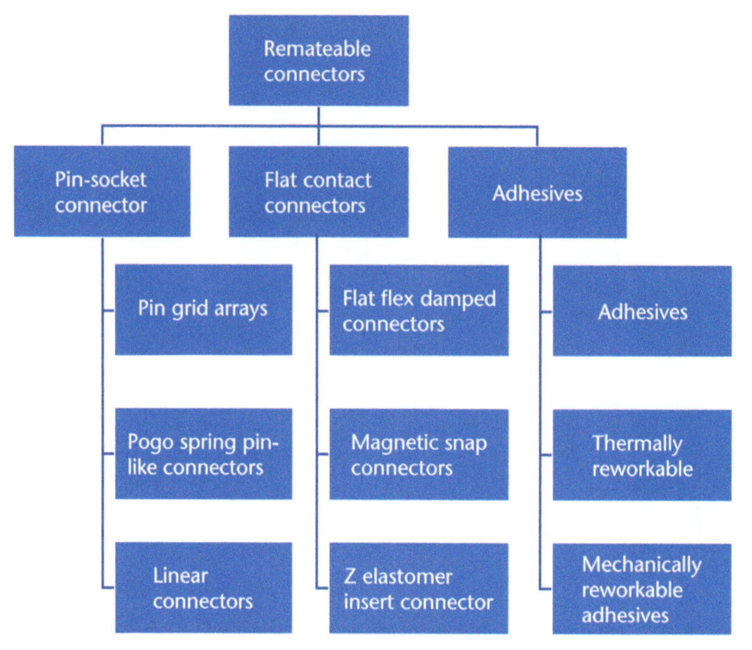

Figure 3.25 Classification of remateable connector technologies and prominent examples.

3.6.1 Pin-Socket Connectors

Mainstream remateable connectors in the wearable technology space use traditional pin and socket designs. Prominent commercial examples are shown in Figure 3.26. These connectors achieve desired characteristics such as low insertion forces and stable contact with gold-coated contacts. These can be in the area-array format such as pin-grid array (PGA) or linear array format such as Bal Seal® connectors that are more common in implantable devices. PGAs are widely adapted for computing applications to provide remateable assembly between the process pin-grid package and the socket. State-of-the-art remateable connectors that are based on bulky PGAs and sockets are not easily scalable and flex compatible [3]. Utilizing a pin-grid array setup solves the engineering constraint by offering multiple pins and sockets within a localized area. One example of a pin-grid array type system is the RK01 series of connectors by Japan Aviation Electronics™, where the design is similar to snap buttons. However, the connectors are featured as an array of pins on a plug that are pushed into receptacles with minimal force to connect and disconnect [48]. Snap buttons are frequently used for clothing, making it an appropriate

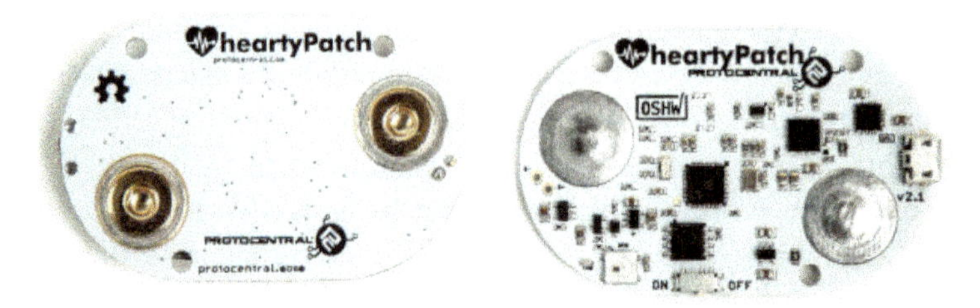

Figure 3.26 Pin-socket connectors in wearable medical devices [48].

design for wearable devices. Such connectors are limited in their ability to increase the number of contact points due to miniaturization constraints.

Mag-Net is a rectangular magnetically coupled connector developed by TT Electronics™ that features a plug with an array of pogo spring pinlike contacts and receptacle with corresponding flat contact points [49]. The plug contacts are based on a spring-loaded head that has a geometry of a pin that peeks out of a body. When mating the connectors, the spring is compressed down as the head of the connector is pressed against the contacts of the receptacle. The key principle behind the mating of such receptacle and pin-arrays are shown in Figure 3.27. The plug and receptacle have magnets within the structures meant to align them automatically, the contact points are arranged to align in any orientation. The receptacle has blind holes, and the plug has corresponding alignment pins. This along with a compressible gasket serves to give the connector an airtight seal. An extension of the area-array connectors is the widely used linear spring-loaded pin type such as the Bal Seal® connectors. Such linear pin and socket connectors involve a pin with ring contacts along the length of its body with spring-loaded contacts in the socket [50]. Linear remateable connectors are only limited to 4–8 interconnects through spring-contact such as in Bal Seal® connectors, while area-array fuzz button connectors, which are also based on compressible spring inserts, are not easily scalable.

3.6.2 Flat Connectors

Flat connectors (FCs) have mating pads that can be pressed against each other. These types of connectors support flexible cables and are applicable for wearable medical devices as they conform to arbitrary geometries of the human anatomy. With small geometries, they address some of the miniaturization needs that wearable devices have. Round connectors require more space and clearance to make a connection, but the flat designs of these flexible connectors reduce the amount of clearance necessary. The smaller pitch distances between contact points further contributes to their miniaturization. Current products have pitches as low as ~0.25

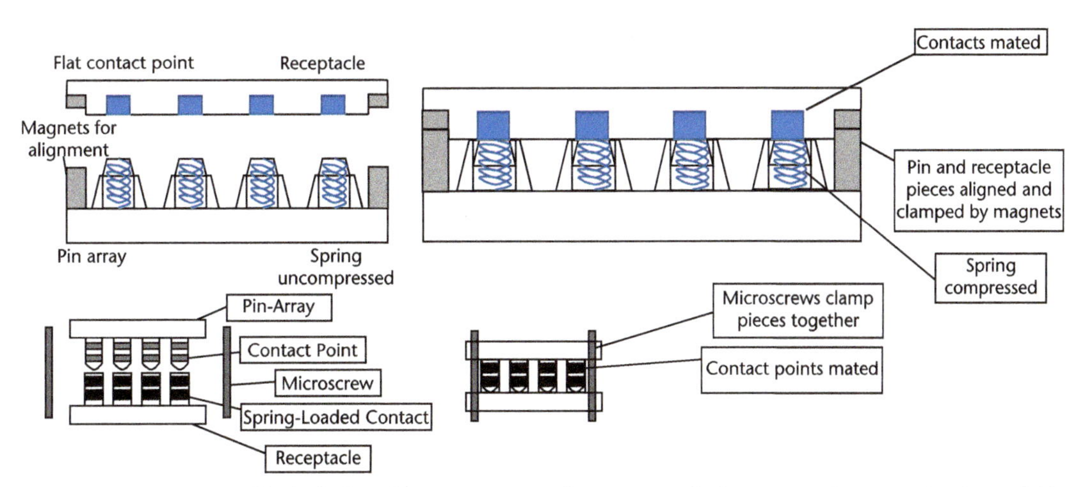

Figure 3.27 Top row: Vertical pin-grid array cross section: unmated pin-array and receptacle pieces (left), connector mated (right). Bottom row: pogo pin connectors: unmated pin and receptacle pieces (left) and mated connector where the spring is compressed for pogo pins (right).

mm and are often used in high-density packaging for electronic systems with cable-to board and cable-to-cable assembly. The engineering design is similar to a traditional pin and receptacle, but their flat geometry allows them to achieve their low profile. Construction of FCs generally consist of a thin plastic film with protruding contact pins or receptacle sockets housed at either end [48]. Flexible printed circuits have contact traces on them that mate with the receptacles of other rigid or flexible connectors. They are constructed by printing conductive traces on a flexible polymer substrate and sealing it with a protective polymer coating [3]. They are exceptionally thin and flexible with very low profiles enabling seamless assembly and integration. The pitch can be as small as ~0.25 mm and they can be inserted at various angles, thus requiring less clearance space in front of the receptacle to make the connection. TE Connectivity offers different types of housing options for flexible connectors including zero insertion force (ZIF) types [51]. Locking mechanisms known as actuators where the inserting section is placed within the receptacle and the actuator is latched. The minimum insertion forces allow for a large number of mating cycles as friction-related erosion that occurs between contact points during regular insertion processes is reduced.

3.6.2.1 Magnetic Snap-On

Magnetic connectors are commonly used in broad consumer electronics such as laptops or phone chargers and are now making their way into medical devices. They offer a plug-and-play type connector option for wearable devices allowing for easy connection and disconnection. They also allow for greater range and distance between electrodes, being especially suited for stretchable circuits and creating flexible wearable devices through stretchable modules using an elastomer matrix. The electronic components would be encapsulated in the elastomer matrix giving it the ability to stretch. The magnets are also encapsulated in the matrix and create the connection between different components. These design ideas were implemented to create a prototype for a wristband where the magnet-based connectors created a signal pathway from a pulse sensor to a power supply and Bluetooth module in a wearable health monitoring device. The sections of the wristband housing magnets and electrical components are rigid. These are illustrated in Figure 3.28.

Magnetic connectors require little to manual alignment, naturally falling into correct orientation with low to potentially zero mating forces. Magnetic connector designs consistently prevent the circuits from overstretching by releasing the connection before the elastomer matrix and electrodes experience flex strains large enough to cause mechanical damage. Overstretching causes plastic deformation of the electrodes and lead to the formation of large cracks. As a result, the resistance of the electrodes increases and leads to their failure. The elastomer and magnet have to be chosen taking the stress and strain characteristics of the elastomer and the magnetic force of the magnetic material into consideration. These properties should be calibrated so that the magnetic connector disconnects before the electrodes are overstretched. Recent findings show the prestrain for the substrate-electrode combination was set at greater than 80% strain [52]. The magnetic connection should disconnect below this level of prestrain and the magnetic force is influenced by the distance between the connecting magnets. The intermagnetic distance is thus selected such that the magnetic force is greater than that of elastic

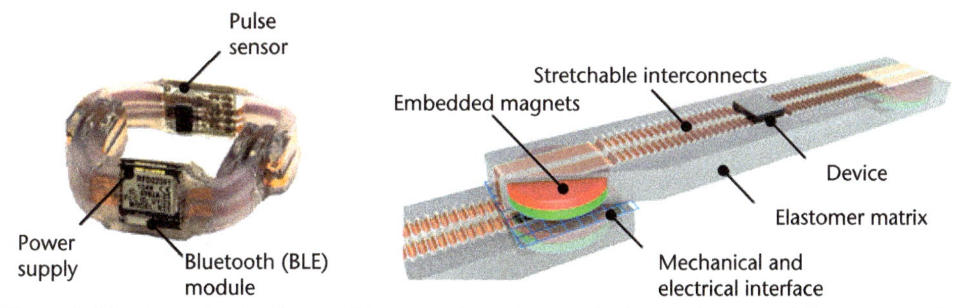

Figure 3.28 Example of soft wearable sensor for pulse monitoring and two modules connected by magnetic connectors adapted from [52].

force up to the prestrain, ensuring that the connection is maintained up to the prestrain beyond which it is severed. Reliability tests were performed with several different intermagnetic distance arrangements to test the snap-off effects of the connector at different levels of strain. Suggestions show that connector design performed a 10,000 connection and disconnection cycles without failure of operation. The design is illustrated in Figure 3.28. This resiliency of this operation suggests that this is a viable design as a remateable connector for wearable medical devices.

3.6.2.2 Conducting Elastomer Inserts

Area-array connections with small pitch distances allow for many conducting points of contact where signals and power can be transferred in a relatively small area, which reduces the overall bulk of the design and makes it easier to use in wearable devices. A version of area-array type connectors are pin-grid arrays where multiple pins align with corresponding sockets to make connections, as described in the previous section. This type of connection design avoids the problems of increasing length for linear inline connectors such as Bal Seal® connectors to include more contact points. An established method of designing area-array vertical interconnects is using a printed circuit board (PCB) substrate to act as the medium. Elastomer inserts are a variant of this approach and makes use of elastomer interposers for an array of pin-to-pin contacts, as shown in Figure 3.29 [53]. This eliminates the necessity for a PCB substrate leading to less material usage and lower manufacturing costs. The design suggests constructing an array of conductive elastomer interconnections through a nonconductive medium. This generic design has several variations where conductive elastomer interconnections are formed using the same material but with conductive particles or conductive fibers. Each remateable interface has a mating pad and the conductive elastomer acts as the interposer between the two pads. The connection can also be made between a PCB and an IC ball grid array. The nonconductive layer has compression limiters to limit compression to about 10% to 40% of the original height. Using an elastomer to interconnect two connection points avoids the need for a PCB substrate with feedthroughs, which saves on cost.

3.6.2.3 Z-Elastomer Interconnects

Considering the future needs for scalability and remateability of miniaturized electronics, a new class of connectors was developed by FIU Packaging Lab. Area-array

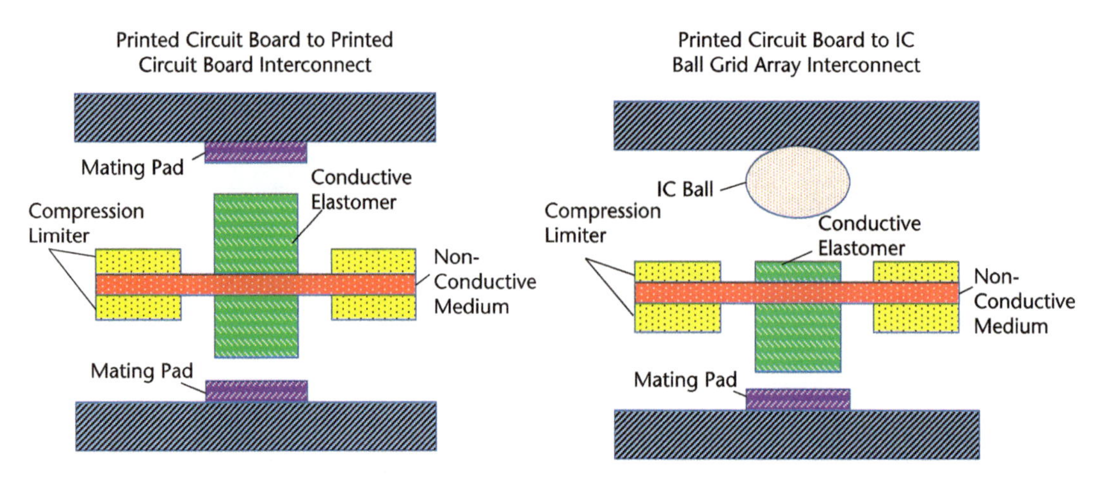

Figure 3.29 Conductive elastomer inserts with mating pads for PCB to PCB and PCB to IC ball interconnections adapted from [53].

interconnects were designed to allow remateability of the interposers on an LCP substrate [54]. Compressible z-interconnects were utilized on an elastomer insert that allow interconnections to feedthroughs with four clamped microscrews. Titanium films were utilized for rigidity of the structure to provide compressive forces on the z-interconnects to form contact on the traces. Figure 3.30 presents a cross-sectional schematic of the deformable z-elastomer interconnect design. The diameter of the interconnects wasdesigned at 200 μm with 800-μm pitch. Different silver inks were evaluated under mechanical and electrical testing to ensure high reliability of the proposed design. Compressible silver ink interconnects were characterized by bending over 8-mm radius and corresponding resistance was measured. Negligible resistance changes were realized indicating a novel potential for flexible, deformable, and compressible interconnects.

3.6.3 Reworkable Adhesives

ACAs provide an avenue for reworkable connectivity suitable in flex-to-flex interconnections often seen in wearable devices. Since these require some postprocessing such as chemical cleaning, they are not considered remateable as defined this chapter. ACAs are matrices of nonconductive adhesive materials with low concentrations

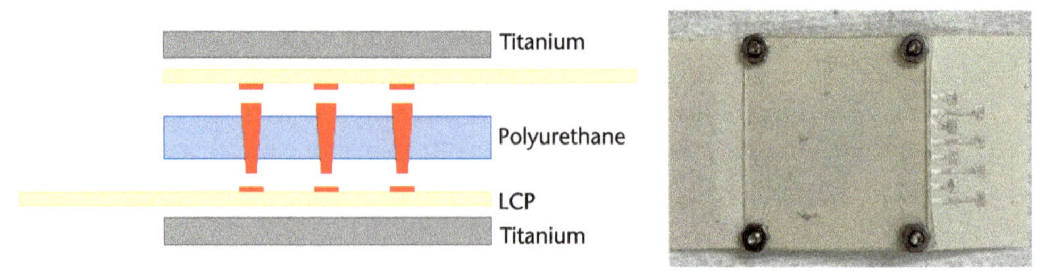

Figure 3.30 Cross section of z-elastomer interconnects with titanium films and polyurethane interposer adapted from [43].

of conductive particles suspended within. They come in the form of films (ACFs) and pastes (ACPs). Adhesives can be used instead of soldering for chip-to-board or chip-to-chip interconnections, working as a reworkable alternative to permanent bonding approaches. ACAs are also more forgiving in terms of the necessary working conditions than soldering or wire-bonding type approaches, requiring lower temperatures to work, less complicated overall processing, and finer pitch. Depending on the application ACAs may also allow for more favorable overall geometry with thinner bond lines, making the design less bulky, a highly desirable trait in wearable and implantable medical devices.

Not all ACAs possess the reworkability desired for chip removal and reassembly. Commercially established ACAs tend not to be easily reworkable due to the presence of a thermosetting adhesive matrix. A potential ACA blend includes thermosetting epoxy and a thermoplastic polymer, an example being a mixture of low-viscosity bisphenol F resin and a thermoplastic polymer (polysulfone (PSU)) with low molecular weight. This combination creates a reworkable ACP prepared for the experiments by Nguyen and others [55]. The experimental ACP was tested on the patterned flex substrates and a commercial ACA was tested on the plain PI films for comparison. The concentration of PSU in the ACP adhesive affects the peel strength. Epoxy-based adhesives have a much higher peel strength than epoxy with PSU. The greater the concentration of PSU, the lower the peel strength [41]. Adhesion primers cause slight increases in peel strength. The epoxy-PSU based adhesive is far more reworkable than the commercial ACF chosen in the experiment. The epoxy-PSU adhesive is far easier to remove using solvents in comparison to the commercial ACF, which proved far more difficult. On the other hand, the commercial ACF outperforms the epoxy-PSU ACP in terms of peel strength. This suggests there is a trade-off between adhesion strength and reworkability. This trade-off can be navigated based on the application.

3.7 Conclusion

The widescale adoption of flexible and wearable electronics will need innovative solutions towards high-density integration and remateability. These emerging solutions are extensively adapted from advanced packaging technologies. The challenges are mostly associated with heterogeneous component integration for RF-power data processing functions. Power telemetry is addressed through innovative inductive link designs and multiferroic telemetry interfaces. Chip embedding with fan-out interconnects will eventually replace the solder assembly technologies for higher routing densities in thinnest form factors. To provide adequate modularity in the system, remateable connectors are developed with fine-pitch area-array interconnects instead of the existing pin-socket or linear connectors. Printed flexible and wearable dry electrodes for monitoring ECG signals, without skin preparation and use of wet gel, have been recently developed. Key advances in package integration technologies for component embedding, advanced component designs, and materials for electrodes will enable next-generation wearable electronics.

References

[1] Hauke, A., P. Simmers, Y. Ojha, et al., "Complete Validation of a Continuous and Blood-Correlated Sweat Biosensing Device with Integrated Sweat Stimulation," *Lab on a Chip,* Vol. 18, 2018, pp. 3750–3759.

[2] Piro, B., G. Mattana, and V. Noël, "Recent Advances in Skin Chemical Sensors," *Sensors,* Vol. 19, 2019, p. 4376.

[3] Kim, J., G. A. Salvatore, H. Araki, et al., "Battery-Free, Stretchable Optoelectronic Systems for Wireless Optical Characterization Of The Skin," *Science Advances,* Vol. 2, 2016, p. e1600418.

[4] Palavesam, N., S. Marin, D. Hemmetzberger, C. Landesberger, K. Bock, and C. Kutter, "Roll-to-Roll Processing of Film Substrates for Hybrid Integrated Flexible Electronics," *Flexible and Printed Electronics,* Vol. 3, 2018, p. 014002

[5] Kraft, T. M., P. R. Berger, and D. Lupo, "Printed and Organic Diodes: Devices, Circuits and Applications," *Flexible and Printed Electronics,* Vol. 2, 2017, p. 033001.

[6] Tummala, R. R., *Fundamentals of Device and Systems Packaging: Technologies and Applications*, New York: McGraw-Hill Education, 2019.

[7] Yu, Z., L. Tetard, L. Zhai, and J. Thomas, "Supercapacitor Electrode Materials: Nanostructures from 0 to 3 Dimensions," *Energy & Environmental Science,* Vol. 8, 2015, pp. 702–730.

[8] Song, B., K.-S. Moon, and C.-P. Wong, "Recent Developments in Design and Fabrication of Graphene-Based Interdigital Micro-Supercapacitors for Miniaturized Energy Storage Devices," *IEEE Transactions on Components, Packaging and Manufacturing Technology,* Vol. 6, 2016, pp. 1752–1765.

[9] Gogotsi, Y., and R. M. Penner, "Energy Storage in Nanomaterials–Capacitive, Pseudocapacitive, or Battery-Like?," *ACS Nano,* Vol. 12, No. 3, 2018, pp. 2081–2083, https://doi.org/10.1021/acsnano.8b01914.

[10] Chen, Q., X. Li, X. Zang, et al., "Effect of Different Gel Electrolytes on Graphene-Based Solid-State Supercapacitors," *RSC Advances,* Vol. 4, 2014, pp. 36253–36256.

[11] Khan, Y., M. Garg, Q. Gui, et al., "Flexible Hybrid Electronics: Direct Interfacing of Soft and Hard Electronics for Wearable Health Monitoring," *Advanced Functional Materials,* Vol. 26, 2016, pp. 8764–8775.

[12] Yoo, J. J., K. Balakrishnan, J. Huang, et al., "Ultrathin Planar Graphene Supercapacitors," *Nano Letters,* Vol. 11, 2011, pp. 1423–1427.

[13] DeYoung, A. D., S.- W. Park, N. R. Dhumal, Y. Shim, Y. Jung, and H. J. Kim, "Graphene Oxide Supercapacitors: A Computer Simulation Study," *The Journal of Physical Chemistry C,* Vol. 11, 20148, pp. 18472–18480.

[14] Shahane, N., P. M. Raj, C. Nair, V. Smet, C. Buch, and R. Tummala, "Nanopackaging for Component Assembly and Embedded Power in Flexible Electronics: Heterogeneous Component Integration for Flexible Systems," *IEEE Nanotechnology Magazine,* Vol. 12, 2018, pp. 6–18.

[15] Wu, Z. S., K. Parvez, X. Feng, and K. Müllen, "Graphene-Based In-Plane Micro-Supercapacitors with High Power and Energy Densities," *Nature Communications,* Vol. 4, 2013, pp. 1–8.

[16] Grover, J., and C. Irvine, "Printed Electronics for Medical Devices."

[17] Bhardwaj, S., R. Pulugurtha, and J. L. Volakis, "High Density Electronic Integration for Wearable Sensing," in *Antenna and Sensor Technologies in Modern Medical Applications,* V. Rahmat-Samii, and E. Toksakal (eds.), Hoboken, NJ: John Wiley & Sons, 2021, pp. 435–467.

[18] Lee, S. P., G. Ha, D. E. Wright, et al., "Highly Flexible, Wearable, and Disposable Cardiac Biosensors for Remote and Ambulatory Monitoring," *NPJ Digital Medicine,* Vol. 1, 2018, pp. 1–8.

[19] "Two Zio Monitors. All The Data You Need," https://www.irhythmtech.com/providers/zio-service/zio-monitors.

[20] Thompson, D. C., O. Tantot, H. Jallageas, G. E. Ponchak, M. M. Tentzeris, and J. Papapolymerou, "Characterization of Liquid Crystal Polymer (LCP) Material and Transmission Lines on LCP Substrates from 30 to 110 GHz," *IEEE Transactions on Microwave Theory and Techniques,* Vol. 52, 2004, pp. 1343–1352.

[21] Li, W., M. Vaseem, S. Yang, and A. Shamim, "Flexible and Reconfigurable Radio Frequency Electronics Realized by High-Throughput Screen Printing of Vanadium Dioxide Switches," *Microsystems & Nanoengineering,* Vol. 6, 2020, pp. 1–12.

[22] Li, W., S. Yang, and A. Shamim, "Screen Printing of Silver Nanowires: Balancing Conductivity with Transparency While Maintaining Flexibility and Stretchability,“ *NPJ Flexible Electronics,* Vol. 3, 2019, pp. 1–8.

[23] Khaleel, H. (ed.), *Innovation in Wearable and Flexible Antennas*, Southampton, United Kingdom: Wit Press, 2014.

[24] Lee, H., C. H. J. Lim, M. J. Low, N. Tham, V. M. Murukeshan, and Y.-J. Kim, "Lasers in Additive Manufacturing: A Review," *International Journal of Precision Engineering and Manufacturing-Green Technology,* Vol. 4, 2017, pp. 307–322.

[25] Arutinov, G., R. Hendriks, and J. Van Den Brand, "Photonic Flash Soldering on Flex Foils for Flexible Electronic Systems," *2016 IEEE 66th Electronic Components and Technology Conference (ECTC)*, 2016, pp. 95–100.

[26] Lee, S.- H., and K.-W. Paik, "A Study on the Nanofiber-Sheet Anisotropic Conductive Films (NS-ACFs) for Ultra-Fine-Pitch Interconnection Applications," *Journal of Electronic Materials,* Vol. 46, 2017, pp. 167–174.

[27] Han, Y., B. Zhang, P. Zhu, et al., "Preparation of Highly Conductive Adhesives by Insitu Incorporation of Silver Nanoparticles," *2016 17th International Conference on Electronic Packaging Technology (ICEPT)*, 2016, pp. 434–438.

[28] Eid, A., J. Hester, Y. Fang, et al., "Nanotechnology-Empowered Flexible Printed Wireless Electronics: A Review of Various Applications of Printed Materials," *IEEE Nanotechnology Magazine,* Vol. 13, 2018, pp. 18–29.

[29] Vital, D., M. M. Monshi, S. Bhardwaj, P. M. Raj, and J. L. Volakis, "Flexible Ink-Based Interconnects for Textile-Integrated RF Components," *2020 IEEE International Symposium on Antennas and Propagation and North American Radio Science Meeting*, 2020, pp. 151–152.

[30] Sayeed, S. Y. B., D. Wilding, J. S. Camara, D. Vital, S. Bhardwaj, and P. Raj, "Deformable Interconnects with Embedded Devices in Flexible Fan-Out Packages," *International Symposium on Microelectronics*, 2019, pp. 000163–000168.

[31] Jeong, E. G., J. H. Kwon, K. S. Kang, S. Y. Jeong, and K. C. Choi," A Review of Highly Reliable Flexible Encapsulation Technologies Towards Rollable and Foldable OLEDs," *Journal of Information Display,* Vol. 21, 2020, pp. 19–32.

[32] Casson, A. J., R. Saunders, and J. C. Batchelor, "Five Day Attachment ECG Electrodes for Longitudinal Bio-Sensing Using Conformal Tattoo Substrates," *IEEE Sensors Journal,* Vol. 17, 2017, pp. 2205–2214.

[33] Ha, T., J. Tran, S. Liu, et al., "A Chest-Laminated Ultrathin and Stretchable E-Tattoo for the Measurement of Electrocardiogram, Seismocardiogram, and Cardiac Time Intervals," *Advanced Science,* Vol. 6, 2019, p. 1900290.

[34] Acar, G., O. Ozturk, A. J. Golparvar, T. A. Elboshra, K. Böhringer, and M. K. Yapici, "Wearable and Flexible Textile Electrodes for Biopotential Signal Monitoring: A Review," *Electronics,* Vol. 8, 2019, p. 479.

[35] Li, G., J. Wu, Y. Xia, et al., "Towards Emerging EEG Applications: A Novel Printable Flexible Ag/AgCl Dry Electrode Array for Robust Recording of EEG Signals at Forehead Sites," *Journal of Neural Engineering,* Vol. 17, 2020, p. 026001.

[36] Ahmad Tarar, A., U. Mohammad, and S. K Srivastava, "Wearable Skin Sensors and Their Challenges: A Review of Transdermal, Optical, and Mechanical Sensors," *Biosensors,* Vol. 10, 2020, p. 56.

[37] Annaidh, A. N., K. Bruyère, M. Destrade, M. D. Gilchrist, and M. Otténio, "Characterization of the Anisotropic Mechanical Properties of Excised Human Skin," *Journal of the Mechanical Behavior of Biomedical Materials,* Vol. 5, 2012, pp. 139–148.

[38] Silver, F. H., J. W. Freeman, and D. DeVore, "Viscoelastic Properties of Human Skin and Processed Dermis," *Skin Research and Technology,* Vol. 7, 2001, pp. 18–23.

[39] Potts, R. O., E. M. Buras Jr, and D. A. Chrisman Jr, "Changes with Age in the Moisture Content of Human Skin," *Journal of Investigative Dermatology,* Vol. 82, 1984, pp. 97–100.

[40] Meziane, N., J. Webster, M. Attari, and A. Nimunkar, "Dry Electrodes for Electrocardiography," *Physiological Measurement,* Vol. 34, 2013, p. R47.

[41] Qin, Q., J. Li, S. Yao, C. Liu, H. Huang, and Y. Zhu, "Electrocardiogram of a Silver Nanowire Based Dry Electrode: Quantitative Comparison with the Standard Ag/Agcl Gel Electrode," *IEEE Access,* Vol. 7, 2019, pp. 20789–20800.

[42] Yokus, M. A., and J. S. Jur, "Fabric-Based Wearable Dry Electrodes for Body Surface Biopotential Recording," *IEEE Transactions on Biomedical Engineering,* Vol. 63, 2015, pp. 423–430.

[43] Park, H., P. Takmakov, and H. Lee, "Electrochemical Evaluations of Fractal Microelectrodes for Energy Efficient Neurostimulation," *Scientific Reports,* Vol. 8, 2018, pp. 1–11.

[44] Sivia, J. S., A. P. S. Pharwaha, and T. S. Kamal, "Analysis and Design of Circular Fractal Antenna Using Artificial Neural Networks," *Progress in Electromagnetics Research,* Vol. 56, 2013, pp. 251–267.

[45] Fan, J. A., W.-H. Yeo, Y. Su, et al., "Fractal Design Concepts for Stretchable Electronics," *Nature Communications,* Vol. 5, 2014, pp. 1–8.

[46] Ho, M. D., Y. Liu, D. Dong, Y. Zhao, and W. Cheng, "Fractal Gold Nanoframework for Highly Stretchable Transparent Strain-Insensitive Conductors," *Nano Letters,* Vol. 18, 2018, pp. 3593–3599.

[47] Zhai, Y., J. Li, X. Chu, et al., "Preparation of Au-MoS 2 Electrochemical Electrode and Investigation on Glucose Detection Characteristics," in *2016 IEEE International Conference on Manipulation, Manufacturing and Measurement on the Nanoscale (3M-NANO),* 2016, pp. 287–290.

[48] Smart Textile Connector "RK01" Series Has Been Launched, https://www.jae.com/en/topics/detail/id=93333.

[49] Auto-aligning and Self-Coupling Textile-Mounting Garmentconnector, https://www.technicaltextile.net/news/tt-electronics-unveils-mag-net-soldier-textile-connector-191928.html.

[50] SYGNUS® Implantable Contact System, https://www.balseal.com/contact/sygnus/.

[51] 0.25mm FPC Connectors, https://www.te.com/usa-en/product-CAT-F839-C76264025.html.

[52] Kettlgruber, G., D. Danninger, R. Moser, et al., "Stretch-Safe: Magnetic Connectors for Modular Stretchable Electronics," *Advanced Intelligent Systems,* Vol. 2, 2020, p. 2000065.

[53] Martin, C. W., C. M. Beers, and J. V. Russell, Method and Structure for Conductive Elastomeric Pin Arrays Using Conductive Elastomeric Interconnects and/or Metal Caps through a Hole or an Opening in a Non-Conductive Medium, U.S. Patent No. 10,886,653, issued January 5, 2021.

[54] Camara, J. F. S., S. Soroushiani, D. Wilding, et al., "Remateable and Deformable Area-Array Interconnects in 3D Smart Wireless Sensor Packages," *2020 IEEE 70th Electronic Components and Technology Conference (ECTC)*, 2020, pp. 671–676.

[55] Nguyen, H.-V., H. Kristiansen, K. Imenes, and K. E. Aasmundtveit, "Peel Adhesion and Reworkability of Anisotropic Conductive Adhesive for Flex-to-Flex Assembly in Medical Devices," *2020 IEEE 8th Electronics System-Integration Technology Conference (ESTC)*, 2020, pp. 1–7.

CHAPTER 4

Wearable Antennas

Jonathan Lundquist, McKenzie Piper, Umar Hasni, and Erdem Topsakal

4.1 Introduction

Wearable antennas have gained great popularity in the past decade due to their ease of fabrication, use of flexible materials, and low costs. Such antennas are perfect candidates for wireless medical telemetry and data acquisition for preventative health [1]. Thus far, most of the research in wearable antennas has been in healthcare applications; however, they have potential in industrial and defense settings [2]. Some of these antennas for the industrial, scientific, and medical (ISM) band below 6 GHz are already on the market [3, 4]. These are all applications where an increased range of motion, better comfort, and a more conformal surface enhance the user's ability to complete their tasks or live their daily lives.

In recent years, several studies appeared in the literature on wearable RFID antennas. In [5], an RFID chip was placed underneath a fabric antenna separated by a polyimide layer. The device was sewn into various other clothing and used to monitor skin temperature and hydration levels via other integrated sensors in conjunction with the RFID tag [5]. In [6], a wearable strain sensor based on knitted RFID technology was used for respiration monitoring. This wearable device consisted of a dipole antenna constructed of conductive yarns knitted using tuck stitches that are produced when the knitting needle holds the original yarn loop up while receiving the new yarn. This produces better connection between the antenna and the RFID chip. The RFID chip was then placed behind the antenna on a flexible printed circuit. Some of the earliest wearable antennas made use of woven copper to construct patch antennas [7]. It was realized early in the development of wearable antennas that the ground plane of patch antennas made them a convenient topology in this space [7].

Wearable antennas also find application for use by industry, first responders, and military personnel. Octane currently markets wearable antennas ranging from 2 MHz to 10 GHz for these purposes [4]. These antennas are intended to be connected to radios worn by the user for communication and telemetry [4]. Several researchers have made attempts to classify wearable antennas into various categories [8]. In this chapter we will classify them by construction methodologies that include embroidered textile antennas, screen-printed ink antennas, and inkjet-printed ink antennas (see Figure 4.1) and focus on fully integrated textile antennas. There are different advantages and disadvantages to each technique and they each have a different role to play in the future of wearable textile antennas.

Each of these construction methods comes with their own set of challenges, but there are some issues that are common to all wearable antennas. Antenna performance and resonant frequency will change when the antenna is flexed, due to the change in topology [9]. Depending on how the antenna is intended to be used this can both be a challenge and an opportunity.

In this chapter, we will discuss different methods of construction and the challenges, advantages, and applications specific to each type of wearable fabric antenna. We will discuss current research as well as progress made on the types of materials used to construct these antennas. There are several important considerations before determining which construction method is best to use for a specific antenna. How conformal does the antenna need to be? Are raised surfaces acceptable and to what extent? How mechanically strong does the antenna need to be? How flexible is flexible enough? What kind of efficiency is needed for the application? What resolution is needed for the antenna? The answers to these questions will inform the type of flexible antenna an engineer will want choose for their application. It may be that a researcher needs to stretch the limits of a specific method to get the results they need and this is where materials and tools will play a critical role. Some conductive inks are stronger and more flexible than others. The same is true of fibers that can be used for embroidery. Also, while inkjet printing achieves excellent resolution, it may be possible to achieve similar results with an embroidered or screen-printed antenna if the fiber size is small enough or the screen is high resolution enough.

4.2 Embroidered Antennas

Embroidery is a commonly used method to develop wearable electronic technology and antennas are no exception to this. Several methods and materials have been introduced over the years to make more efficient, durable, and high-resolution

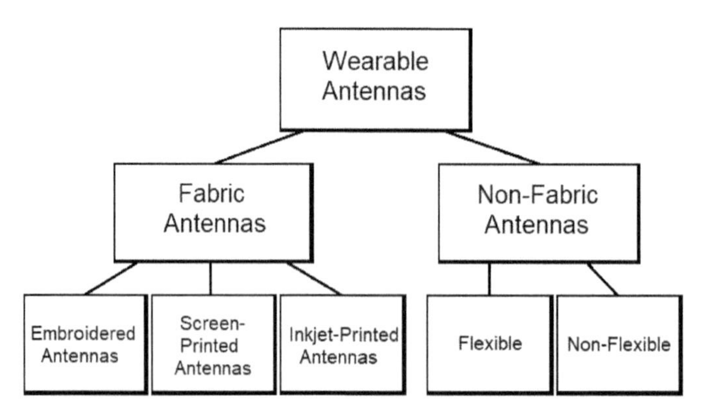

Figure 4.1 Classification of wearable antennas.

embroidered antennas. Embroidery is an effective way to achieve an all-textile antenna, but there are important design considerations required to use it effectively. Not all threads/fibers are the same, and this is potentially the most important factor when it comes to embroidered antennas. However, important decisions also need to be made in the substrate the antenna is printed on, the direction of stitch, and the location on the body that it will be attached to. We will discuss how all of these decisions impact the performance, durability, and flexibility of the antenna design.

Early attempts to design embroidered antennas have seen the use of conductive thread as an option for the antenna design. These materials have been found to have high losses, limited flexibility, and limited durability [10]. These earlier material choices led to antennas that were less conductive and had difficulty surviving machine washing. Given that reusability of these expensive wearable devices is highly desirable, and washing is absolutely necessary in patient care settings, this was very disadvantageous [11]. E-fibers have garnered much recent attention in this space, due to their ability to solve most of these problems. E-fibers are metallicized polymers with ideal mechanical characteristics such as flexibility and durability that are also highly conductive. They experience little to no permanent deformation after repetitive mechanical exercise [12]. Researchers at Ohio State University were able to use e-fibers to build embroidered antennas with a resolution of 0.1 mm using Elektrisola-7 threads and printing optimization with a single thread layer. This was an improvement over advancements they had already made with a 0.3-mm process [13]. The importance of a single thread layer is that it allows for a significantly more conformal antenna, which has been a challenge for embroidered antennas in past. Thread is not the only material consideration for embroidered antennas. All textile antennas have to consider the substrate being used. This can be an application-specific choice, and ultimately the antenna design may be required to work with the substrate for the given application. Occasionally, engineers may have the option to pick a specific fabric for the substrate, or use another material as an added layer to change the substrates mechanical or electrical properties.

Many aspects of a chosen fabric substrate can affect its electrical properties, such as the type and composition of fibers used, the weave of the fabric, and the density. The direction the antenna is placed on the material will also have an impact as the dielectric constant is often not isotropic in these materials [14]. To make matters more difficult, information is not always readily available for a given fabric and characterization in this case must be done by the antenna designer. When possible, it is best to find a material that provides the optimal electrical properties for a given antenna. This, however, is often not the only priority impacting design. For many medical applications a wearable device is limited to the type of garment required for the data being collected. Stretchability, comfort, and weight all factor into the material selection process and the electrical properties may need to be characterized for the material that is needed for the task. Motion capture systems that use wearable antennas should not limit the wearers range, which may interfere with the data being collected. In this case it is important that the material used is also very flexible [15]. However, fabrics used for military and first responders may prioritize a fabric that is fire-retardant [16].

One of the most obvious advantages of the embroidery process is that it is not temperature-dependent. This increases the number of textiles that can be used as the substrate when compared with screen printing and ink-jet printing that have a curing process that is detrimental to certain fabrics. Embroidery also may be easily automated with computer-driven embroidery machines. One of the biggest advantages is that it allows for a method of connecting the antenna to e-textile circuitry that is seamlessly integrated. Embroidered antennas also tend to be more wash-resistant than other methods of manufacturing fabric antennas [17]. Many of the disadvantages of embroidery can be mitigated with the use of modern materials and informed design. Embroidery machines are already commonly used in fabrication of textile products, which make the production equipment and trade skills readily available. This makes embroidery a promising field of research for wearable antenna fabrication.

Disadvantages of such antennas are more a function of material selection and construction methodologies than they are inherent in the design of embroidered antennas. Embroidered antennas that use conductive thread may require several layers to increase conductivity and have lower precisions than other methods. This reduces flexibility of the antenna and renders it less conformal than thinner designs. Furthermore, conductive thread itself is not a mechanically resilient material. Another issue with embroidered antennas that is not seen in screen-printed or ink-jet-printed antennas is that stitch direction can be an important factor depending on the thread density and the thread type. Researchers in the United Kingdom have found that the antennas perform best when the stitch is aligned in the direction of current flow in the antenna. This makes understanding the topology being used vitally important to good design [18]. The antenna array shown in Figure 4.2 shows the stitching in the direction of current flow for each patch [19].

4.2.1 Design and Construction

As with any antenna, the purpose of the antenna drives much of the design process. The purpose will inform the desired frequency band and radiation pattern, and these along with the constraints of the construction materials and antenna placement will drive the chosen topology. Once a substrate and thread have been chosen for the antenna, these materials need to be characterized so that they can be appropriately simulated. The conductivity of the thread at the frequency of interest needs to be characterized for the type density and direction of stitch. The substrate also needs to be characterized so that its relative permittivity is known. This information will be critical to simulation of the antenna [20]. While methods vary, researchers at the University of California, Los Angeles developed a copper patch antenna on the

Figure 4.2 Embroidered array. (From [19].)

substrate and used this to match the permittivity [21]. This method has also been used for characterizing screen-printed antenna substrates [22].

Several methods have been used to simulate embroidered antennas, but most at minimum tend to include a high conductivity serpentine element with the long lines in the direction of stitch, with a lower conductivity element as in-fill [21–23] (see Figure 4.3). These methods appear to have yielded reasonable results in predicting antenna performance and radiation patterns.

It is important to work within the constraints of the equipment available such as the resolution of both the embroidery machine and the thread. The process of embroidering can be somewhat of an art form, but in general a smaller more conductive thread will allow for a more accurate representation of the design topology and higher thread density while additional layers will result in higher conductivity [24].

4.3 Screen-Printed Antennas

Prints have been used to pattern fabrics since ancient times. Much of this original printing was done with block stamps. In the early twentieth century, screen printing began to become popular as a new printing technology [25]. One of the reasons printing is so popular is that it makes it relatively easy to make the same pattern over and over again. This is also true when using this technique to print antennas onto fabrics.

Screen printing consists of patterning an image on a screen using a photographic method. The negative space on the screen will act similar to a stencil. A squeegee is used to push ink through this pattern and onto and/or into the fabric materials (Figure 4.4). The ink is then usually cured through heating.

Like other forms of wearable antennas, screen-printed antenna performance is heavily affected by the materials chosen to work with. Different types of substrates will have different types of weave or may be more or less smooth. Also,

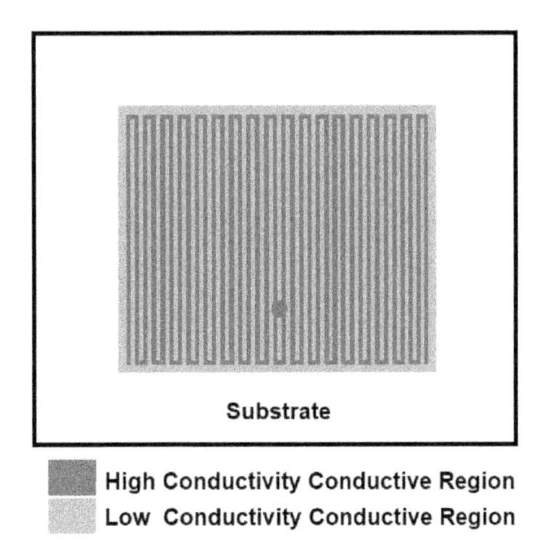

Figure 4.3 Typical surface regions for simulated embroidered antennas [25, 26].

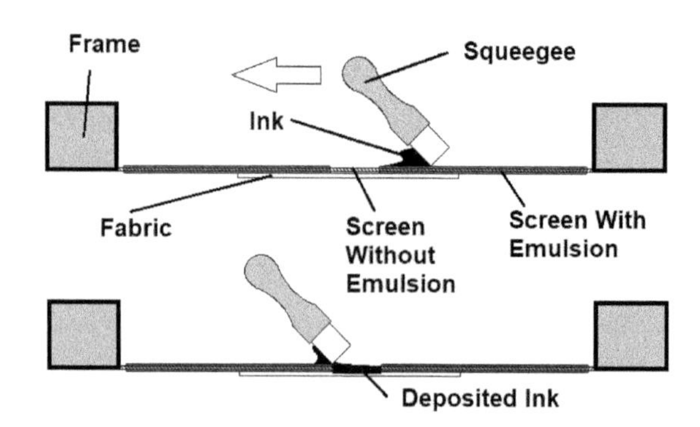

Figure 4.4 Application of ink to substrate.

the substrate material will have to survive whatever curing process is used for the chosen ink. Table 4.1 shows an assortment of vendor inks and their cure times and temperatures.

It is important to ensure that the chosen substrate can survive the cure temperature of the ink to be used for the antenna. Each of these inks listed above has its own advantages and disadvantages, with the cure temperature being the most obvious immediate distinction.

The Novacentrix inks in Table 4.1 were characterized on 21 different fabrics in a study reported in [22]. The study looked at the sheet resistance of each of the inks as well as defects in the prints before and after laundering. The researchers found that while different substrates would yield different sheet resistances for a given ink, the variation between inks was more significant (see Figure 4.5 and Table 4.2). Other factors such as failure modes and print surface were either directly a function of the substrate printed on or were heavily affected by it [22].

On flat, smooth substrates, Novacentrix FLX9 and FLX15 suffered very few defects in print and little to no defects after washing. These two inks could be stretched without defect on fabrics that FG181 and FG57B could not. One of the biggest takeaways from this study is that the variation in ink, for substrates that were appropriate to print on, was more impactful to performance and survivability than the substrate itself. A screen-printed patch was used to determine the relative permittivity of the most suitable ink-substrate combination, which was satin and

Table 4.1 Screen-Printing Inks and Associated Cure Temperatures and Times

Ink Name	*Cure Time*	*Temperature*	*Reference*
Electroninks, El-918	20 min	<100°C	[26]
Novacentrix, HPS-FG57B	30–60 min	120°C	[22]
Novacentrix, HPS-FG181*	30–60 min	120°C	[22]
Novacentrix, HPS-FLX9*	5–10 min	160°C	[22]
Novacentrix, HPS-FLX15*	5–10 min	160°C	[22]
Creative Materials, Extremely Conductive Elastomeric Ink, 127-07	3–5 min	175°C	[27]
Creative Materials, Screen-Printable Electrically Conductive Ink, 127-48	15–30 min	125°C	[28]

*This is a development ink and may not be available for general purchase.

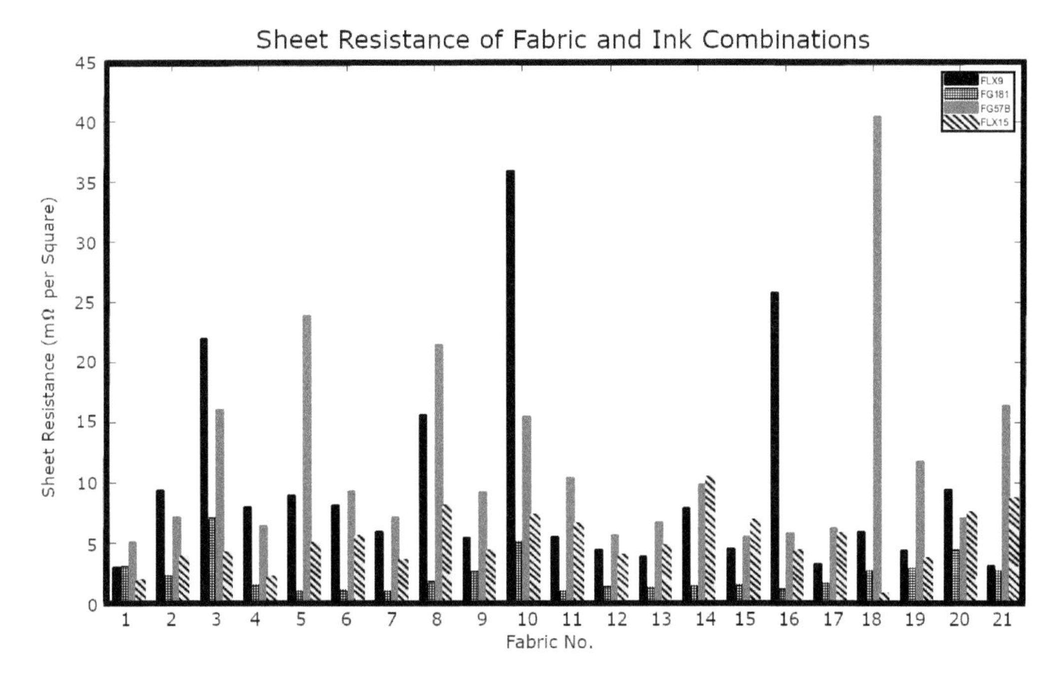

Figure 4.5 Sheet resistance of fabric and ink combinations [22].

Table 4.2 Composition of Fabrics in Figure 4.2

Fabric No.	*Fabric Content (%)*										
	Viscose	*Silk*	*Cotton*	*Spandex*	*Rayon*	*Nylon*	*Polyester*	*Acrylic*	*Lyocell*	*Linen*	*Wool*
1	80	20	0	0	0	0	0	0	0	0	0
2	0	0	97	3	0	0	0	0	0	0	0
3	0	0	5	0	4	5	38	10	0	0	38
4	0	0	100	0	0	0	0	0	0	0	0
5	0	0	100	0	0	0	0	0	0	0	0
6	0	100	0	0	0	0	0	0	0	0	0
7	0	0	100	0	0	0	0	0	0	0	0
8	0	35	0	0	65	0	0	0	0	0	0
9	0	58	0	0	0	42	0	0	0	0	0
10	0	100	0	0	0	0	0	0	0	0	0
11	40	0	60	0	0	0	0	0	0	0	0
12	0	0	60	0	0	0	40	0	0	0	0
13	0	0	100	0	0	0	0	0	0	0	0
14	0	0	100	0	0	0	0	0	0	0	0
15	0	0	10	0	16	0	0	0	49	25	0
16	0	100	0	0	0	0	0	0	0	0	0
17	0	65	0	0	0	35	0	0	0	0	0
18	0	0	0	0	0	0	0	0	0	0	100
19	0	0	100	0	0	0	0	0	0	0	0
20	0	0	0	0	45	0	0	0	0	55	0
21	0	0	100	0	0	0	0	0	0	0	0

From: [22].

FLX15, due to wash resistance, lack of defects, flexibility, and ease of determining ink thickness (Figure 4.6) [22].

A wearable microstrip patch antenna without any metallic particles using PEDOT:PSS conductive polymer ink, making the antenna completely organic, was proposed in [29].

Similar to embroidered antennas, attempts have been made to use different coatings or other materials to increase the wash resistance of screen-printed antennas. A breathable TPU was used as a protective material for wearable antennas to improve stability in wash cycles [30]. While inks and coatings are important, the impact of substrate cannot be discounted. It is critical to successful screen printing that the substrate be nontextured and as flat as possible for proper printing. This allows for the best print quality [31].

The first noticeable advantage of screen printing is that it is almost perfectly conformal to the fabric substrate. Screen printing is an expedient method of manufacturing on large print areas and dealing with large volumes of the same product design [31]. The process for screen printing is easy and cost-effective. Screen printing can easily achieve a higher resolution by using a higher mesh count for the screen [32]. The ability to make high-resolution patterns cheaply, quickly, and easily makes this an appealing method of construction. Some disadvantages of screen printing can be mitigated by the use of specific materials [22] or applying a coating material over the printed pattern [30]. In this way it is similar to embroidery, in that material selection and characterization can do a lot to improve the end product.

Although there are many advantages to this technique, there are also some disadvantages. Earlier ink blends had very limited wash resistance and this is still a struggle for manufacturers of conductive screen-printing inks [33]. However, with recent advances in ink manufacturing new inks are becoming more wash-resistant [22]. In particular, working with polymeric binders to make silver screen-print inks stronger has been researched [34] and can be found in inks currently on the market [22]. Normally, when significantly stretched a screen-printed silver ink will begin to crack or deform. It does eventually reach a point where breakdown due to mechanical stress reaches a plateau [35]. This can be mitigated by choosing a substrate

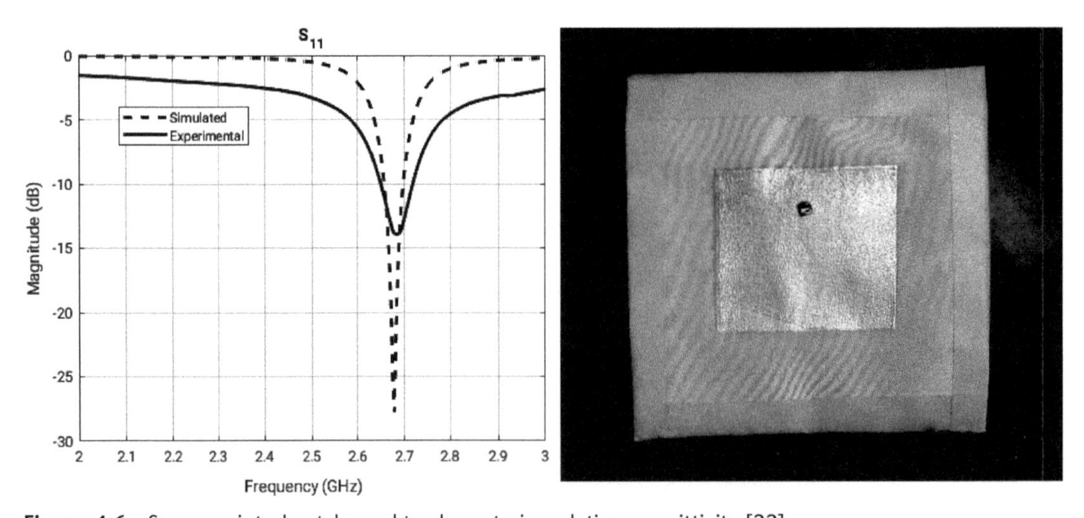

Figure 4.6 Screen-printed patch used to characterize relative permittivity [22].

that has a flexibility within the range the ink is designed for. Furthermore, different inks will have different stretch characteristics. This can often be found in the data sheet [27].

4.3.1 Design and Construction

The first concern when designing screen-printed antennas is determining whether the application is a good fit for this technology. Screen-printed antennas may not be ideal for textiles worn in heavy industrial, military, or first responder applications that experience significant wear and tear. This is because they do not retain their characteristics under conditions with high mechanical stress [34–36]. These antennas are also not optimal for use on substrates that are heat sensitive due to the curing process using high temperatures [26–28].

The choice of ink to use may depend on the substrate itself. Different inks have different curing temperatures, flexibility, and wash resistance. Using an ink not suited to the flexibility of the substrate could result in defects such as cracks, as seen in Figure 4.7 [22]. While the substrate could be chosen to fit the desired ink, often the substrate is determined by the application itself and the desire to have the antenna fully integrated into the garment. The relative permittivity of the substrate can be characterized using a microstrip patch resonator and matching the resonant frequency [37].

The chosen ink will need to be characterized for average sheet resistance if not already available on the data sheet. This can be done by using several samples and a four-point ohmmeter [22]. Most ink data sheets contain information for cure time, temperature, stretchability, and may have a recommended mesh size [26–28]. A higher mesh count will give more resolution but may result in a lower conductivity. This data can be used to simulate the antenna.

The fabrication process usually involves using contact photography to create an image stencil with UV-sensitive emulsion on the screen. The screen can then be used to paste the image onto the cloth. It will then need to be cured in accordance with the manufacturer's specifications. Attaching electronic components to the screen-printed antenna can be problematic and one common approach is to

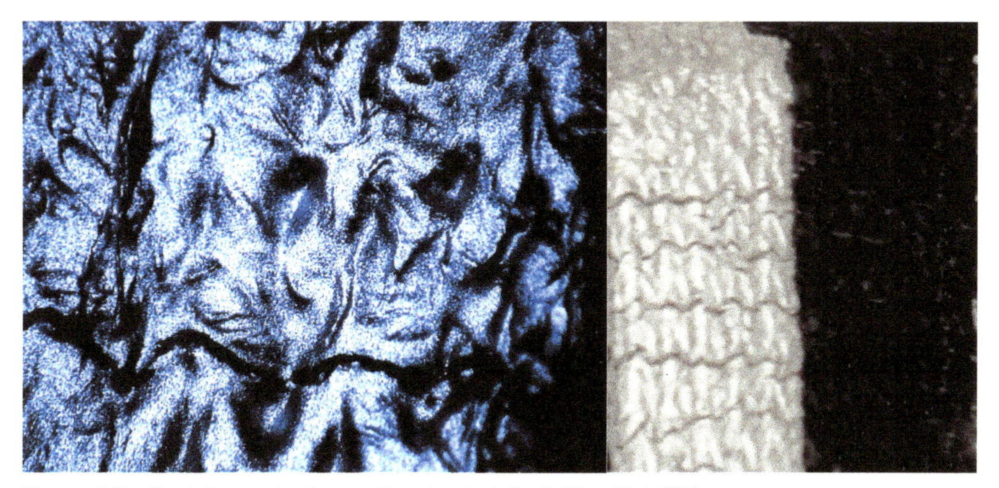

Figure 4.7 Crack in conductive portion due to inflexibility of ink [22].

use a conductive adhesive. Conductive thread can also be used to sew the components to the garment. If more wash resistance and mechanical strength is required, a protective layer, such as a polyurethane coating, has been shown to be effective [38]; however, this needs to be factored into the antenna design and simulation to achieve the correct resonant frequency. The relative permittivity of the protective layer will affect the design parameters [39].

4.4 Inkjet-Printed Antennas

Inkjet-printed antennas are constructed using inkjet printers with metal nanoparticle inks. Whereas screen printing uses an imaged stencil to place nanoparticle inks, inkjet printers can place very specific amounts of ink in specific locations and have no need for the construction of a stencil. This can allow for rapid prototyping of an antenna but is slow and expensive at high volume [40]. Silver nanoparticle inks are commonly used to fabricate the antennas. The antenna resolution is only limited by the resolution of the printer, which is usually more precise than screen-printing or embroidery. Interface layers can be screen-printed onto the fabric to make the jetting surface smoother, leading to even better results [40, 41]. Like screen-printed inks, inkjet inks must be cured onto the fabric, usually in a process that requires an elevated temperature for a specific amount of time [42]. For this reason, this technology is not suitable for all fabrics.

Inkjet-printed antennas use low viscosity conductive nanoparticle inks suitable for commercial inkjet printers. Substrates can be a unique challenge for inkjet-printed antennas, because the conductive nanoparticles themselves are not typically flexible [42, 43]. Also, the substrate may not lend itself well to ink jetting if it is textured, woven loosely, or has significant surface roughness. Researchers often combat this by preparing the surface with a screen-printed interface layer to print the antenna on. Polyurethane is often used for this purpose [42, 43]. When choosing the substrate, it is also important to choose one that is compatible with the curing process of the chosen ink [44]. The combination of materials or material used as the substrate will have to be characterized for relative permittivity as with any other antenna.

Different inks will have different particle sizes, viscosities, and surface tensions. It is important to ensure that the chosen ink works with the print cartridge being used to print the antenna. Large particle sizes can easily clog cartridge nozzles, resulting in expensive print failures [44]. Table 4.3 lists the specifications for some common inkjet inks.

Depending on the printer used, inkjet printing can be cheap [44], but the primary motivation behind the use of inkjet printed antennas is that it is easy to achieve a higher resolution of print than other methods [42–44]. While being similar to screen printing in that the process deposits conductive nanoparticles onto the fabric, a complex lithographic process is not required to make the print [43], which is the primary reason that inkjetting is easier to implement. Inkjet printers and supplies for them can be considerably expensive compared to screen printing, depending on the printer used. These machines can also be more complex to maintain than screen printing equipment and require consistent care and maintenance. Failures that may occur include air in the cartridge, fluids collecting on the nozzles, nozzle

Table 4.3 Specifications for Silver Inkjet Inks

Ink Brand and Model	*Concentration (wt %)*	*Viscosity (cP)*	*Particle Size (nm)*	*Surface Tension (dyne/cm)*	*Resistivity (Ω-cm)*	*Reference*
Novacentrix, Metalon, JS-A101A	40	5–7	30–50	19–30	<3.1E-05*	[45]
Novacentrix, Metalon, JS-A102A	40	8–12	30–50	19–30	<3.1E-05*	[46]
XTPL, Ag Nanoink IJ36	32–36	26–30	35–50	30	3.95E-06	[47]
Sigma Aldrich, 907022	18–20	10–18	90–110	40	Not specified	[48]

*When thermally cured at 140°C for 10 minutes (nonthermal curing methods available) [48, 49].

clogs, and misconfigurations, to name a few. All of these failures may prevent a successful print [49]. Finding the correct print settings for a particular ink can be a challenge if there are no recommended jet settings from the manufacturer. If the resolution is set too high, the pattern may bleed into the fabric and if it is set too low it defeats the purpose of using the inkjet altogether [42]. Some of these issues can be mitigated by using cartridges with different drop sizes. Some printers such as the Fujifilm Dimatix printers offer cartridges with different drop volumes [49].

Another issue with inkjet-printed antennas is that it can be difficult to properly print directly onto the fabric [44]. This can, however, be easily mitigated by using a screen-printed or painted-on interface layer that allows for a smoother printing surface [42–44]. Our group is currently researching using iron-on transfer paper as an interface layer for inkjet-printed antennas (see Figure 4.8).

4.4.1 Design and Construction

Choosing the substrate will be important because its roughness and the tendency of the ink to bleed in it will determine whether an interface layer is needed to refine the print [42–44]. It may be desirable to use an interface layer in any event depending on the ink being used, because it may reduce the number of layers needed to achieve the desired efficiency of the antenna. Also, obtaining a conductive layer using inkjet

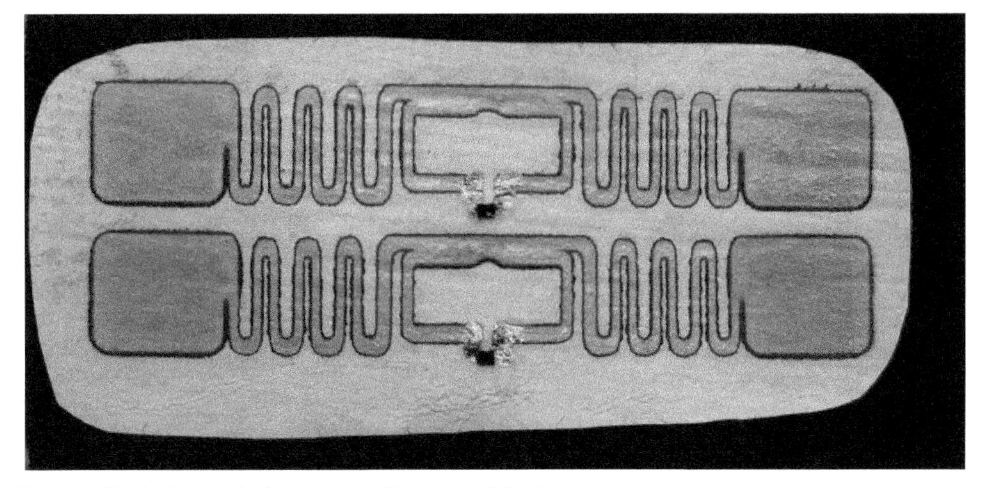

Figure 4.8 Inkjet-printed antenna with iron on interface layer.

printing is difficult on fabric and may require the interface layer in order to achieve this on a given material [50]. There are multiple choices that can be made for the interface layer and the choice may impact the resolution at which you can achieve using the inkjet printing process. Surface tension of the interface layer or substrate can impact the way the ink spreads on the surface when dropped and thus impact the resolution of the print [51]. As discussed in previous sections, the substrate and interface layer, if there is one, will need to be characterized to determine its relative permittivity for antenna design, simulation, and modeling. While conductive silver inks are very common there are several kinds of inks that can be used for inkjet printing antennas, each of which vary in resistivity, surface tension, viscosity, and particle size, not only with material, but with brand and model as well [45–48]. A given ink will only be compatible with a printer if it falls in the range of specification for that printer [49], which may be a limiting factor in selection. Other types of ink include inks made from graphene silver composites, carbon nanotubes, and conductive polymers [52–54]. Some of these inks may not be ideal for textile antenna fabrication, but may have benefits as an additional layer.

After selecting an ink that is appropriate for both the substrate and printer, it is important to appropriately tune the print settings (if possible) for the printer being used to achieve optimal resolution and uniformity. This can be done by choosing a cartridge with an appropriate drop volume, as well as changing the drop distance and delay time [51].

Drops spacing and cartridge nozzle diameter will impact resolution along with the surface tension of the substrate [51]. This results in a multivariate problem impacting uniformity and resolution. It is important to do a detailed characterization of the print quality as a function of settings on any new material with a given ink before deciding which settings to use to proceed, with the idea being to maximize resolution, uniformity, and conductivity. Once materials and print settings have been characterized, simulation and fabrication of the antenna can be performed. It is important to make sure the print surface is as flat as possible on the printer prior to print and that thickness is known so the z-axis height can be set appropriately [49]. Once the print is complete the ink must be cured using the manufacturer recommended procedure.

4.5 Material Considerations: Fabrics, Inks, and Threads

Material characteristics are the most important part of designing wearable fabric antennas. The chosen materials will make large differences in the antenna efficiency, elasticity, durability, and wash resistance. As a result, most of this chapter has focused on the material properties of various construction materials for each type of wearable antenna. In this section we will dive a little deeper into the materials that are used to construct these antennas.

4.5.1 Fabrics

If the fabric substrate material is capable of stretching beyond what the conductor is able to handle without cracking or deforming, then this may not be an ideal substrate for the chosen construction method, or it may not be an ideal construction method for the chosen fabric substrate. While factors such as stretchability are important, they are trivial to determine. Electrical properties such as relative permittivity and loss tangent are not as simple [55]. Relative permittivity is not simply a matter of fiber composition either as it is affected by humidity, density of the weave, and any interface layer printed onto the substrate. One can relatively easily extract a close relative permittivity of a substrate using a patch antenna by fabricating the patch on the substrate. The equations for a patch antenna then can be solved at the measured resonant frequency and simulations used to refine the calculated permittivity [55]. This however does not directly provide information for loss tangent. Researchers in Belgium used a weighted cost function over a set of frequencies that determined where the error in measured reflection was minimized versus a simulation for a range of loss tangent and relative permittivity [56]. This allowed both relative permittivity and loss tangent to be extracted. Permittivity can be impacted not only by humidity but also other environmental factors such as ambient temperature and the salinity of any water absorbed by an antenna. A comprehensive characterization and mitigation of these effects can be found in [57].

Another method used to determine the relative permittivity and loss tangent of a fabric substrate is to use a microstrip transmission line. Researchers in Finland and China used the least squares method and maximum power gain to find relative permittivity, loss tangent, and conductivity. This required measuring S_{11}, S_{12}, and S_{21} to extract the necessary data [58]. The heat resistance of the substrate may be important depending on the method of fabrication. The "Fibers, 1. Survey" by Fritz Schultze-Gebhardt and Karl-Heinz Herlinger is a thorough resource for this information, along with several other material properties of fabrics [59]. It is important to ensure that the heat resistance of the fabric substrate used is higher than that of the process being used to make the print.

4.5.2 Conductive Fibers

Generally, conductive fibers can be made either by embedding or coating a fabric in a conductive material such as metal or conductive polymers, or by using thin strands of metals as the fiber [60]. Each of these techniques results in a different conductivity, flexibility, and strength for antennas they are used in.

Mechanical strength and flexibility are important aspects for designing a long-lasting antenna that can hold up under rugged conditions. Modern e-fibers made from metallicized polymers have been shown to perform well in both of these categories [13]. Research has shown that coating these fibers with a protective polymer, such as PDMS, can improve their stretchability and allow them to be flexed a greater number of times without degradation [61]. Research has also demonstrated that a precision process with small threads can achieve a relatively high resolution [13]. It is also important to understand the conductivity of a given fiber in order to accurately simulate an antenna during the design process. Conductivity of a given thread can be determined using a waveguide cavity method [60]; however, an antenna designer is more likely to be concerned with the conductivity value that will result as both a function of the thread and weave used. This can be determined by model fitting of a microwave transmission line made from the thread [58].

4.5.3 Conductive Inks

The concerns facing engineers using conductive ink are similar for those using conductive fibers. Mechanical strength and flexibility are highly desirable to minimize impact to the wearer and conductivity is important for antenna efficiency. The transmission line characterization model discussed for conductive thread apply to any combination of substrate and conductor [58]. For some inks, manufacturers supply information on the typical volume resistivity of the printed ink in their data sheets [45–47].

Mechanical strength, wash resistance, and flexibility of an ink is dependent on the ink being used, method of application, the substrate it is printed on, and whether an interface layer or coating layer was used. Several different conductive materials are used to make conductive inks, including carbon, silver, copper, and conductive polymers [62, 63]. Materials made from nanoflakes tend to have good characteristics for flexibility and conductivity [64]. Often manufacturers will place the nanoparticle shape on the data sheet of a particular ink [47]. While most inks are cured by heating or drying, UV-cured inks are now available. These UV-cured inks allow for use of textiles that would otherwise be sensitive to the heat curing process used with most inks [65]. Still, UV curing systems are expensive and not all inks that use heated curing use the same temperature, and it may be possible to find a conductive ink that has a curing temperature lower than the chosen textile substrate can tolerate [22].

4.6 Applications

A major challenge facing wearable antenna designers is the human body itself. The human body is very lossy and has a higher permittivity than most of the substrates an antenna will be printed or sewn onto. As a result, an antenna designed in free space will see frequency shift when placed against the human body and will likely see a decrease in efficiency. These changes may not be consistent because clothing that hangs loosely on the body will be impacted by varying degrees as the antenna moves closer and further from the skin [66, 67]. This can be mitigated to some extent by using an antenna backed with a ground plane [66], which has the added effect of minimizing specific absorption rate (SAR). For best results however, the antenna should be designed with on-body use in mind. Human body phantoms and simulations that include human tissue can provide a good indication of how the antenna behaves on or near skin. It may be desirable to use a wideband antenna to reduce the impact of the frequency shift [67].

Tissue-mimicking gels (see Figure 4.9) that match the electrical properties of materials in the human body are commonly used to make phantoms for the purpose of testing on-body antennas. These gels are cheap and easy to make and can be tailored to the varying tissues and thicknesses in the human body. The gels are soft, easy to cut, and can be layered to form more complex structures that mimic the layers of skin, adipose tissue, muscle, bone, or other biological tissue [68].

Another important consideration for designing on-body antennas is minimizing SAR. Device manufacturers need to adhere to standards and regulations regarding SAR and the antenna design can significantly contribute to this [69]. Using

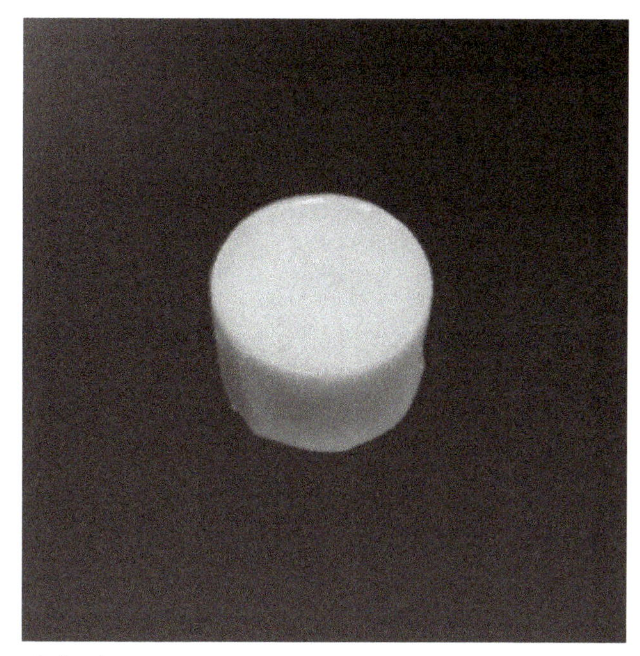

Figure 4.9 Skin gel phantom.

directional antennas that radiate away from the body and backing omnidirectional antennas with ground planes can help with this. Research is actively being conducted in the areas of antenna topology, modeling, design, and materials for wearable antennas. In this section we will discuss recent papers published in these areas and the impact future research could have on wearable antennas.

A new coplanar keyhole topology that has a bidirectional radiation pattern (see Figure 4.10) for wearable medical applications is demonstrated in [22]. The gain is highest in both directions normal to the surface of the skin. This allows for transmission both inside and outside of the body. One of the purposes of this design was

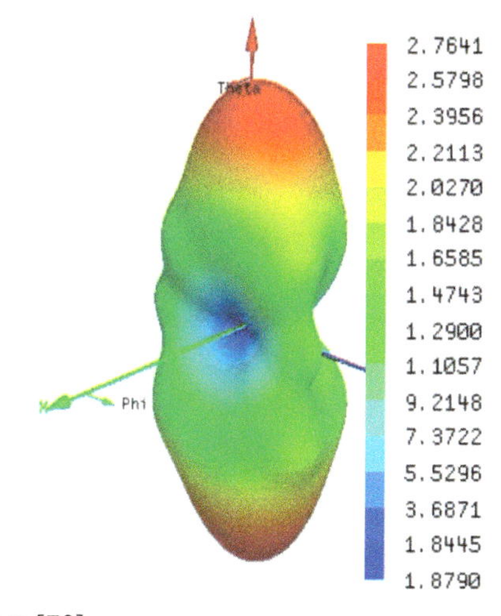

Figure 4.10 Fabric coplanar keyhole antenna and radiation pattern [70].

to allow the antenna to communicate with implanted medical devices and relay that information to a more distant receiver [70].

This antenna was fabricated by the screen printing of Novacentrix HPS FG-32 conductive silver ink onto a nylon/spandex blend. An interface layer was added to modify permittivity and increase the antenna size for ease of fabrication. A wash study was conducted, and some frequency shift was noted after washing (Figure 4.11). The antenna was tested with a radio transmitter and receiver and was found to have no packet loss up to a distance of 20 ft and showed a received power of -50 dBm at this distance [70].

A screen-printed RFID was also presented in this study (Figure 4.12). This antenna used the newly characterized inks and fabrics from [22] to inform its design. The antenna was a meandering dipole topology and was simulated in on-body and off-body conditions in Ansys HFSS (Figure 4.13) [70].

Materials advancements have been made by researchers in the United Kingdom who have developed a planar inverted cone ultrawideband antenna made of graphene on fabric. Chemical vapor deposition (CVD) was used to plate the graphene onto nickel. The graphene was then transferred to a material that could be cut on a vinyl cutter using an adhesive. The vinyl cutter was used to cut the antenna shape, which was then transferred to the fabric made of cellulose fibers using a laminator. This process can be seen in Figure 4.14 [71].

The cone antenna seen in Figure 4.15 was designed to have a resonance between 3 and 9 GHz. The measured and simulated S_{11} values on and off body support this (Figure 4.16). This wide bandwidth was seen both in free space and in on-body measurements. A liquid-filled phantom designed to mimic the electrical properties of a human was used to perform the on-body measurements in an anechoic chamber. The antenna was simulated in CST with and without human tissue. A wash study conducted showed the antenna survived at least 50 wash cycles [71].

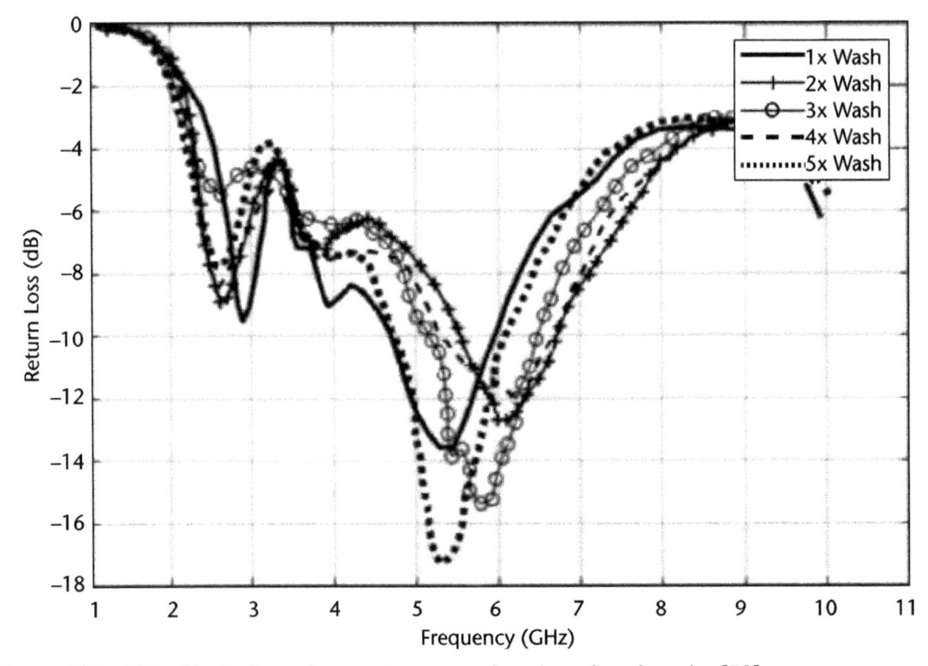

Figure 4.11 S11 of keyhole coplanar antenna as a function of wash cycles [70].

Figure 4.12 Screen-printed RFID using Novacentrix FLX15 [70].

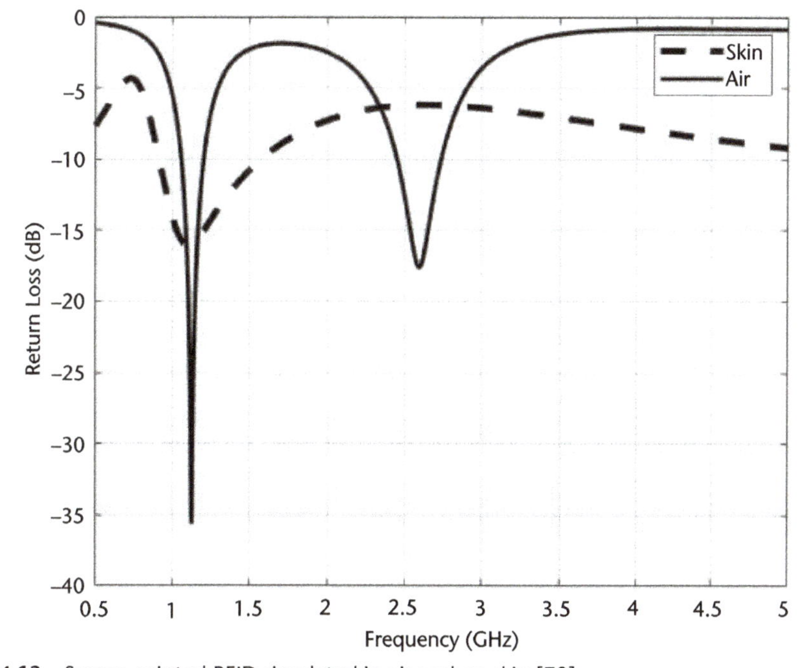

Figure 4.13 Screen-printed RFID simulated in air and on skin [70].

Figure 4.14 Manufacturing process for graphene antenna. (©2020 Elsevier [71].)

In [72, 73], advancements in design, materials, and topology have been demonstrated with the design and theoretical models for biomatched antennas (BMAs)

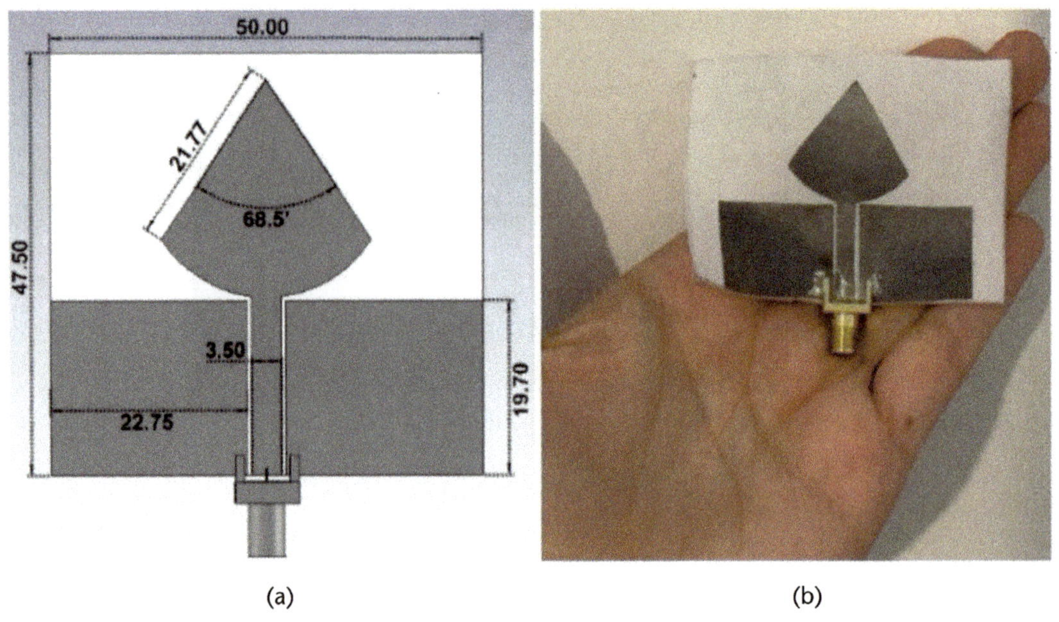

(a) (b)

Figure 4.15 Graphene cone antenna on fabric. (©2020 Elsevier [71].)

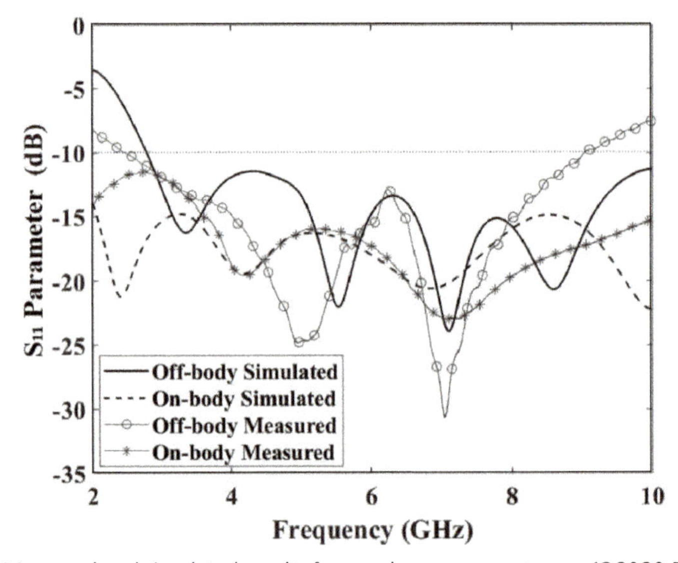

Figure 4.16 Measured and simulated results for graphene cone antenna. (©2020 Elsevier [71].)

using periodic dielectric structures. It is important in medical applications for some wearable antennas to be able to transmit into the body [72].

These researchers have overcome challenges regarding antenna performance by using these periodic water and plastic structures and flared antenna topologies to develop a wideband biomatched horn for this application. Water was used because of the high water content in human tissue. The researchers found that anisotropic structures provided the best performance for these antennas [72]. Figure 4.17 shows one of these biomatched antennas from a previous study by these researchers and Figure 4.18 shows the measurement results.

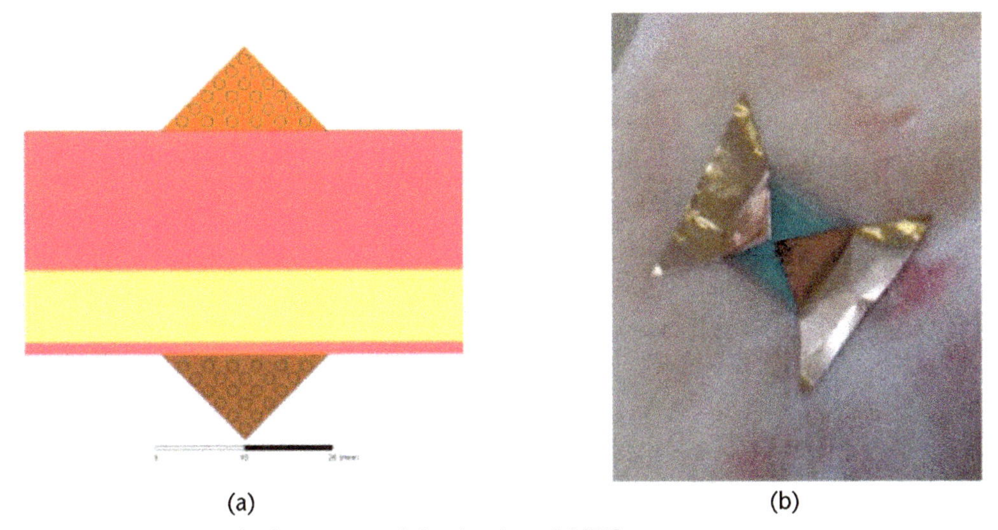

Figure 4.17 Biomatched antenna and simulated model [73].

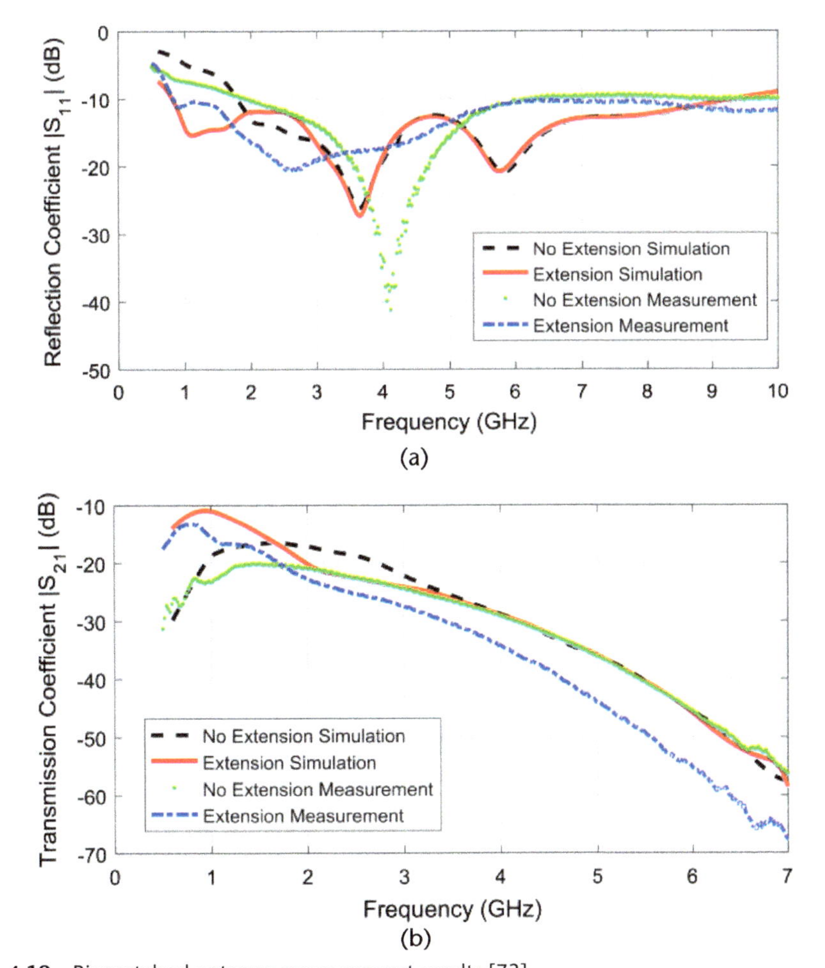

Figure 4.18 Biomatched antenna measurement results [73].

The structures were found to have a cutoff frequency where the permittivity significantly increases as the size of the structures become comparable to the wavelength. The antenna itself was found to have a low frequency cutoff when its edge length is about half the wavelength when modeled as a quasi-conical antenna. The paper provides equations for these cutoff frequencies [72].

References

[1] Umar, H., *Multimode Wearable Antennas for IOT Applications*, Dissertation, Virginia Commonwealth University, 2020.

[2] Salvado, R., et al., "Materials for the Design of Wearable Antennas: A Survey," *Sensors*, No. 12, 2012, pp. 15841–15857.

[3] Liu, B., "Application Specification: ISM 433MHz Flex Antenna," Molex, Technical Data Sheet, Revision A, 2017.

[4] Octane Wireless, "Developer-Manufacturer of Wearable Antenna Products," *Octane Wireless,* June 10, 2021, https://www.octanewireless.com/products/wearable-antennas/.

[5] Wen, S., et al., "A Wearable Fabric-Based RFID Skin Temperature Monitoring Patch," *2016 IEEE SENSORS*, Orlando, FL, 2016, pp. 1–3.

[6] Liu, Y., et al., "An Improved Design of Wearable Strain Sensor Based on Knitted RFID Technology,“ *2016 IEEE Conference on Antenna Measurements & Applications (CAMA)*, Syracuse, New York, 2016, pp. 1–4.

[7] Salonen, P., and H. Hurme, "A Novel Fabric WLAN Antenna for Wearable Applications," *IEEE Antennas and Propagation Society International Symposium Digest Held in Conjunction with: USNC/CNC/URSI North American Radio Sci. Meeting (Cat. No 03CH37450)*, Vol. 2, 2003, pp. 700–703.

[8] Kirtania, S. G., et al., "Flexible Antennas: A Review," *Micromachines,* No. 11, 2020, pp. 847.

[9] Bai, Q., and R. Langley, "Crumpling of PIFA Textile Antenna," *IEEE Transactions on Antennas and Propagation*, Vol. 60, No. 1, 2012, pp. 63–70.

[10] Kim, G., et al., "Design of a UHF RFID Fiber Tag Antenna with Electric-Thread Using a Sewing Machine," *2008 Asia-Pacific Microwave Conference*, Hong Kong, China, 2008, pp. 1–4.

[11] Koski, K., et al., "Durability of Embroidered Antennas in Wireless Body-Centric Healthcare Applications," *2013 7th European Conference on Antennas and Propagation (EuCAP)*, Gothenburg, Sweden, 2013, pp. 565–569.

[12] Wang, Z., L. Zhang, and J. L. Volakis, "Textile Antennas for Wearable Radio Frequency Applications," *Textiles and Light Industrial Science and Technology (TLIST) Journal,* Vol. 2, No. 3, July 2013, www.tlist-journal.org.

[13] Kiourti, A., C. Lee, and J. L. Volakis, "Fabrication of Textile Antennas and Circuits with 0.1 mm Precision," *IEEE Antennas and Wireless Propagation Letters*, Vol. 15, 2016, pp. 151–153.

[14] Salvado, R., et al., "Textile Materials for the Design of Wearable Antennas: A Survey," *Sensors*, Vol. 12, No. 11, 2012, pp. 15841–15857.

[15] Hasni, U., and E. Topsakal, "Multi-Mode Smart Wearable Fabric Antennas for Augmented Touch Tracking and Motion Detection on Human Skin," *USNC-URSI National Radio Science Meeting*, 2019.

[16] Tian, M., et al., "Investigating the Thermal-Protective Performance of Fire-Retardant Fabrics Considering Garment Aperture Structures Exposed to Flames," *Materials*, Vol. 13, No. 16, 2020, p. 3579.

[17] Gharbi, E., et al., "A Review of Flexible Wearable Antenna Sensors: Design, Fabrication Methods, and Applications," *Materials*, Vol. 13, No. 17, 2020, p. 3781.

[18] Seager, R., et al., "Effect of the Fabrication Parameters on the Performance of Embroidered Antennas," *IET Microw. Antennas Propag.*, Vol. 7, No. 14, 2012, pp. 1174–1181.

[19] Wang, Z., J. L. Volakis, and A. Kiourti, "Embroidered Antennas for Communication Systems," in *Electronic Textiles: Smart Fabrics and Wearable Technology*, T. Diaz (ed.), Cambridge, UK: Woodhead Publishing, 2015, pp. 201–237.

[20] Moradi, E., et al., "Characterization of Embroidered Dipole-Type RFID Tag Antennas," *2012 IEEE International Conference on RFID-Technologies and Applications (RFID-TA)*, Nice, France, 2012, pp. 248–253.

[21] Song, L., D. Zhang, and Y. Rahmat-Samii, "Towards Embroidered Textile Antenna Systematic Design and Accurate Modeling: Rectangular Patch Antenna Case Studies," *2018 IEEE International Symposium on Antennas and Propagation & USNC/URSI National Radio Science Meeting*, Boston, MA, 2018, pp. 1609–1610.

[22] Lundquist, J., et al., "Characterization of Screen-Printed Fabric and Ink Antennas," *USNC-URSI National Radio Science Meeting*, 2021.

[23] Tsolis, A., et al., "Embroidery and Related Manufacturing Techniques for Wearable Antennas: Challenges and Opportunities," *Electronic*, Vol. 3, No. 2, 2014, pp. 314–338.

[24] Kiourti, A., and J. L. Volakis, "High-Geometrical-Accuracy Embroidery Process for Textile Antennas with Fine Details," *IEEE Antennas and Wireless Propagation Letters*, Vol. 14, 2015, pp. 1474–1477.

[25] Dawson, T. L., and, C. J. Hawkyard, "A New Millennium of Textile Printing," *Review of Progress in Coloration and Related Topics*, Vol. 30, No. 1, 2000, pp. 7–20.

[26] EI-918, Electroninks, Technical Data Sheet, 2021.

[27] 127-07 Extremely Conductive Elastomeric Ink, Creative Materials, Technical Data Sheet, Revision C, 2019.

[28] 127-48 Screen-Printable Electrically Conductive Ink, Creative Materials, Technical Data Sheet, Revision A, 2019.

[29] Li, Z., et al., "All-Organic Flexible Fabric Antenna for Wearable Electronics," *Journal of Materials Chemistry C*, Vol. 8, No. 17, 2020.

[30] Scarpello, M. L., et al., "Stability and Efficiency of Screen-Printed Wearable and Washable Antennas," *IEEE Antennas and Wireless Propagation Letters*, Vol. 11, 2012, pp. 838-–841.

[31] Lee, J., et al., "Flat Yarn Fabric Substrates for Screen-Printed Conductive Textiles," *Advanced Engineering Materials*, Vol. 22, No. 12, 2020, p. 2000722.

[32] Alharbi S., R. M. Shubair, and A. Kiourti, "Flexible Antennas for Wearable Applications: Recent Advances and Design Challenges," *12th European Conference on Antennas and Propagation (EuCAP 2018)*, London, UK, 2018, pp. 1–3.

[33] Virkki, J., et al., "Reliability of Washable Wearable Screen-Printed UHF RFID Tags," *Microelectronics Reliability*, Vol. 54, No. 4, 2014, pp. 840–846.

[34] Hsu, C. P., et al., "Effect of Polymer Binders in Screen Printing Technique of Silver Pastes," *Journal of Polymer Research*, Vol. 20, No. 10, 2013, Article 227.

[35] Virkki, J., et al., "The Effects of Recurrent Stretching on the Performance of Electro-Textile and Screen-Printed Ultra-High-Frequency Radio-Frequency Identification Tags," *Textile Research Journal*, Vol. 85. No 3. 2015.

[36] Shahariar, H., et al., "Porous Textile Antenna Designs for Improved Wearability," *Smart Materials and Structures*, Vol. 27, No. 4, 2018.

[37] Sankaralingam, S., and B. Gupta, "Determination of Dielectric Constant of Fabric Materials and Their Use as Substrates for Design and Development of Antennas for Wearable Applications," *IEEE Transactions on Instrumentation and Measurement*, Vol. 59, No. 12, 2010, pp. 3122–3130.

[38] Kazani, I., et al., "Dry Cleaning of Electroconductive Layers Screen Printed on Flexible Substrates," *Textile Research Journal*, Vol. 83, No. 14, 2013, pp. 1541–1548.

[39] Balanis, C. A., *Antenna Theory: Analysis and Design,* Hoboken, NJ: John Wiley & Sons, Inc. 2016.

[40] Whittow, W. G., et al., "Inkjet-Printed Microstrip Patch Antennas Realized on Textile for Wearable Applications," *IEEE Antennas and Wireless Propagation Letters*, Vol. 13, 2014, pp. 71–74.

[41] Karimi, M. A., and A. Shamim, "A Flexible Inkjet Printed Inverted-F Antenna on Textile," *2016 IEEE Middle East Conference on Antennas and Propagation (MECAP)*, Beirut, Lebanon, 2016, pp. 1–2.

[42] Li, Y., et al., "Inkjet Printed Flexible Antenna on Textile for Wearable Applications," *2012 Textile Institute World Conference*, 2012.

[43] Kao, H. L., C. H. Chuang, and C. L. Cho, "Inkjet-Printed Filtering Antenna on a Textile for Wearable Applications," *2019 IEEE 69th Electronic Components and Technology Conference (ECTC)*, Las Vegas, NV, 2019.

[44] Mohamadzade, B., et al., "Recent Advances in Fabrication Methods for Flexible Antennas in Wearable Devices: State f the Art," *Sensors*, Vol. 19, No.10, 2019, p. 2312.

[45] Metalon JS-A101A," Novacentrix, Technical Data Sheet, 2018.

[46] Metalon JS-A102A, Novacentrix, Technical Data Sheet, 2018.

[47] Technical Data Sheet: Ag Nanoink IJ36, XTPL, Technical Data Sheet, 2021.

[48] Product Specification: PRD.0.ZQ5.10000102531, Sigma Aldrich, Technical Data Sheet, 2021.

[49] Fujifilm Dimatix Materials Printer DMP-2850 Users Manual, Fujifilm, 2016.

[50] Chauraya, A., et al., "Inkjet Printed Dipole Antennas on Textiles for Wearable Communications," *IET Microw. Antennas Propag.*, Vol. 7, 2013, pp. 760–767.

[51] Gao, M., L. Li, and Y. Song, "Inkjet Printing Wearable Electronic Devices," *J. Mater. Chem. C*, Vol. 5, No. 12, 2017, pp. 2971–2993.

[52] Deng, D., S. Feng, M. Shi, and C. Huang, "In Situ Preparation of Silver Nanoparticles Decorated Graphene Conductive Ink for Inkjet Printing, Vol. 28, No. 20, 2017, pp. 15411–15417.

[53] Tortorich, R. P., and J. W. Choi, "Inkjet Printing of Carbon Nanotubes," *Nanomaterials*, Vol. 3, No. 3, 2013, pp. 453–468.

[54] Wilson, P., C. Lekakou, and J. Watts, "A comparative assessment of surface microstructure and electrical conductivity dependence on co-solvent addition in spin coated and inkjet printed poly(3,4-ethylenedioxythiophene):polystyrene sulphonate (PEDOT:PSS)," *Organic Electronics*, Vol. 13, No. 3, 2012, pp. 409-418.

[55] Sankaralingam, S., and B. Gupta, "Determination of Dielectric Constant of Fabric Materials and Their Use as Substrates for Design and Development of Antennas for Wearable Applications," *IEEE Transactions on Instrumentation and Measurement*, Vol. 59, No. 12, 2010, pp. 3122–3130.

[56] Declercq, F., et al., "Environmental High Frequency Characterization of Fabrics Based on a Novel Surrogate Modelling Antenna Technique," in *IEEE Transactions on Antennas and Propagation*, Vol. 61, No. 10, 2013, pp. 5200–5213.

[57] Lilja, J., P. Salonen, T. Kaija, and P. de Maagt, "Design and Manufacture of Robust Textile Antennas for Harsh Environments," in *IEEE Transactions on Antennas and Propagation,* Vol. 60, No. 9, 2012, pp. 4130–4140.

[58] Le, D. et al., "Microstrip Transmission Line Model-Fitting Approach for Characterization of Textile Materials as Dielectrics and Conductors for Wearable Electronics," *Int. J. Numer. Model*, Vol. 32, No. 6, 2019, e2582.

[59] Frank, E., V. Bauch, F. Schultze Gebhardt, and K. H. Herlinger, "Fibers, 1. Survey," in *Ullmann's Encyclopedia of Industrial Chemistry,* Seventh Edition, Weinheim, Germany: Wiley-VCH Verlag GmbH, 2011.

[60] Ouyang, Y., and W. J. Chappell, "High Frequency Properties of Electro-Textiles for Wearable Antenna Applications," *IEEE Transactions on Antennas and Propagation*, Vol. 56, No. 2, 2008, pp. 381–389.

[61] Kiourti, A., and J. L. Volakis, "Stretchable and Flexible E-Fiber Wire Antennas Embedded in Polymer," *IEEE Antennas and Wireless Propagation Letters*, Vol. 13, 2014, pp. 1381–1384.

[62] He, P., et al., "Screen-Printing of a Highly Conductive Graphene Ink for Flexible Printed Electronics," *ACS Appl. Mater. Interfaces*, Vol. 11, No. 35, 2019, pp. 32225–32234.

[63] Copprint LF-371, Copprint, Technical Data Sheet, 2020.

[64] Tai, Y., and Z. Yang, "Preparation of Stable Aqueous Conductive Ink with Silver Nanoflakes and Its Application on Paper-Based Flexible Electronics," *Surface and Interface Analysis*, Vol. 44, No. 5, 2011, pp. 529–534.

[65] Hong, H, J. Hu, and X. Yan, "UV Curable Conductive Ink for the Fabrication of Textile-Based Conductive Circuits and Wearable UHF RFID Tags," *ACS Appl. Mater. Interfaces*, Vol. 11, No. 30, 2019, pp. 27318–27326.

[66] Abbas S. M, et al., "On-Body Antennas: Design Considerations and Challenges," *2016 URSI International Symposium on Electromagnetic Theory (EMTS), Espoo, Finland*, 2016, pp. 109–110.

[67] Rahman, A., et al., "Analysis on the Effects of the Human Body on the Performance of Electro-Textile Antennas for Wearable Monitoring and Tracking Application" *Materials*, Vol. 12, No. 10, 2019, p. 1636.

[68] Rahmat-Samii, Y., and E. Topsakal, "Antenna and Sensor Technologies in Modern Medical Applications," in *In Vitro and In Vivo Testing of Implantable Antennas*, R.B Green, M.V Smith, and E. Topsakal, Hoboken, NJ: John Wiley & Sons, 2020, pp. 145–189.

[69] Zhang, H., et al., "Design of Low-SAR Mobile Phone Antenna: Theory and Applications," *IEEE Transactions on Antennas and Propagation*, Vol. 69, No. 2, 2021.

[70] Hasni, U., et al., "Screen Printed Fabric Antennas for Wearable Applications," *IEEE Open Journal of Antennas and Propagation*, Vol. 2, 2021.

[71] Ibanez-Labiano, I., et al., "Graphene-based Soft Wearable Antennas," *Applied Materials Today*, Vol. 20, 2020.

[72] Blauert, J., and A. Kiourti, "Theoretical Modeling and Design Guidelines for a New Class of Wearable Bio-Matched Antennas," *IEEE Transactions on Antennas and Propagation*, Vol. 68, No. 3, 2020, pp. 2040–2049.

[73] Blauert, J., and A. Kiourti, "Bio-Matched Antennas with Flare Extensions for Reduced Low Frequency Cutoff," *IEEE Open Journal of Antennas and Propagation,* Vol. 1, 2020, pp. 136–141.

CHAPTER 5

Wearable Sensors

Shubhendu Bhardwaj, Dieff Vital, and John Volakis

5.1 Sensing with Wearables

While human-sensor interfaces have advanced, current state-of-the-art healthcare devices still rely on obtrusive and battery-based electronics (smart watch, flex circuit-board patches, and electronic cuffs) to collect and relay medical data. As a result, these interfaces have not been actively employed or accepted by the medical community for long-term data collection. The latter does not serve our elderly population who require the highest level of comfort. Concurrently, there is an imminent need for privacy-preserving artificial intelligence/machine-learning (AI/ML) techniques that take advantage of low-cost computations. Notably, continuous medical monitoring has direct relevance to cardiac and diabetic complications, affecting 26.5% of the U.S. elderly population [1]. Further, the correlation between diabetes and cardiovascular conditions is well-known [2, 3].

The potential of continuous health-tracking and associated data learning is yet to be realized. This is because of the challenges associated with the *wearability* of modern sensing systems; namely, the sensing systems are not comfortable for daily or prolonged use. Indeed, state-of-the-art sensors use obtrusive electronics with appendages near the sensing site (see Figure 5.1, [4–10]). Obtrusive electronics lead to limited user experience, especially for elderly people and children who may require the highest level of comfort. Past attempts [11, 12] have focused on the use of plastic electronic substrates. But the technology for integrating electronics into fabric surfaces is missing. Flexible monolithic electronics, where the transistors and didoes are grown on flexible substrates, has also been considered [5, 13–15], but actual devices have yet to be translated to GHz-RF electronics and require innovations in material sciences.

Obtrusive electronics and closed systems have been intrinsically a part of the medical sensing technology but suffer from limited wearability and comfortable sensing. To address this, integration of the electronics with the fabric substrate with expandability features is a key target technology. Also, simple wireless power and data transfer modalities have become apparent, and further, to create an open

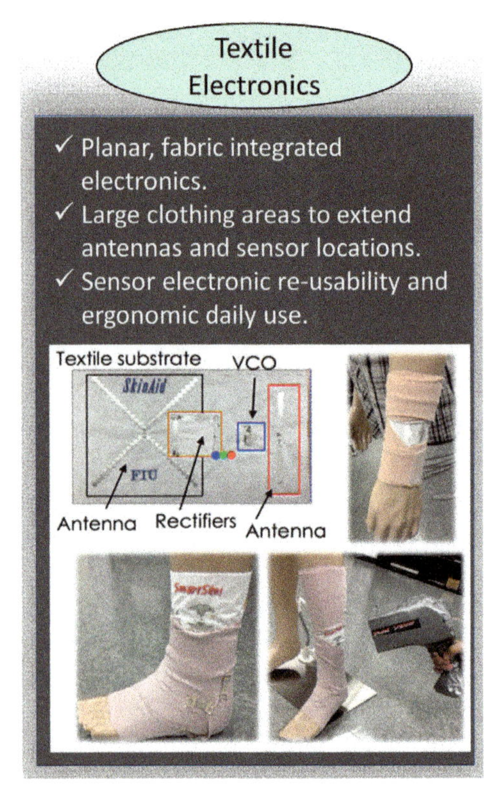

Figure 5.1 Motivations for integrating sensor systems to textile surfaces.

system, remateability of interconnections on planar electronic is required, which will be discussed in this chapter.

5.2 Wearable Electronics for Biomarker Extraction

Future smart health and connectivity will be based on smart bandages and skin sensors that have minimalistic electronics and can provide seamless monitoring of biomarkers. Remote detection of electrochemical sensor data is a challenge that requires special attention. Electrochemical sensors and the associated data are quite useful and are routinely used for detection of uric acid for wound monitoring applications, detection of cortisol of stress, detection of glucose for diabetes, and many other modalities. A typical detection system would wireless powering method and on-fabric simplistic circuitry to realize smart sensing systems with low-power consumption. If developed, this improvement would help patients overcome their financial burdens and healing process (fast recovery), and obtain a better quality of life.

However, often sensors and electronics uses Li-ion batteries to power the circuitry and operate the bandages. The latter routinely also features complex and rigid microcontroller circuits that cannot be integrated into clothing. To make that possible, a voltage-controlled oscillator (VCO)-based wound health monitoring circuit is described in this chapter. The sensor system uses flexible, fabric-integrated, and wireless electrochemical sensing. The system is shown for wound monitoring and pH monitoring cases. Using this system, the characterization of the electrochemical fluid can be done with specificity and selectivity, which are desired features in such systems. For integration into the bandage, the circuit is developed on a fabric substrate, which allows possible bandage integration for comfort, flexibility, and wearability of the sensor. Furthermore, in the shown concept, the sensor system features are designed to be powered from an incident radio frequency signal remotely given by a nearby interrogator, which will enable transmit-reflect type operation of the sensor. This operation features a very critical circuit, the voltage-controlled oscillator that converts the sensor's DC output current into an RF signal whose frequency can be modulated based on the output current information.

5.3 Wound Monitoring RFID Bandage on Textile Surface

A system schematic and operational view of the envisioned wireless wound monitoring system is shown on the left side in Figure 5.2. The proposed system will use an interrogator, which will transfer wireless RF power to be received by the fabric-based antenna integrated in the bandage. An RF to DC rectifier circuit, which is based on a flexible textile substrate, is used to convert the received RF power into DC voltage. The obtained DC power is used to power a uric acid sensor and voltage-controlled oscillator to modulate the DC output from the sensor into an RF frequency. The frequency of the RF signal from the VCO will be a function of the wound fluid's uric acid concentration. The generated RF signal will be transmitted back to the interrogator device transferring the wound data back to the interrogator. In this chapter, we will demonstrate the feasibility and operation of a textile-integrated VCO-based wireless sensor, which may be used for developing the system shown in Figure 5.1. In this work, uric acid is used as the biomarker and its characterization is done by electrochemical sensing (see Figure 5.2, right (2)). The sensor used is a textile-based enzymatic sensor [16]. The sensor is powered by the textile-based rectifier and after sensing the uric acid concentration and sends the wound information in the form of an electric signal that is fed in the textile-based oscillator, which at the end modulates the wound-data in the form of frequency that

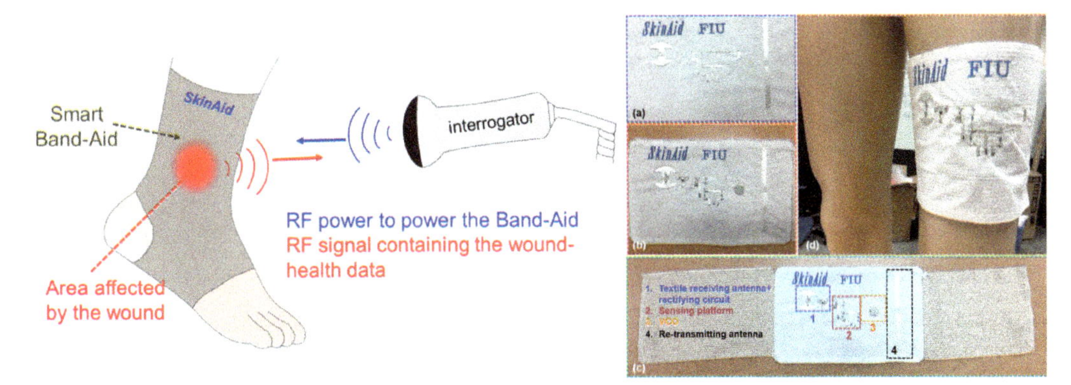

Figure 5.2 Left: Smart bandage RFID concept relies on interrogator power in transmit-reflect modality where the data-signal containing the wound-health data is modulated and returned. Right: A practical implementation of the circuit on a flexible clothing surface.

is communicated back to the interrogator. In the next sections, the electrochemical sensor, the VCO, and a benchtop version of the smart bandage will be presented and discussed.

5.4 Textile Based Voltage-Controlled Oscillator

The key component in creating a medical monitoring RFID for electrochemical sensor is a VCO block, which is capable of converting a DC current signal from a sensor to an RF frequency, which can be modulated by the DC current signal's value. The electrochemical sensors have three electrodes: a reference electrode, a working electrode, and a counter or auxiliary electrode. The DC bias is required to be applied at the working electrode with respect to the auxiliary electrode, while the output voltage is measured between the working electrode and the reference electrode. In the proposed RFID system, the output is connected to a VCO to convert the DC signal into a communicable RF signal.

The VCO circuit was designed using a Crystek CVCO055BE and realized on a flexible fabric substrate by using embroidery of conductive thread (Elektrisola-7) onto a gauze fabric ($\varepsilon_r = 1.67$ and $\tan\delta = 0.07$) and can be seen in Figure 5.3. The conductive surfaces were made using embroidery on a Brother sewing machine, which uses an automated CAD design method to implement a desired pattern of conductive traces. The voltage-controlled oscillator chip, a Crystek CVCO055BE, was used to convert a DC output level obtained from the sensor to an oscillating RF frequency for data modulation. The capacitance $C_1 = 1$ nF, and $C_2 = 10$ nF, and the $V_{CC} = 3.14$ V (and 10.479-mA current) were used to calibrate the IC for operation based on the desired output and input. To test the IC's performance on a fabric substrate, we varied the control voltage from $V_{tuning} = 0$ to 6.2V (and 4 μA current) in order to emulate a feedback of the electrochemical sensor due to changing

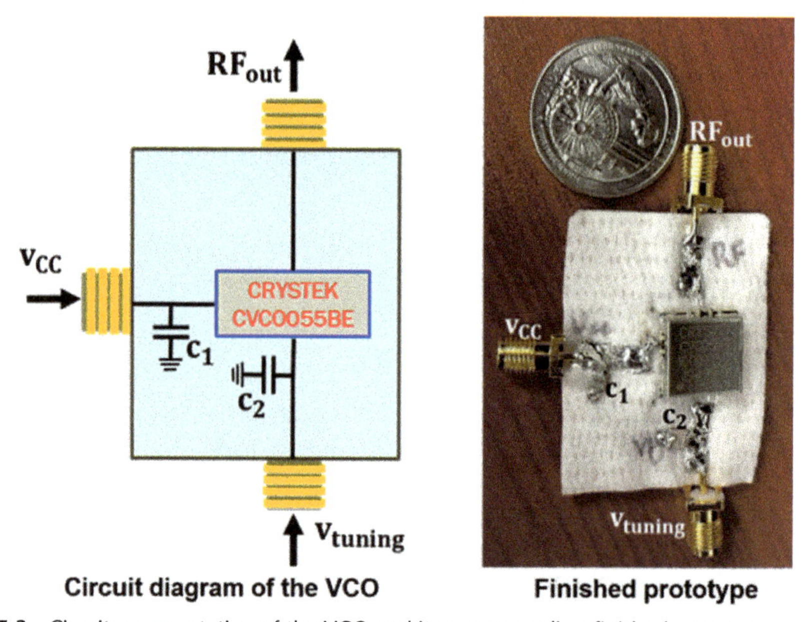

Figure 5.3 Circuit representation of the VCO and its corresponding finished prototype.

concentration of uric acid present in the wound fluid. The DC voltage was given using a Keysight E36312A power supply (see Figure 5.4, right). The current drawn by the IC was recorded to be 10.5 mA. Therefore, power requirements of the operation were within 33 mW. A comparison of the output frequency obtained in experiments with the IC's normal operation based on the equation

$$f_{VCO} = 836 \times \left(1 + \operatorname{asinh}\left(\frac{2C_1}{C_2} \times V_{CTRL}\right)\right) \tag{5.1}$$

where the related results and measurements are provided in Figure 5.5, left. The V_{tuning} was varied from 0 to 6.2V and the output modulated frequency (f_{VCO}) was measured to be between 836 MHz and 1.712 GHz. The spectra associated to these frequencies were measured using a Keysight PXA signal analyzer N9030B and were between 0.15 dBm and 3.8 dBm (see Figure 5.5, right). There is an agreement between the theoretical and measured values showing an expected operation of the VCO for the sensor-integration application. The experiment shows that flexible substrates may be used for integrating commercially available VCO for operation, proving the utility for a textile-based RFID.

5.5 Wound Assessment Using Data Modulation

The feasibility of a VCO-based sensor for wound monitor data modulation for RF transfer is tested by conducting a benchtop experiment (shown in Figure 5.6), in which a DC power source is connected to a TI BQ25504 power management system, which would provide a constant input to the uric acid sensor while also powering the VCO IC. The DC source emulates the power received from a rectenna powered from a wireless power transmission source. In addition to the PMC, it contains the sensor, the VCO, and a pair of 915-MHz microstrip patch antennas distanced by 25 cm, used for power reflection back to the interrogator, a Keysight

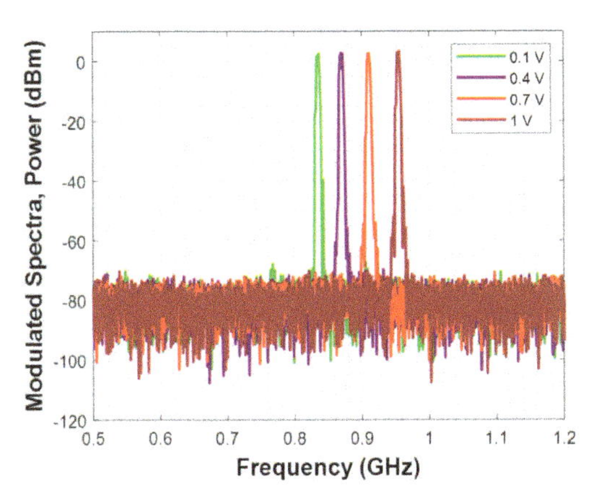

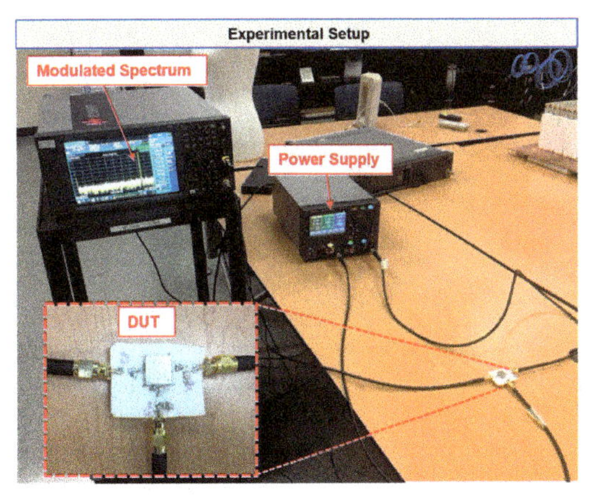

Figure 5.4 Experimental setup and results from the VCO characterization. Left: modulated spectra from frequencies obtained from tuning the VCO from 0.1 to 1V. Right: the experimental setup.

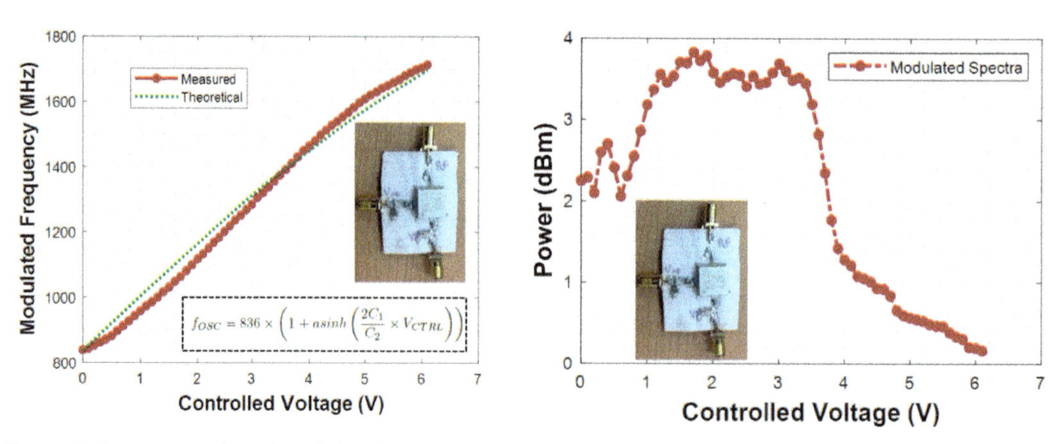

Figure 5.5 Measured results of the characterization of the VCO. Left: power level of different spectra recorded from frequency modulation. Right: measured output modulated frequency of the VCO.

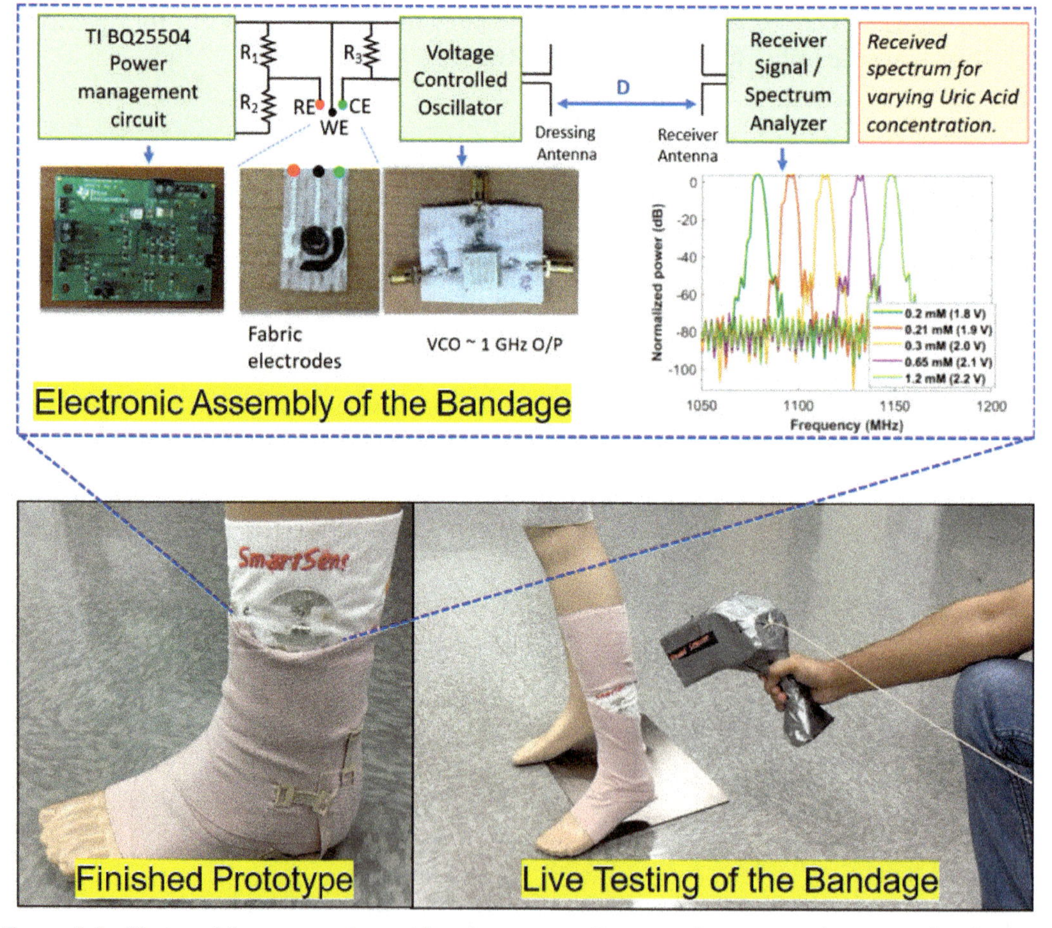

Figure 5.6 Photos of the proposed smart bandage responding to an interrogator by responding back to the interrogator with a modulated signal of the wound-data (bottom left) finished prototype of the bandage, and (bottom right) live testing of the bandage based on the layout (zoomed-out on top). (©2021 IEEE, [17].)

PXA signal analyzer. This is the equivalent of a testing system where the proposed smart bandage is illuminated by a scanner (interrogator). The tests are conducted by using uric acid solutions of different molar concentrations to understand the sensor's modulation capability in relation to practical molar values in wound. Figures 5.6 and 5.7 show the experimental setup and measurement results for this wound assessment test. In this setup, the TI BQ25504 ultralow-power boost converter uses a voltage divider to assure a constant voltage of -0.6V (as suggested in [16]) is being delivered to the sensor. This power management circuit is to be replaced by a constant output rectifier in future work in this direction. The latter used a bias power of 10 mW from a Keysight E36312A triple output programmable DC power supply. The sensor is dipped into a container of uric acid molar solution at different concentrations. The VCO was supplied 0.38 mW by the power management circuit (BQ25504). The frequency modulation was evaluated for uric acid of concentration ranging from 200 μM to 1 mM. The feedback voltage from the output of the VCO ranged from 1.875 V to 2.15V following the equation

$$V_{tuning}(V) = 2.175 + 0.105 \times \log_2\left(0.235 \times \left(C_{UA}\right)^{0.205}\right) \tag{5.2}$$

The feedback of the sensor was also recorded using a Keysight U1242B true RMS multimeter. Measured results can be seen in Figure 5.8. The wound-data provided by the VCO is captured using a Keysight PXA signal analyzer N9030B showing the spectrum for uric acid level and the corresponding equation is

$$f_{Bandage}(MHz) = 1121.256 + 12.365 \times \log_2\left(0.965 \times C_{UA}\right) \tag{5.3}$$

The output return transmitted spectra are shown in Figure 5.8(c) and (d), which shows the translational shifting of the spectral peak with changing levels of uric

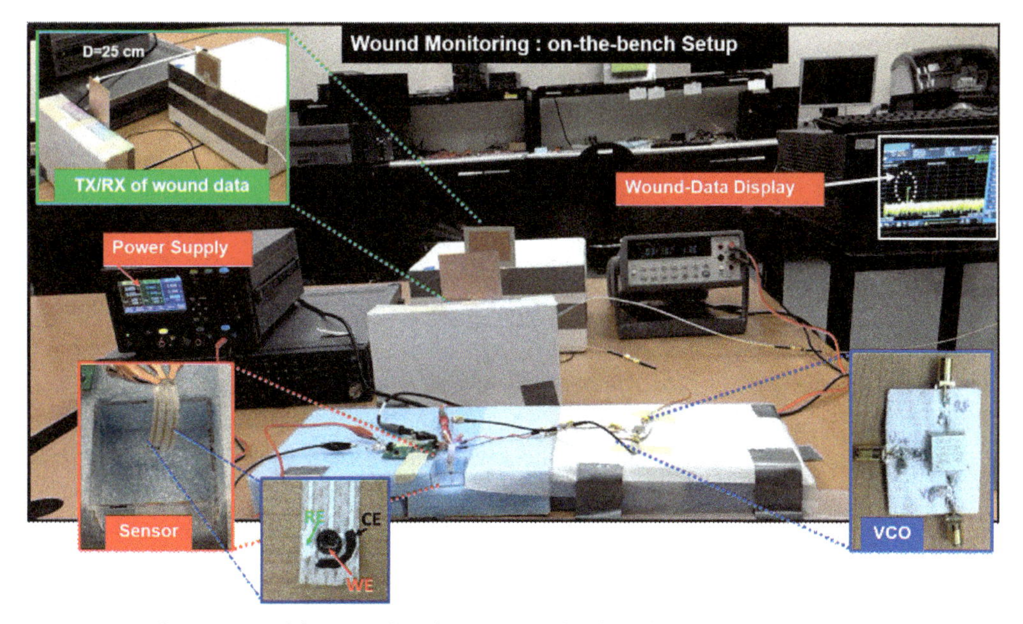

Figure 5.7 Benchtop setup of the smart bandage system developed with the integration of on-fabric circuits emulating the wound healing assessment. (©2020 IEEE [18]. ©2021 IEEE [17].)

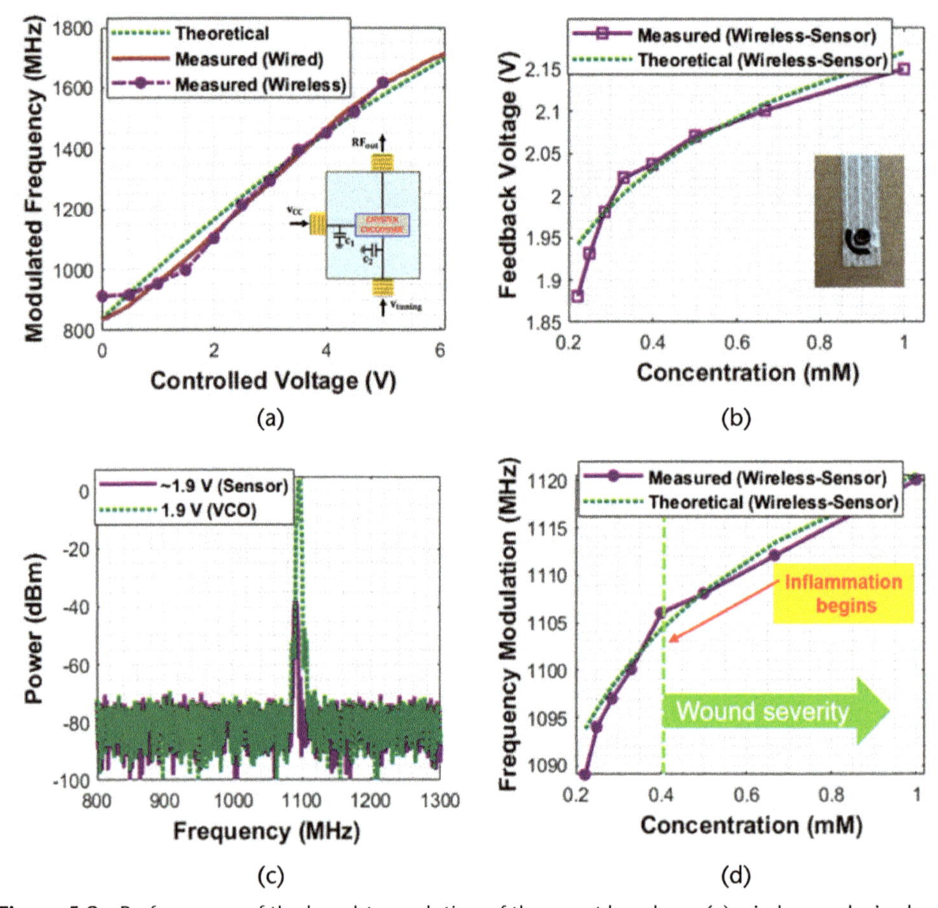

Figure 5.8 Performance of the benchtop solution of the smart bandage: (a) wireless and wired performance of the textile-based VCO compared with its theoretical model, (b) feedback response of the electrochemical sensor, (c) spectra representing the output of the VCO along with that of the sensor + VCO, and (d) measured and theoretical models of the wound assessment using the smart bandage.

acid concentration. The variation of the VCO's output frequency spectral peaks are shown in Figure 5.8(b) with changing molar concentration (CUA) levels. The signals from the modulated frequency were retransmitted back to the interrogator at 25 cm using a pair of 915-MHz antennas. The signals can also be received by any remote receiver, such as a phone, tablet, or computer.

An important parameter is the sensitivity of modulated data with changing uric acid concentration, which is calculated to be 44 MHz/μM. This sensitivity is adjustable by changing the ratio of resistors C1 and C2. The total power consumed by the sensor alone is 24 μW. To parameterize the assessment of a wound healing status based on an obtained frequency, an equation was developed by curve fitting of the experimental data. Two models were developed. The first model represents the relation between uric acid concentration CUA in millimolars (mM) with the applied voltage VCTRL in volts (V), as can be seen in (5.2). The second represents the modulated frequency ($f_{Bandage}$) in MHz as a function of the variable uric acid concentration (CUA) and displayed in (5.3). These equations can be used for quick and reliable assessment of the wound. Because of these unique attributes, this smart bandage should be appealing to customized and personalized care because of its cost-efficiency, ease to fabricate, and the theoretical model used for quick and reliable assessment.

5.6 Smart Bandage Integration for Practical Measurements

Previously published electrochemical monitoring wearable solutions are battery-powered and/or consisted of transceiver ICs for transmittance of the modulated signal to a remote receiver. These solutions lead to bulky and uncomfortable-to-wear electronics [19–21] on the human body. New wearable devices will require textile integration with minimalistic electronics (with the ability to be integrated into a single chip).

Therefore, a potential solution should cater to challenges of (1) wireless illumination, (2) wireless data transfer, and (3) conformable and low-profile clothing-integrable electronics. To do so, a robust frequency modulation-based RFID modality is proposed (consisting of a transmit-reflect module), which directly modulates the sensor-data and reflects frequency-modulated signals (with the help of a voltage-controlled oscillator) to a remote location. A remotely powered, smart dressing solution is used to reduce battery requirements, while frequency data-modulation and signal reflection are used for robust electrochemical assessment of chronic wounds (see Figure 5.9). In this new experiment, we demonstrate the performance of a battery-free electrochemical sensing platform integrated on fabric surfaces with wireless power transfer and harvesting for health assessment. We

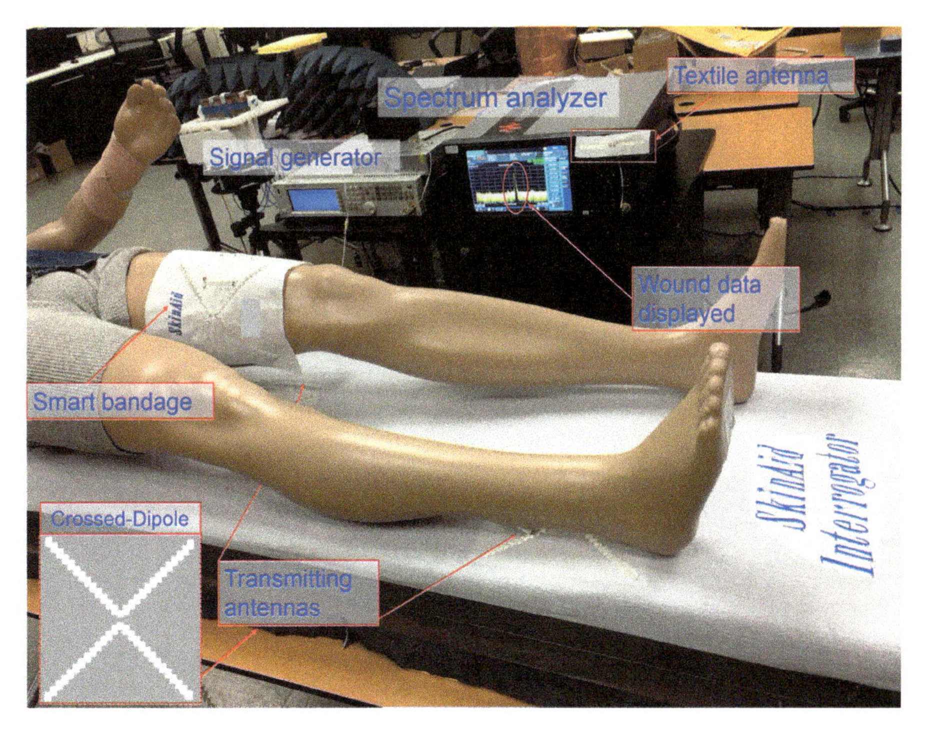

Figure 5.9 Complete wirelessly powered system representing the smart bandage with all the different components. Wound data is collected at more than 3 feet away from the bandage by the textile antenna (attached to the spectrum analyzer).

present a fully functional setup for wound monitoring consisting of an interrogator, a bandage consisting of a textile-based rectifier, a textile VCO, and a textile-based antenna for data transfer. A remote receiver emulated by a spectrum analyzer is used to capture the wound data. We emulate the electrochemical sensor's DC signal output through a resistive load in the circuit.

Figure 5.9 shows the blocks of the system that operate as follows: (1) the power transmitter module (e.g., integrated in a bedsheet) consists of a transmitting antenna to provide RF power to a similar receiving antenna located in the bandage, using frequencies between 350 and 573 MHz. (2) The bandage features a textile-substrate rectifier to convert the RF power into DC to power an electrochemical sensor and a VCO. (3) A trielectrode sensor with typical power usage of 24 μW senses uric acid (emulating chronic wound fluid) concentration to provide a DC current as a function of the concentration. (4) The VCO converts the DC signal into an RF signal, providing a simple data modulation for transmittance through a second textile-based data-antenna (operating at 915 MHz). The modulated output signal, based on its frequency, will provide the uric acid concentration extracted from the wound fluid. Based on the value of the concentration, the wound-health status can be assessed (about the severity or restoration of damaged tissue) by obtaining a unique frequency from the remote receiver.

5.7 Wireless Power Telemetry Link

5.7.1 Near Field Power Transfer Using a Corrugated Crossed-Dipole Antenna

The system in Figure 5.9 illustrates a bedridden patient with a bandage on their left leg. The bandage circuit is powered through a transmitter antenna integrated in the bed. The proposed antennas were designed and fabricated using conductive fiber embroidery [13]. The shape of the antenna (see Figure 5.9, inset), a corrugated crossed dipole, is based on prior-proposed misalignment resilient anchor-shaped antennas [14]. Misalignment resilience helps patients who may be subject to intermittent movements, which could interrupt the effective transfer of power from the bed to the bandage. The topology is based on prior works [14], and features corrugations known for modifying the electromagnetic (EM) fields of antenna apertures and enable axial symmetry [15]. It is anticipated that the performance of the system will be stable on an average under lateral, diagonal, and angular misalignments. As can be seen in Figure 5.10, the peak power transfer efficiency (PTE) is 80%, when the misalignment is introduced along the direction of stronger electric fields and 60% along the direction of stronger magnetic fields. It can be inferred that the PTE is approximately constant under all misalignments. This is an indication that the system will be continuously powered regardless of the movement of the bearer.

5.7.2 Textile-Based Rectifier

Figure 5.11 shows the rectifier featured in the proposed smart bandage. This single-diode rectifying circuit is a modified version of the first textile-based single-diode rectifier using a $\lambda/8$ shorted stub [21]. This circuit uses an extra $\lambda/8$ stub before

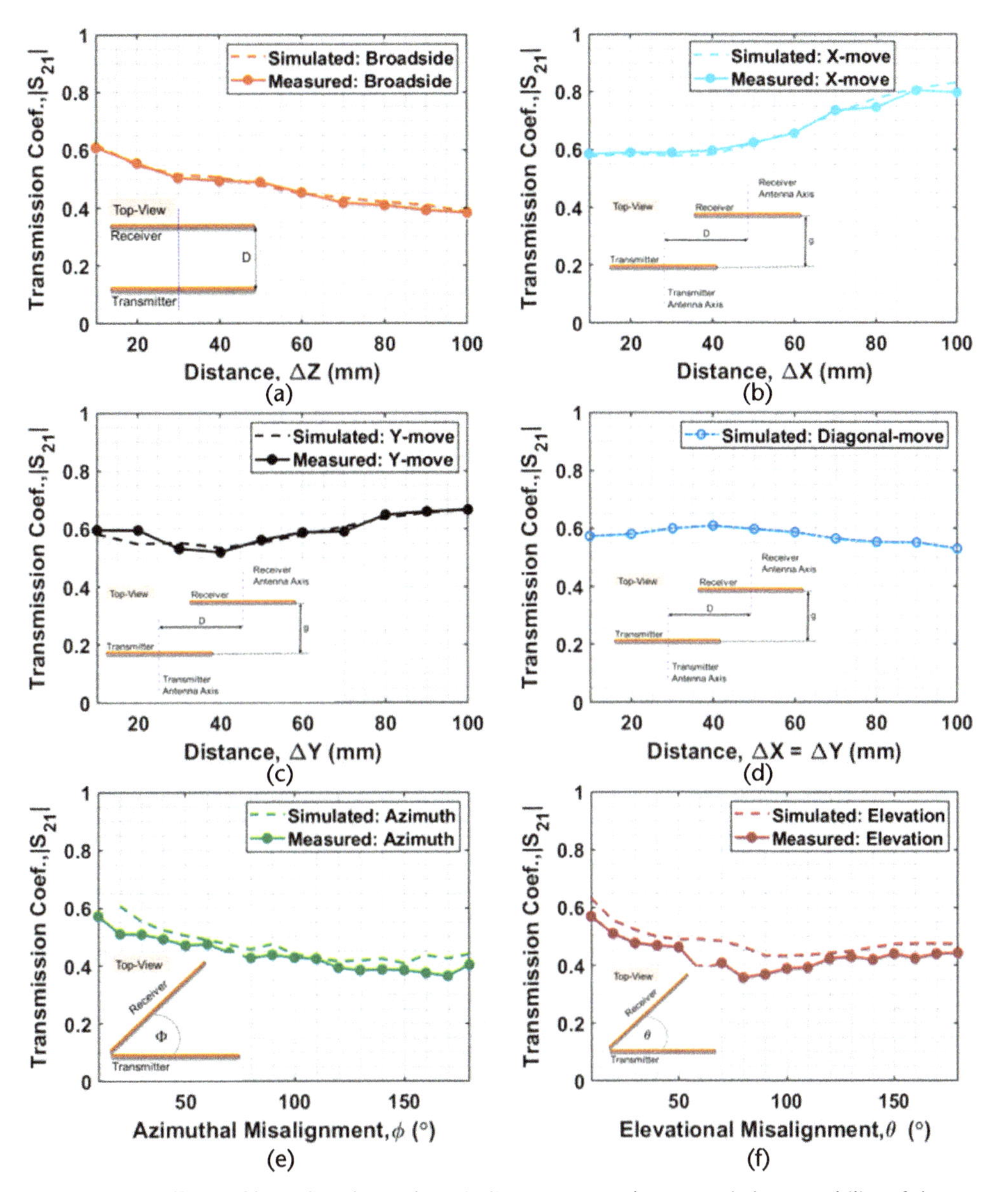

Figure 5.10 Effects of lateral and angular misalignments on the transmission capability of the antenna: (a) broadside direction when D varies from 1 to 10 cm, (b) and (c) lateral misalignments where g = 1 cm and DX, DY varies from 1 to 10 cm, (d) diagonal misalignment where g = 1 cm and DX = DY varies from 1 to 10 cm, (e) elevational misalignment, and (f) azimuthal misalignment.

the diode to compensate for the capacitive effect introduced by the diode. The impedance of the circuit is

$$Z_{circuit} = \frac{R_e}{1+\left(R_e C_{total}\omega\right)^2} - \frac{jR_e^2 C_{total}\omega}{1+\left(R_e C_{total}\omega\right)^2} \tag{5.4}$$

where $C_{total} = C_e + C_2$, C_e is the capacitance of the diode, and C_2 the capacitance of the small capacitor added in parallel with the diode. The imaginary part of (5.4)

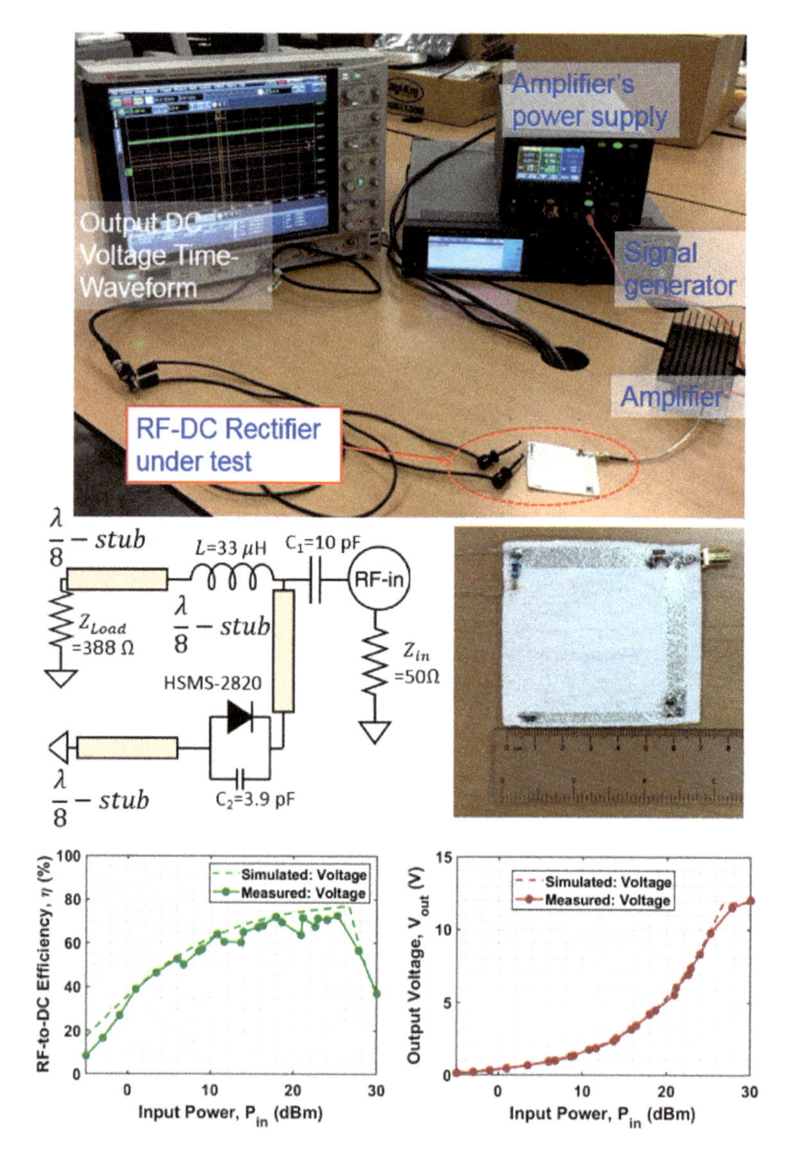

Figure 5.11 Textile-based rectifier. Top: measurement setup for the rectifying circuit. Middle left: circuit diagram of the rectifier with all the lumped components. Middle right: finished prototype of textile rectifier. Bottom left: simulation and measurement results of RF-to-DC conversion efficiency. Bottom right: simulated and measured collected DC voltage.

shows that the rectifier is primarily capacitive. For high-efficiency achievement, the capacitive part of the impedance should be suppressed by conjugate matching.

The goal here is to eliminate the presence of the imaginary part of $Z_{circuit}$ and minimize the real part as much as possible. The conjugate matching is achieved by equating the magnitude of the impedance of the $\lambda/$ stub to the $I(Z_{circuit})$. The rectifier was designed to operate at around 500 MHz. A full-wave simulation was realized to estimate the RF-to-DC conversion efficiency as well as the collected DC voltage using ADS. The circuit was later fabricated using automated embroidery of Elektrisola-7 onto gauze fabric and tested. It was found that the textile rectifier exhibited an RF-to-DC conversion efficiency of 76% at 26 dBm input and the DC voltage collected at that power level was 10V. This efficiency level is comparable or better than the current state of the art reported in [20].

5.8 Measurement Setup Realized to Emulate In Vivo Electrochemical Sensing and Monitoring Scenarios

To emulate real-life settings, we assembled a measurement system (apparatus) that comprised of all the subsystems mentioned above. The apparatus included a smart bed (Figure 5.12), the smart sensing system (Figure 5.13), and a point-of-collection of wound data provided by the modulated small RF signal given by the textile-based VCO. The RF signal is received by an omnidirectional antenna (Figure 5.14). The resulting measurement setup can be seen in Figure 5.15. The smart bed is illuminated by an RF signal generator and power amplifier. The RF signal transmitted by the smart bed is captured by the smart dressing solution (smart bandage) that converted it into DC to power the sensor and bias the VCO.

The sensor, upon dipping in the wound-fluid (here emulated by placing a resistor at the sensor location) senses the concentration of uric acid present in the wound fluid and the resulting DC signal is fed to the VCO. The VCO modulates the DC signal, and the obtained RF signal is sent to the 915-MHz dipole to retransmit it wirelessly to the remote receiver (spectrum analyzer). Upon reception of the modulated RF signal, the frequency is used for assessment. Using (5.3) we can determine the corresponding uric acid level that was detected by the sensor. The uric acid level that resulted for applying the assessment equation (5.3) will be

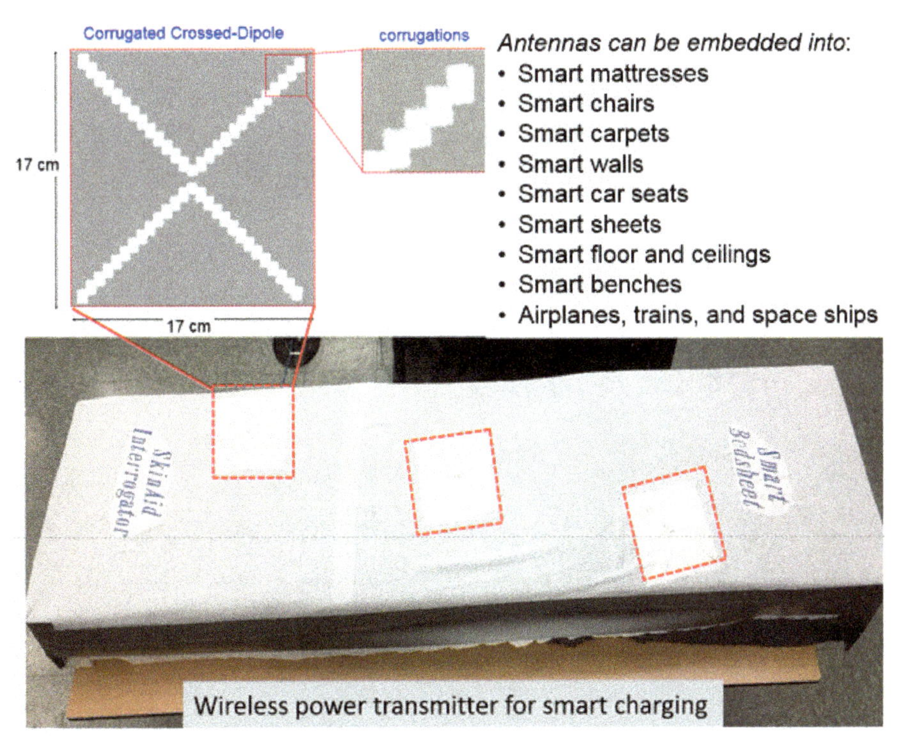

Figure 5.12 Wireless transmitting RF module to illuminate any receiving system for a smart charging, monitoring, and sensing platform.

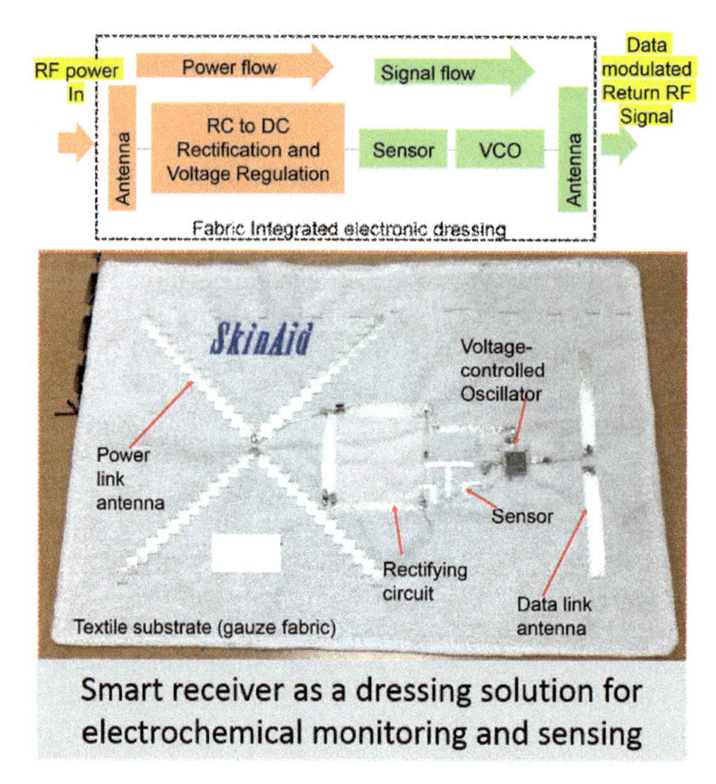

Figure 5.13 Top: A close-up of the smart bandage diagram showing the powering method and the data collection method. Bottom: The finished product representing the circuits corresponding to the various parts of the diagram made of conductive threads embroidered onto fabric substrates.

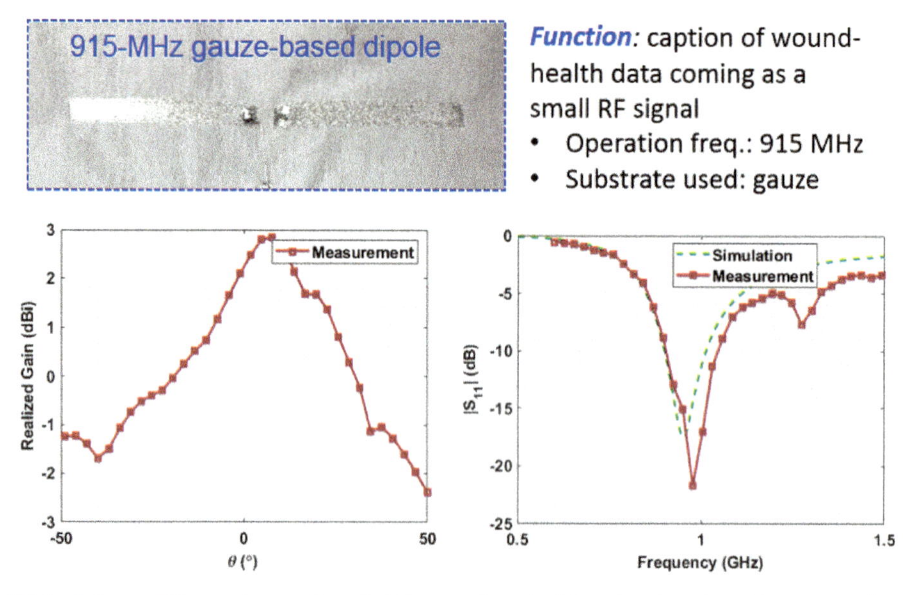

Figure 5.14 Retransmitting antenna placed at the input-port of the signal analyzer to capture the small RF signal resulted from the modulation of the wound-health data.

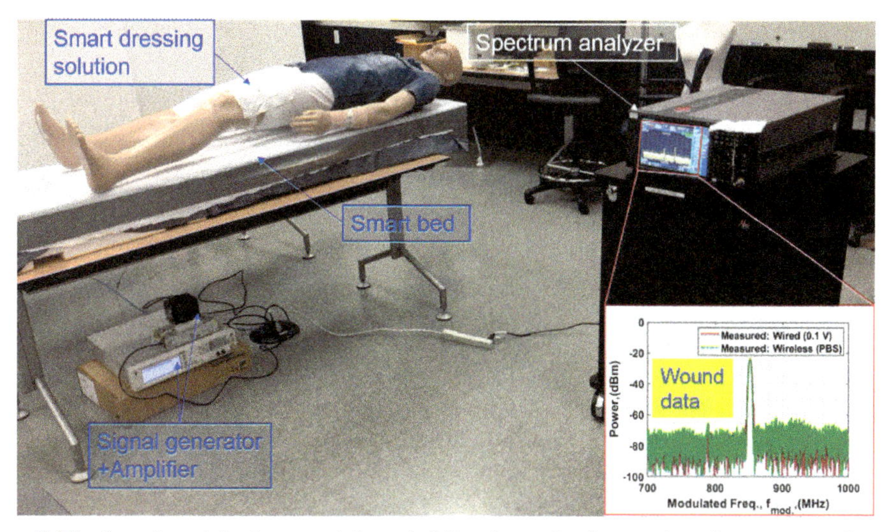

Figure 5.15 Experimental setup consisting of all the elements of power transfer, smart sensing, and data transmittance to achieve reliable electrochemical assessment of chronic wounds. This apparatus can be used as a reference to real-life scenarios.

compared with the threshold, which is 0.4 mM. The system was tested (see Figure 5.15) where the transmitter was the bed separated from the bandage receiver by 6 inches and the wound fluid was emulated by a resistive load. As can be seen in Figure 5.15, the output signal sent by the sensor had a voltage level of 0.1V and the corresponding frequency modulation was at 845 MHz.

This signal was captured at a distance of more than 3 feet away from the bandage. The corresponding concentration of the wound fluid is 0 mM < 0.4 mM. This result suggests that the wound is healed.

5.9 Conclusion

In this chapter, we proposed smart bandages that are textile-integrated, self-powered, and provide new methods to modulate the electrochemical sensor data for transmission. This was presented through two generations of smart bandages. Direct frequency modulation is achieved using a VCO-based DC-to-RF modulation. The bandages feature a textile-based rectifier and VCO along with misalignment-resilient textile-based antenna (anchor and corrugated crossed dipole) for power telemetry and a dipole antenna for wound-data communication link. In summary, wireless powering and wireless data-extraction using frequency modulation was established with excellent agreement with theoretical expectations of the rectifier and VCO. To the authors' knowledge, this is the first time a wirelessly powered battery-less smart bandage system, with power telemetry and wireless link, has been demonstrated with frequency modulation. The proposed system could positively affect personalized, connected care by providing comfortable wearability and long-term usage. It is demonstrated that robust and reliable assessment of chronic wounds can be done using a reduced electronic and textile-based voltage-controlled oscillator. The results can be used as a reference experiment for in vivo electrochemical monitoring of chronic wounds.

References

[1] *National Diabetes Statistics Report, 2020 (Estimates of Diabetes and Its Burden in the United States),* US Department of Health and Human Services (Centers for Disease Control and Prevention), https://www.cdc.gov/diabetes/library/features/diabetes-stat-report.html.

[2] Krittanawong, C., H. J. Zhang, Z. Wang, M. Aydar, and T. Kitai, "Artificial Intelligence in Precision Cardiovascular Medicine," *Journal of the American College of Cardiology*, Vol. 69, No. 21, May 30, 2017, pp. 2657–2664, doi: 10.1016/j.jacc.2017.03.571.

[3] Ariza, M. A., V. G. Vimalananda, and J. L. Rosenzweig, "The Economic Consequences of Diabetes and Cardiovascular Disease in the United States," *Reviews in Endocrine and Metabolic Disorders*, Vol. 11, No. 1, March 27, 2010, pp. 1–10, doi: 10.1007/s11154-010-9128-2.

[4] Mostafalu, P., et al., "Smart Bandage for Monitoring and Treatment of Chronic Wounds," *Small*, Vol. 14, No. 33, August 2018, p. 1703509doi: 10.1002/smll.201703509.

[5] Heikenfeld, J., et al., "Wearable Sensors: Modalities, Challenges, and Prospects," *Lab on a Chip*, Vol. 18, No. 2, January 21, 2018, pp. 217–248, doi: 10.1039/c7lc00914c.

[6] ECG Sensor|TAGecg Wearable ECG Sensor, https://www.welchallyn.com/en/products/categories/cardiopulmonary/wearable-ecg-monitors/tagecg-wearable-ecg-sensor.html.

[7] Dieffenderfer, J., et al., "Low-Power Wearable Systems for Continuous Monitoring of Environment and Health for Chronic Respiratory Disease," *IEEE J. Biomed. Heal. Informatics*, Vol. 20, No. 5, September 2016, pp. 1251–1264, doi: 10.1109/JBHI.2016.2573286.

[8] Umasankar, Y., M. Mujawar, and S. Bhansali, "Towards Biosensor Enabled Smart Bandages for Wound Monitoring: Approach and Overview," *Proceedings of IEEE Sensors*, Vol. 2018, October 2018, doi: 10.1109/ICSENS.2018.8589786.

[9] Wearable Self-Powered Adaptive Platform—Center for Advanced Self-Powered Systems of Integrated Sensors and Technologies (ASSIST), https://assistcenter.org/self-powered-ecg-shirt/.

[10] Wound Healing—Center for Advanced Self-Powered Systems of Integrated Sensors and Technologies (ASSIST), https://assistcenter.org/wound-healing/.

[11] Zhang, Y., et al., "Battery-Free, Lightweight, Injectable Microsystem for In Vivo Wireless Pharmacology and Optogenetics," *Proc. Natl. Acad. Sci. U. S. A.*, Vol. 116, No. 43, October 2019, pp. 21427–21437, doi: 10.1073/pnas.1909850116.

[12] Gutruf, P., et al., "Wireless, Battery-Free, Fully Implantable Multimodal and Multisite Pacemakers for Applications in Small Animal Models," *Nat. Commun.*, Vol. 10, No. 1, December 2019, pp. 1–10, doi: 10.1038/s41467-019-13637-w.

[13] Chung, S., K. Cho, and T. Lee, "Recent Progress in Inkjet-Printed Thin-Film Transistors," *Adv. Sci.*, Vol. 6, No. 6, March 2019, p. 1801445doi: 10.1002/advs.201801445.

[14] Xu, L., et al., "3D Multifunctional Integumentary Membranes for Spatiotemporal Cardiac Measurements and Stimulation Across the Entire Epicardium," *Nat. Commun.*, Vol. 5, No. 1, February 2014, pp. 1–10, doi: 10.1038/ncomms4329.

[15] Semple, J., D. G. Georgiadou, G. Wyatt-Moon, G. Gelinck, and T. D. Anthopoulos, "Flexible Diodes for Radio Frequency (RF) Electronics: A Materials Perspective," *Semicond. Sci. Technol.*, Vol. 32, No. 12, 2017.

[16] Zeng, F. G., S. Rebscher, W. Harrison, X. Sun, and H. Feng, "Cochlear Implants: System Design, Integration, and Evaluation," *IEEE Rev. Biomed. Eng.*, Vol. 1, 2008, pp. 115–142, doi: 10.1109/RBME.2008.2008250.

[17] Vital, D., S. Bhardwaj, J. L. Volakis, P. Bhushan, and S. Bhansali. "Smart bandage for electrochemical monitoring and sensing using fabric-integrated data modulation." U.S. Patent No. 11,065,164, issued July 20, 2021.

[18] Vital, D., S. Bhansali, J. L. Volakis, and S. Bhardwaj, "Electronic Wound Monitoring Using Fabric-Integrated Data Modulation," *2020 IEEE International Symposium on Antennas and Propagation and North American Radio Science Meeting,* IEEE, July 2020, pp. 1351-1352.

[19] Mostafalu, P., et al., "Smart Bandage for Monitoring and Treatment of Chronic Wounds," *Small*, Vol. 14, No. 33, August 2018, p. 1703509, doi: 10.1002/smll.201703509.

[20] Derakhshandeh, H., S. S. Kashaf, F. Aghabaglou, I. O. Ghanavati, and A. Tamayol, "Smart Bandages: The Future of Wound Care," *Trends in Biotechnology*, Vol. 36, No. 12. December 2018, pp. 1259–1274, doi: 10.1016/j.tibtech.2018.07.007.

[21] Kassal, P., et al., "Smart Bandage with Wireless Connectivity for Uric Acid Biosensing as an Indicator of Wound Status," *Electrochem. Commun.*, Vol. 56, pp. 6–10, 2015, doi: 10.1016/j.elecom.2015.03.018.

[22] Vital, D., P. Gaire, S. Bhardwaj, and J. L. Volakis, "An Ergonomic Wireless Charging System for Integration with Daily Life Activities," *IEEE Transactions on Microwave Theory and Techniques,* pp. 69.

CHAPTER 6

Wearable RF Harvesting

Dieff Vital, Shubhendu Bhardwaj, and John L. Volakis

6.1 Part 1: Far-Field Integrated Power Transfer and Harvesting for Wearable Applications

6.1.1 Introduction

Integration of the wireless charging platform on clothing is enabled by the proposed embroidery processes. At the same time, it is hypothesized that clothing surfaces can allow larger power collection due to larger surface available area. We consider the integration of antenna array and rectifiers for far-field power collection in this chapter. In effect, the goal of implementing a novel RF harvesting jacket is realized. We begin with the design, simulation, and fabrication of a textile-based single patch antenna made from the Elektrisola-7 thread embroidered onto organza. This is followed by the design, simulation, and fabrication of a single-diode rectifying circuit resonating at the same frequency. Later, both antenna and rectifying circuit were combined into 2×2 and 2×3 rectenna arrays for power transfer and harvesting. Our rectifier is distinct from other published rectifiers since (1) it features a single-diode topology for the rectifier circuit to achieve an RF-to-DC efficiency of 70% at 8 dBm and (2) does not use a management circuit. The design-optimization and fabrication process of the rectifying circuit was precisely controlled to avoid sacrificing the conversion efficiency. In this chapter we also report the performance of the 2×2 rectenna array in a realistic Wi-Fi environment and the 2×3 rectenna array by amplifying the available Wi-Fi signals and then harvesting them. In effect, we proposed and tested the large area and fully flexible arrays of RF-power transfer and harvesting rectennas (see Figure 6.1). This system uses automated embroidery of Elektrisola-7 thread onto organza fabric. As a result, the prototyped structure is robust, flexible, and washable. In past works, several other flexible electronics approaches have been demonstrated for power transfer and harvesting. These include conductive tapes [2–4], screen printing and inkjet printing [5, 6], and liquid metal alloy [7], along with conductive textiles known as e-threads [8–11] and/or metallized fabrics [12, 13]. In the next sections, a complete design, simulation, and fabrication of the full system is presented and discussed.

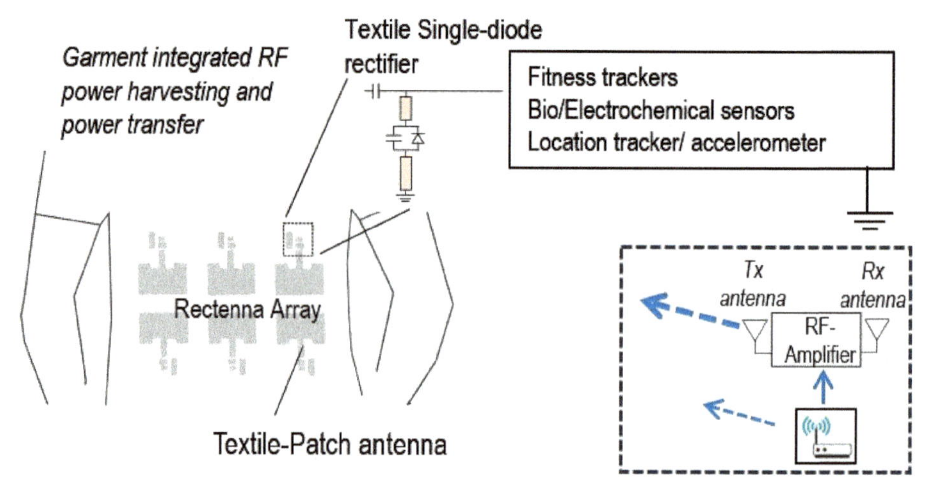

Figure 6.1 Wireless power harvesting system comprising of Wi-Fi router (bottom), rectenna array (left), and various wearable applications (top right). (©2019 IEEE [1].)

6.1.2 Conductive Thread Embroidery-Based Fabrication of Patch Antenna

The steps involved in the fabrication process are shown in Figure 6.2. The conductive surfaces are made of Elektrisola-7 and optimized via circuit or full-wave simulations and then are converted to respective CAD models. The CAD models are then imported to a Brother™ Innovis VM5100 embroidery software, where a pattern of needle-paths (digitization of the CAD model) is generated for the embroidery process.

The digitized model is then fed to the automated embroidery machine for fabrication. The conductive surfaces are generated using automated embroidery of e-threads (Elektrisola-7) on a regular cloth and a second layer of textile embroidery is used to implement the ground plane as shown in Figure 6.3(a). In this work,

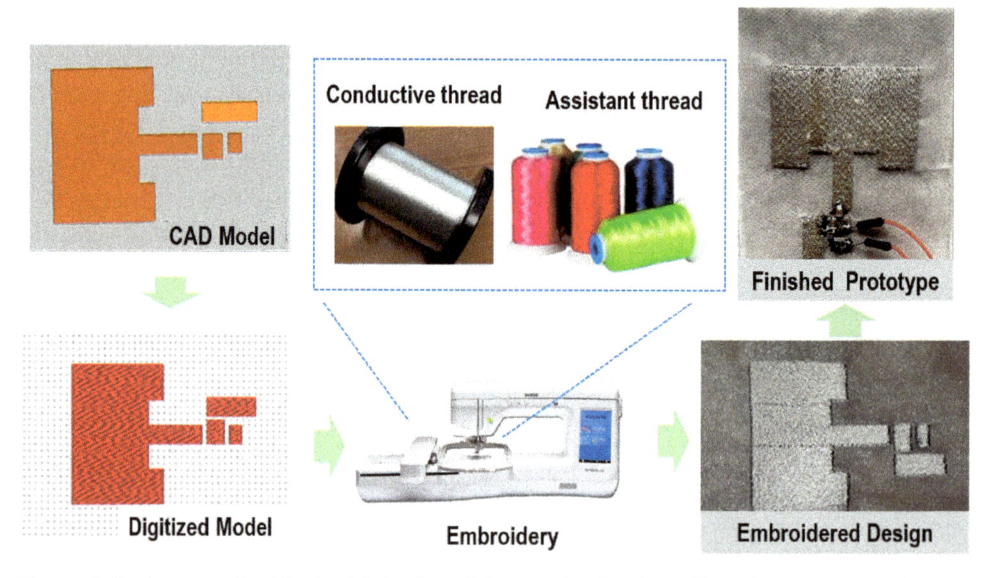

Figure 6.2 Steps involved in the fabrication of the conductive thread-based RF circuits and antennas from the optimized CAD models. (©2019 IEEE [1].)

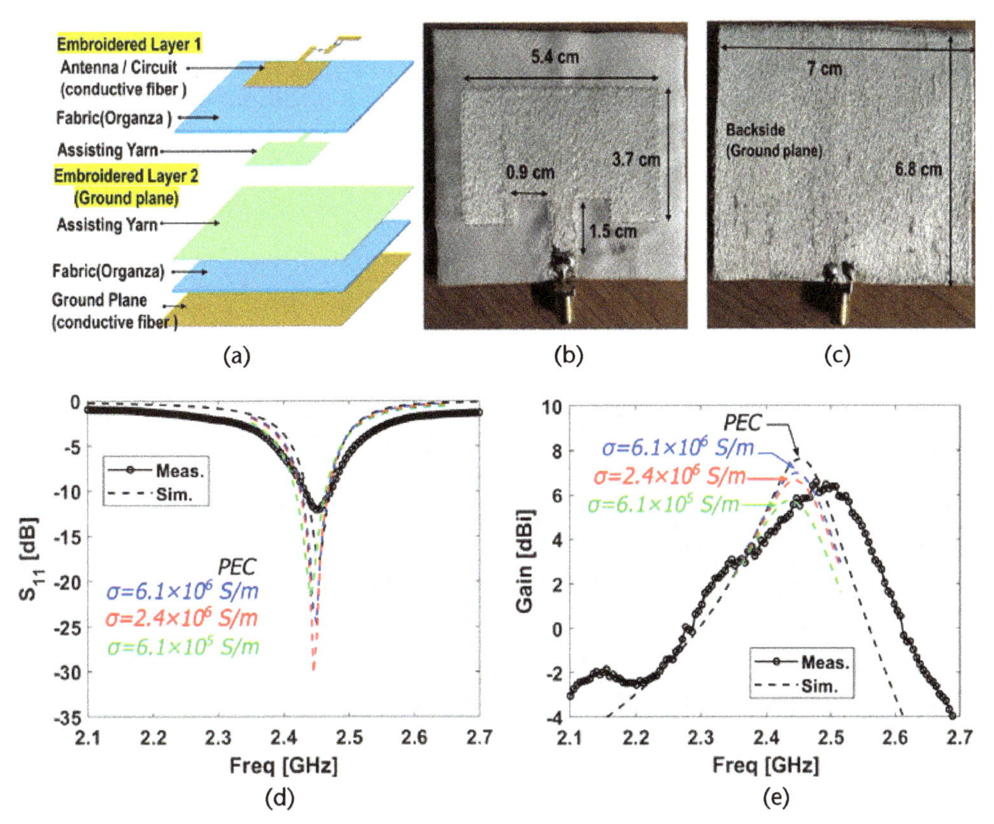

Figure 6.3 A 2.45-GHz patch antenna made from conductive threads embroidered onto fabric substrate: (a) exploded view of various layers used in the design, (b) top front view of the patch antenna, (c) ground plane view of the antenna, (d) measured and simulated (Ansys HFSS) return loss coefficient, and (e) realized gain as measured and simulated (Ansys HFSS) in the broadside direction. (©2019 IEEE [1].)

Elektrisola-7 (supplied by vendor Elektrisola, www.elektrisola.com), consisting of silver-coated copper strands (Cu/Ag50 amalgam), was used to build the conductive interfaces of the RF modules used in the study. In past works, conductive surfaces made of this particular thread have demonstrated loss performances in the same order as observed in rigid substrates (or PCBs) where copper is cladded on the substrates for maximum conductivity [9, 14, 15]. For low conductive losses and maintaining the flexibility of cloth after embroidery, three related parameters are taken into consideration. These are stitching density (number of threads per unit length, stitching pattern (the alignment of the embroidered threads with respect to the direction of the RF current), and thread tension (the precision about how much the top and bottom threads have to pull each other in order to realize a smooth embroidery on both sides of the seam while avoiding thread breakage). As expected, high thread density is required for high conductivity. Therefore, the highest possible stitching density of 14 threads per mm was used in the embroidery. By using this optimal thread density, we envision a compensation of the full conductivity when the structure is subject to mechanical deformations that force the threads to fray away from one another. In addition, stitching patterns have been known to impact the conductivity of RF structures. Selected stitching patterns were optimized to align the threads in the direction of the RF current. In a prior work, such alignment is shown to have minimized losses in transmission lines to approximately 0.3 dB/cm for frequencies up to 3.5 GHz [14]. Finally, thread tension was controlled to ensure the highest possible precision without damaging the threads to maintain continuous conductive surfaces. For this experiment, we chose a thread tension setting of 4 (out of available scale of 0 to 9) for our application.

6.1.2.1 Antenna Design, Prototype, and Measurements

A rectangular patch antenna was considered as per [16], but a crucial unknown was the dielectric constant of the textile substrate. To find the dielectric constant, we measured the resonant frequency (f_r for a wavelength λ_r) of several patch antennas with varying lengths (l) that were made using organza and (sticky) stabilizer substrate onto which conductive surfaces were embroidered. By using the relation $l = \lambda_r / 2\sqrt{\varepsilon_r}$, the effective dielectric constant of the fabric substrate was estimated as $E_r = 2.75$. The effect of fringing fields was ignored in this estimation due to small thickness of the substrate (≈1.5 mm). Using this information, the length of the patch was determined to be $l = \lambda_r / 2\sqrt{\varepsilon_r}$ ≈3.7 cm for 2.45-GHz operation. Other dimensions, shown in Figure 6.3(b) and (c), were appropriately optimized using full-wave simulations to achieve a desired S_{11} bandwidth.

Based on previous conductivity studies [17], it was found that conductive surfaces can be modeled using a perfect electrical conductor (PEC) model during the design's optimization. This behavior can be understood by considering the nature of the conductive thread. At 2.45 GHz, considering copper (conductivity as $\sigma = 5.95 \times 10^7/\Omega\text{m}$) as a conductive material, the corresponding skin depth ($\delta = (\rho/(\pi f_0 \mu_r \mu_0))$) was found to be 1.32 μm. Compared to this, the diameter of a single thread is 280 μm, which is the equivalent of 210 skin depths. Considering this calculation, we predict that the skin-depth effects are not dominant, and for design purposes the embroidered surfaces can be treated as a PEC. As depicted in Figure 6.3(d) and (e), the measured gain and S_{11} agree well with simulated results. Specifically, the S_{11} curve shows a resonance at 2.45 GHz and impedance matching in the Wi-Fi band. The antenna exhibited a maximum gain of 6.5 dBi at 2.47 GHz in the broadside direction, with a gain-frequency profile closely following the simulated one. Next, we pursue efficiency calculations for the fabricated antenna. For reference, we consider the simulation model with PEC conductive surfaces and no dielectric loss that shows 100% radiation efficiency and 98.2% total efficiency that includes the losses due to impedance mismatch. That is, the total efficiency (e_t) and radiation efficiency (e_r) are related by the reflection coefficient S_{11}, as

$$e_t = e_r\left(1 - |S_{11}|^2\right) \tag{6.1}$$

Since simulation model uses PEC surfaces and lossless dielectric materials, e_r = 100% and e_t = 98.3% for S_{11} = –17.59 dB at 2.45 GHz. Assuming measured and simulated directivities of the antenna are identical, reduction is gained from simulation to measurements due to added conductor losses, dielectric losses, and additional mismatch losses. Therefore, the total measured efficiency of the prototype is calculated using

$$e_{tm} = e_t \times 10^{(G_{pm}/10)} / 10^{(G_{ps}/10)} \tag{6.2}$$

where G_{pm} = 6.57 dBi and G_{ps} = 7.61 dBi are the measured and simulated peak gains, respectively. Furthermore, measured radiation efficiency (e_{rm}) is estimated by removing the mismatch losses that were observed in the measurements, as

$$e_{rm} = e_{tm} / \left(1 - |S_{11m}|^2\right) \tag{6.3}$$

where suffix S_{11m} refers to measured reflection coefficient. Comparing peak gain from this simulation (i.e., 7.61 dBi), with measured peak gain of 6.57 dBi, we calculate the total measured efficiency (e_{tm}) of the fabricated antenna as e_{tm} = 77.3%. e_{tm} takes into account all forms of losses (i.e., dielectric losses and conductive losses) as well as mismatch losses at the antenna port. From the knowledge of measured mismatch losses, we also calculate the measured radiation efficiency (which includes only dielectric and conductive losses) of the antenna to be e_{rm} =89.1%.

The differences in measured and simulated gain can be used to estimate the conductivity of the fabricated conductive surfaces. To achieve the latter, we modeled the antenna with varying conductivity values of conductive surfaces as shown in Figure 6.3(d) and (e). Assuming that dielectric losses are small and can be neglected, conductivity of the embroidered surface accounts for all losses in the antenna. In Figure 6.3(e), the peak gain for the case with $\sigma = 2.4\times10^6$ S/m matches the peak measured gain, suggesting that fabricated surfaces should have conductivity $\sigma = 2.4\times10^6$ S/m or better. In reality, this value should be slightly higher, since mismatch losses are higher in the measured prototype, and therefore conductor losses are overestimated in this conductivity comparison. Nevertheless, this analysis provides an approximate value of the effective conductivity of the developed embroidered surfaces.

The effects of the human body on the performance of the textile antenna were investigated using full-wave modeling of the antenna in presence of human body (Figure 6.4(a) and (b)). The electrical properties of different layers used in the simulation are provided in Table 6.1. The mass densities of the layers of skin, fat, and muscle were chosen as ρ = 1.02 g/cm^3, 0.9094 g/cm^3, and 1.059925 g/cm^3, respectively. A small area with mass of 0.2 kg per layer was chosen for this study. The measurement setup was realized in accordance with [18]. The antenna was placed at a representative distance of 5 mm ($\approx \lambda/25$) from the human body, although this distance could be variable. The gain and S_{11} results (Figure 6.4(c) and (d)) suggest that the antenna performance was not affected due to field interactions with the human body. This is primarily because of the ground plane, which provides isolation between the antenna and the human body. Simulations also show that the surface absorption rate (SAR) was limited within a maximum of 0.87 W/kg, which in accordance with the limit of 1.6 W/kg as per IEEE standard C95.1-2005 for safety levels with human exposure [19].

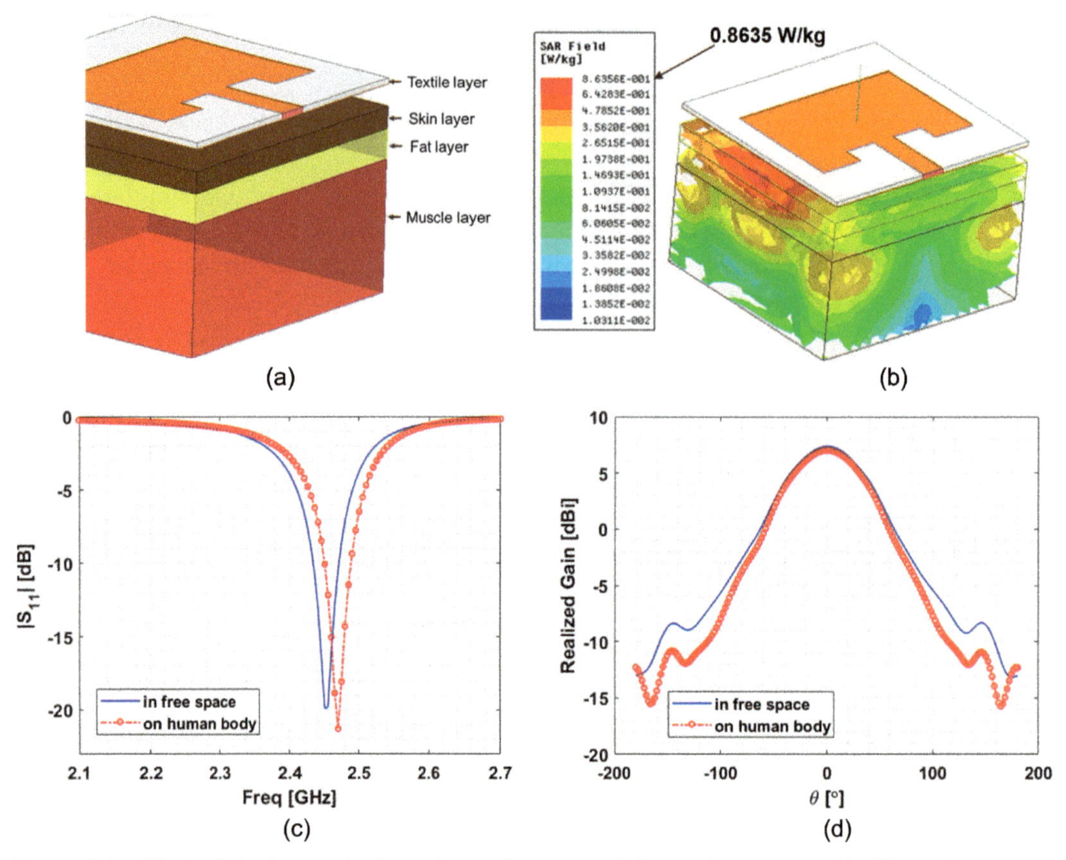

Figure 6.4 Effects of the human body on the performance of the textile antenna: (a) different layers considered for the simulation setup, (b) average SAR found to be 0.8635 W/kg when 1W of power was used, (c) simulated S11, and (d) realized gain of the textile-based antenna when exposed in air and mounted on a human body. (©2019 IEEE [1].)

Table 6.1 Electrical Properties of the Layers in Human Body Simulation*

Layer	*Thickness (mm)*	*Dielectric Constant E_r*	*Bulk Cond. σ (S/m)*
Skin	5	38	1.46
Fat	7	5.3	0.11
Muscle	30	52.7	1.77

*A spacing of 5 mm between the ground plane and the human model was considered.

6.1.3 Textile-Based Single-Diode Rectifier in Wearable Applications

6.1.3.1 Rectifier Design, Prototype, and Measurements

Figure 6.5 depicts the rectifier circuit prototype and measured performances. Achieving high efficiencies at low input power levels (below 0 dBm) and using minimal circuitry is a primary challenge for this design. The operation of a single-diode rectifier involves the use of a resonant transmission line that is shorted at one end. The position of the diode is optimized by tuning the lengths L_1 and L_2 to

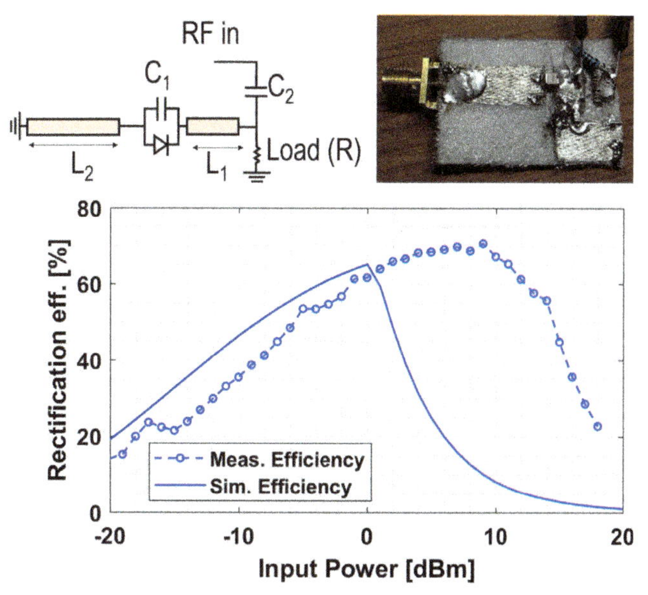

Figure 6.5 Textile-based, wearable rectifier using conductive embroidery onto organza substrate. Top left: schematic of the rectifying circuit using single-diode topology. Bottom left: fabricated prototype on organza-based clothing material. Right: measurement and simulated (Keysight ADS) efficiency at 2.45 GHz. (©2019 IEEE [1].)

increase the efficiency. The idea is to utilize the power stored in a standing wave, which causes higher voltage across diode terminals. As expected, the standing wave amplitude could be up to twice of the propagating wave (full rectification). This allowed us to achieve RF-DC conversion efficiency of 70%, which is greater than the 50% limit of single-diode half-wave rectifiers. Furthermore, the capacitors C_1 and C_2 were added and optimized for impedance matching with incoming RF and the rectifying diode. There is fair agreement between the simulated and measured rectifier efficiency for input power level less than 0 dBm (see Figure 6.5). But the measured results deviate for input power levels greater than 0 dBm. This deviation is a known phenomenon among single-diode rectifiers as also noted in [20], and is due to the nonlinear behavior of the diode. Since a linear RC model was used in ADS circuit simulations, we observe deviations between the simulated and the measured data. This is due to the fact that a linear model was used to design the rectifying circuit, which in reality, should be nonlinear.

The rectifying diode is a crucial component of the harvester (rectifying circuit) and must exhibit low turn-on voltage and low-leakage current to achieve high RF-to-DC conversion efficiency. In general, diodes must provide a compromise between these quantities. In this regard, Schottky diodes provide a good compromise

Table 6.2 Lumped Component Values Placed in the Rectifier Circuit

C_1	2 pF	Matching at 2.45 GHz
C_2	0.4 pF	Matching at 2.45 GHz
R	1 kΩ	Representative of a typical load
L_1	18 mm	Matching at 2.45 GHz and efficiency optimization
L_2	6 mm	Matching at 2.45 GHz and efficiency optimization

due to their smaller turn-on voltage [21–23] and their ability to rectify low-input RF signals without the requirement of a significant biased voltage. For our design, the Skyworks SMS7630 Schottky diode was selected. A comparison of the proposed rectifier to some recent rectifiers is shown in Table 6.3. Our design achieved close to 20% efficiency at –20 dBm input power. By comparison, earlier designs [26, 28, 29] show 10.5%, 2%, and 1% efficiency, respectively. Specifically, our circuit showed an RF-to-DC rectification efficiency between 20% and 60% for input-power level varying from -20 dBm to 0 dBm and this is in close agreement with the simulations. This textile rectifier is combined with the previously discussed antenna to realize an array of rectennas for power transfer and harvesting. The full details are reported in the next sections.

6.1.4 Design and Optimization of Textile Rectenna Array

6.1.4.1 Textile Array Design and Fabrication

Using the rectifier and patch antennas optimized and tested in previous sections, rectenna elements were fabricated (see Finished Prototype in Figure 6.2). The elements were then used to fabricate 2 × 2 and 2 × 3 arrays as shown in Figures 6.6 to 6.9. The interelement distance in each of the arrays was chosen to be 8.5 cm in horizontal and 5.5 cm in vertical directions. This choice was based on ease of fabrication, but further optimization could be pursued to increase area density for optimal power collection. Overall, the 2 × 2 and 2 × 3 arrays occupied an area of 300 cm^2 and 437 cm^2, respectively.

Prior to testing the developed arrays, we characterized the available ambient RF power. The test results are presented in the next section.

6.1.5 RF-Power Availability Tests

Ambient RF power measurement was conducted using a low-gain, broadband spiral antenna with 2.4 dBi gain at 2.45 GHz. The antenna was placed 20 cm from the radiating device (Wi-Fi router or phone) and was connected to a Keysight PXA N9030B Signal Analyzer to record the power-frequency spectrum. Peaks were observed in the spectrum at 700 MHz, 830 MHz, and 2.4 GHz. Tables 6.4 and 6.5 summarize the peak power observed from a cell phone and a Wi-Fi router, respectively.

Table 6.3 Efficiency Comparison of the Textile Rectifier Circuit with Recent Publications

Input Power (dBm)	*Efficiency (%)*	*Rectifying Diode*	*Reference*
–20, 8	18, 70	SMS7630	This work
–13.3	2	SMS7630	[24]
–13	9	SMS7630	[25]
–20.4	10.5	SMS7630	[26]
10	53	HSMS2852	[27]
–20, 3	1, 45	HSMS2860	[28]
–20, 8	2, 73	HSMS2860	[29]

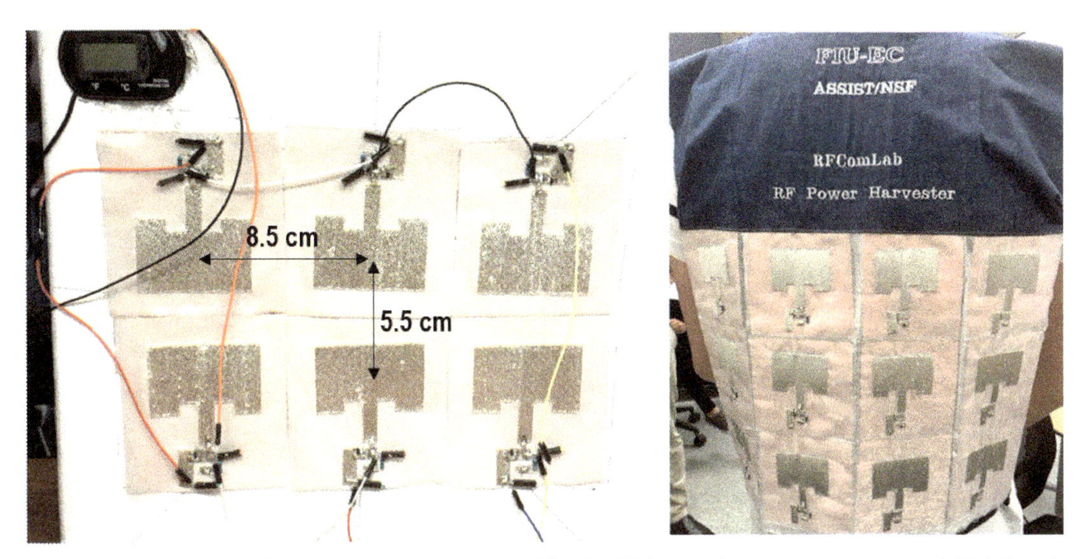

Figure 6.6 (a) The six-element prototype array used for the RF-harvesting measurements, and (b) full-scale jacket with power harvesters demonstrating the wearability. (©2019 IEEE [1].)

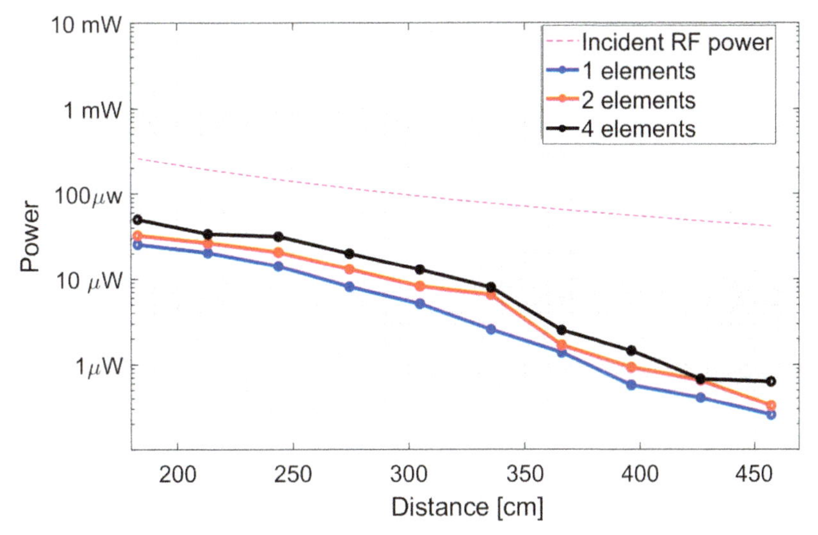

Figure 6.7 Power-harvester performance using a 2 × 2 rectenna array. The average power collected by the 1, 2, and 4-element arrays and corresponding average incident power are shown as a function of distance. (©2019 IEEE [1].)

Table 6.4 shows the emitted peak-power from a cell phone at frequencies of 700 MHz, 830 MHz (Long Term Evolution (LTE) bands), and 2.4 GHz (Wi-Fi band), when a voice or a video call was in session. As noted, the maximum power level of 1.5 dBm was recorded at 700 MHz during the video call. For a Wi-Fi enabled video call, a peak emission of –22 dBm was observed from the cell phone. Likewise, emissions from a Wi-Fi router are shown in Table 6.5. Emissions at 2.4 GHz during a Wi-Fi audio call and Wi-Fi video call were recorded to be –17 dBm and –11 dBm, respectively.

The measurements show that peak power levels of the RF signal could be strong, especially for LTE video calls. Even so, the average values could be smaller.

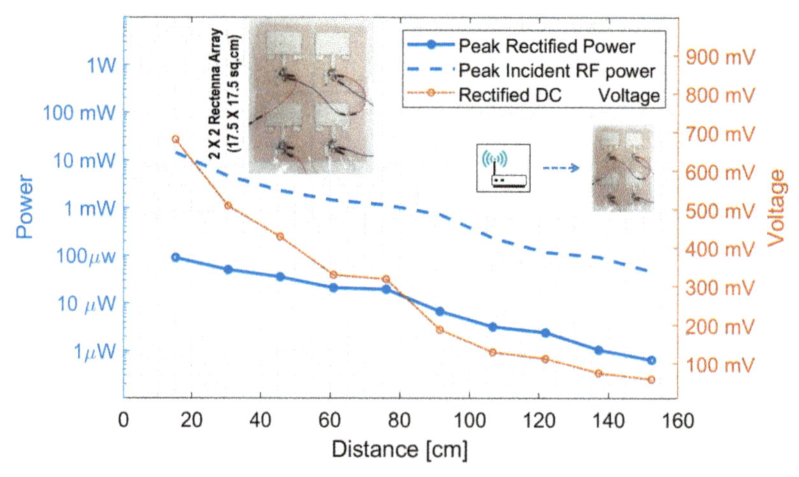

Figure 6.8 Power-harvester performance of a 2 × 3 rectenna array. The DC voltage, its corresponding collected average power collected by the rectifier circuit, and peak incident power are shown as a function of distance. (©2019 IEEE [1].)

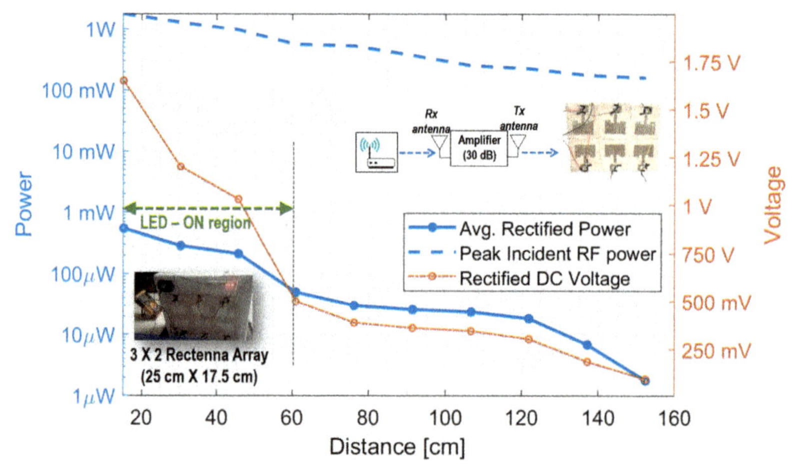

Figure 6.9 Power-harvester performance using a 2 × 2 rectenna array. The DC voltage, its corresponding collected peak power collected by the rectifier circuit, and peak incident power are shown as a function of distance. (©2019 IEEE [1].)

Table 6.4 Recorded Power Levels of the Ambient RF Signal 20 cm from the Respective Devices

Frequency	*Voice Calls*	*Video Call*	*Ambient*
700 MHz	1.5 dBm	0 dBm	–52 dBm
830 MHz	–22 dBm	–7.3 dBm	–60 dBm
2.4 GHz	–27 dBm*	–22.4 dBm*	–60 dBm

*Wi-Fi-enabled call.

We consider two approaches to attain higher levels: (1) increasing the antenna gain by developing rectenna arrays, and (2) boosting the ambient Wi-Fi signal (via amplification) for enhanced power collection. These two approaches are discussed in the next section.

Table 6.5 Peak RF Power Level Measured from a Wi-Fi-Router*

Frequency	*Audio Calls*	*Video Call/Streaming*	*Ambient*
2.4 GHz	−17 dBm	−11 dBm	−60 dBm

*The Wi-Fi router used was a Kasda AC 1200M Dual Band KW6515 and the phone was an iPhone 6S.

6.1.6 Power Harvesting Using Textile Rectenna Arrays

To characterize the rectenna array and the effect of increasing the number of elements, we investigated the power collected by a single rectenna element, 2 × 1 rectenna array, and 2 × 2 rectenna array under a dedicated RF signal with continuous sine waveform (see Figure 6.7). To test the Wi-Fi power-harvesting capabilities under ambient Wi-Fi conditions, we also reported the power collection using 2 × 2 and 2 × 3 rectenna arrays as provided in Figures 6.8 and 6.9. The rectenna elements of the array were connected in series to combine the harvested power from each antenna. Further, we did not combine the RF power to avoid having to deal with incident waves received at different phases, which could eventually decrease the received power level. For ambient Wi-Fi cases, a Wi-Fi-router at 2.45 GHz was used to stream a video. For the 2 × 3 rectenna array (Figure 6.9), we considered the case of amplifying the ambient RF power using a stand-alone receiver-amplifier-transmitter unit (see Figure 6.9, inset). The amplification unit used consisted of a receiver antenna (a horn with a gain of 9 dBi), RF amplifier (mini-circuit coaxial amplifier of gain 30 dB), and retransmitter antenna (a horn with a gain of 9 dBi) and was placed 5 cm away from the Wi-Fi router. The distance between the transmitter horn and the power harvester array was varied, as shown in Figure 6.9.

6.1.6.1 Measurement s and Discussion

First, we characterized the rectenna array and the effects of adding more elements on the rectified DC power, as seen in Figure 6.7. The transmitting power (P_t) should be such that when it is output by the transmitter, its equivalent power ($EIRP_{max}$) concentrated into a smaller area by the receiving antenna should be:

$$EIRP_{\max} = P_t(dBm) + G_t(dB) - P_g(dB) \tag{6.4}$$

P_t is limited to 30 dBm (1W) and the total EIRP ($EIRP_{max}$), 36 dBm (4W) as per the FCC regulations [30]. P_g is minus 1 for each 3 dB of antenna gain over 6 dB [31]. In this work, 500 mW RF power was transmitted using an Agilent N5182A MXG vector signal generator and a minicircuit ZHL-42W+ amplifier connected to a 9-dBi gain horn antenna (Figure 6.7). The transmitting system corresponds to an EIRP of 4 W (36 dBm), which is in compliance with FCC regulations according to [30, 31]. At 5 feet, a single element shows a collection of 25 μW, which is enhanced to 55 μW when four rectenna elements are serially connected. Thus, additional power gain is not directly proportional to the number of elements but provides some enhancements. In the future, appropriate serial or parallel power combination architecture should be explored to achieve more robust DC power collection.

In a realistic Wi-Fi environment, due to the intermittent nature of the signal (bursty), the collected DC power should be characterized using averaged DC or peak DC power. In practical designs, a power management circuit can be used to account for the intermittent nature of the signal [28, 32]. But for our characterization, we simply conducted peak and averaged DC power measurements using a mixed-signal oscilloscope (MSO, Keysight MSOS254A) connected to the output of the rectenna arrays. The peak DC voltage (V_p) was recorded using the MSO and then the corresponding peak power (P_p) was calculated using $P_p = V^2/R$ where R=5.064 kΩ and R=5.295 kΩ as load resistance values. As a reference, total available RF power was also calculated by measuring it using a spiral antenna and then extrapolating for a given harvesting area. For calculation of the peak incident RF power, we first used a reference spiral antenna of gain (G) 2.4 dBi connected to a spectrum analyzer to record peak RF power (P_{RF}) at 2.4 GHz. Then by using the spiral's effective area ($A_e = G\lambda^2/(4\pi)$), the power density of the incident signal ($S = P_{RF}/A_e$) was calculated. The available peak power RF power incident on the rectenna array was then calculated by using the array aperture area A as $P_{inc} = S \times A$, and the corresponding measurements are given by the dashed curves in Figures 6.8 and 6.9.

The harvested peak power using a 2 × 2 array was found to be varying between 100 and 0.8 μW as the distance was varied from 10 to 150 cm from the Wi-Fi router (Figure 6.8) when the load resistance of 5.295 kΩ is measured across the terminals of the array. The corresponding DC voltage level was tested to be between 58 and 680 mV. These power levels may be considered small for many sensors used in health monitoring applications, although, with the use supercapacitors, we can store up to nanowatt-level power to operate a wide range of typical sensors.

The Wi-Fi signal amplification and collection were conducted using the 2 × 3 rectenna array. The results are shown in Figure 6.9. The selected amplifier unit was effective in increasing the incident power by two orders of magnitude, as shown by the dashed curve in Figure 6.9 when compared to the dashed curve in Figure 6.8. For the 2 × 3 array case, we report the averaged DC power, where the averaging was done over a period of 10 minutes. As noted in Figure 6.9, average power levels varied between 1 to 600 μW as the distance was changed from 160 and 10 cm. Specifically, at a distance of 60 cm, a DC power of 80 μW was recorded. Three LEDs were connected in parallel at the output of the array and their lighting is shown for distances up to 60 cm. This means that the collected power levels are quite reasonable for practical applications such as charging supercapacitors or powering sensors for a variety of healthcare and IoT-type systems [33, 34]. Notably, an LED lighting experiment suggests that no power management circuit is needed to use the proposed harvesting system. Figure 6.9 also provides a rough estimate of total radiated power to understand its FCC compliance. As the transmitter antenna was a directive horn and a rectenna array was placed at a close distance of 10 cm, most power emitted from the transmitter was received by the rectenna array. Estimated peak power is found to be 1W for this case, as shown in Figure 6.9. We note that average power could be 10 to 100 times lower, as the Wi-Fi signal comes in bursts of pulses. This suggests that the setup shown is compliant with the FCC's maximum power limit of 1W [31]. Many wearable rectennas have been proposed prior to this work [35–41]. The comparison shows that this work reports the first embroidery-based textile rectenna system, which is completely integrated on fabric.

Furthermore, the use of the embroidery process and a minimalistic rectifier circuit has allowed us to extend the number of elements to 6, which is the highest reported elements for wearable applications. Thus, we have proposed a method to exploit large clothing areas for RF power-harvesting and power-transfer applications.

6.1.6.2 Potential Applications

The clothing-integrated RF-power harvesting circuitry onto textile substrates (see Figure 6.6) comes with several advantages such as flexibility, comfort, light weight, and durability. Such advantages are contingent to the controlled design and fabrication processes. These attributes make it useful for the wearable electronics in defense and space applications, where it can be used to integrate communication interfaces and sensing tactical gears for dismounted personnel and sensor-enabled space suits for space explorers. Within healthcare and general fitness applications, there is interest for smart wear with fitness devices. In childcare, they apply to smart garments for sleep monitoring and location tracking. The developed RF-power harvesting textiles also apply to smart homes, where the RF modules can be embedded in couches, curtains, and carpets to harvest neighboring Wi-Fi signals to power lamps, desk clocks, and other devices. Other wearable applications that require a closer proximity between the transmitter and receiver have also caught the attention of recent research projects. However, when the receiving system is implemented into clothing and then worn (by patients, consumer users, etc.), it is subject to misalignments provided by the intermittent movements of the wearers and can compromise the RF performance of the system. In the next section, we will present and discuss a novel, patent-pending antenna topology used to mitigate the issue of misalignments while keeping excellent RF performance.

6.2 Part 2: Near-Field Integrated Power Transfer and Harvesting for Wearable Applications

6.2.1 Introduction

This section investigates wireless power transfer to wearable devices using near-field power transfer systems integrated on items of clothing for proximity applications. A new antenna design is presented first—an anchor-shaped antenna—which will be shown to be resilient to physical misalignment [42]. Then, we apply the proposed design to realize an ergonomic power transfer system that can use the daily use cases of activities for wireless power supply [43]. We present and discuss a new class of antennas suitable to mitigate the effects of angular and lateral misalignments in near-field power transfer and harvesting. This antenna topology exploits the combination of both electric and magnetic coupling modes within the regime of inductive wireless power transfer to achieve a high PTE and provides resilience to the lateral and angular misalignments. We present a fabric-integrated wearable charging platform that exploits daily life activities, such as sitting on a chair or lying in a bed, to operate wearable devices. This wearable charging feature is relevant in the context of prior studies on sedentary lifestyles, which reports a dominance of sedentary states in our daily activities [44]. This study reported that on average, test subjects spent approximately 57% of the time in activities like lying, reclining, and sitting, and 37% in light activities where they could be in close vicinity of the surfaces (back of a chair, mattress, sheets, etc.) to receive wireless RF power. This prevalent lifestyle provides an opportunity for wireless charging of phones and IoT/IoHT devices (Figure 6.10). In the next sections, the fundamentals are presented and discussed (Section 6.2.2) and followed by the integrated of the antenna structure into clothing for wearable applications (Section 6.3.1).

6.2.2 Anchor-Shaped Antenna: Fundamentals

Electric and magnetic coupling mode discontinuous loops have been studied in microwave filters [45]. The fringe-enabling cavity or gap allows generation of fringing electric fields, leading to an electric-field type coupling between the two loops. This electric field is added to the magnetic field that is inherent to the metallic structure of the loop. As a result, the fringe-enabling cavity can enable a magnetoelectric coupling, where double resonance and frequency bifurcation effects take place [46]. The proposed structure is referred to as an anchor-shaped structure that uses the mixture of electric and magnetic couplings for power transfer. Specifically, electric coupling modes are enabled, when anchor-shaped antennas are misaligned from their normal, perfectly aligned (broadside) position.

6.2.2.1 The Proposed Anchor-Shaped Antenna

To demonstrate the new functionality of the structure, we compare its properties with the two basic types of resonators: a loop resonator (magnetic resonator) and a dipole resonator (electric resonator). We summarize the current distributions on the three resonators at their respective resonant frequencies as shown in Figure 6.12. As can be seen in this figure, a dimension L = 15 cm was chosen to create similar footprints of the three resonators in order to realize a fair comparison. As shown in Figure 6.12, the loop antenna shows one-wavelength-long current distribution along its total length and the dipole shows a half-wavelength-long current

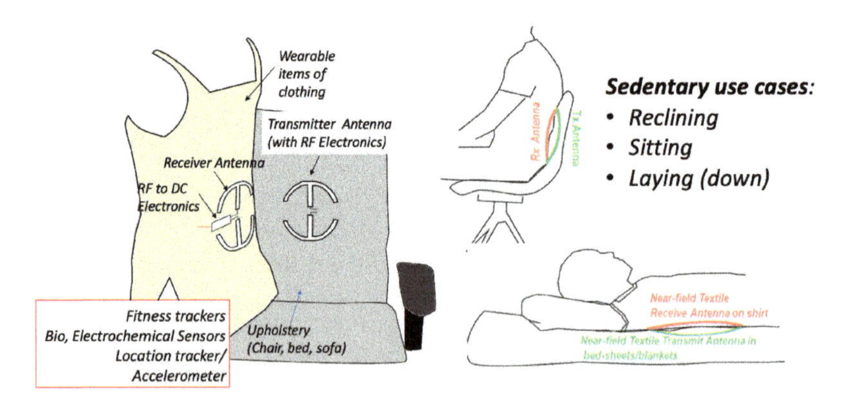

Figure 6.10 Integration of the proposed wireless power transfer and harvesting system consisted of a textile-based anchor-shaped antenna (operating at 360 MHz) into clothing and upholstery: (a) implementation of the system into a dress and chair with potential sensing applications, and (b) illustration of different sedentary use cases. (©2021 IEEE [43].)

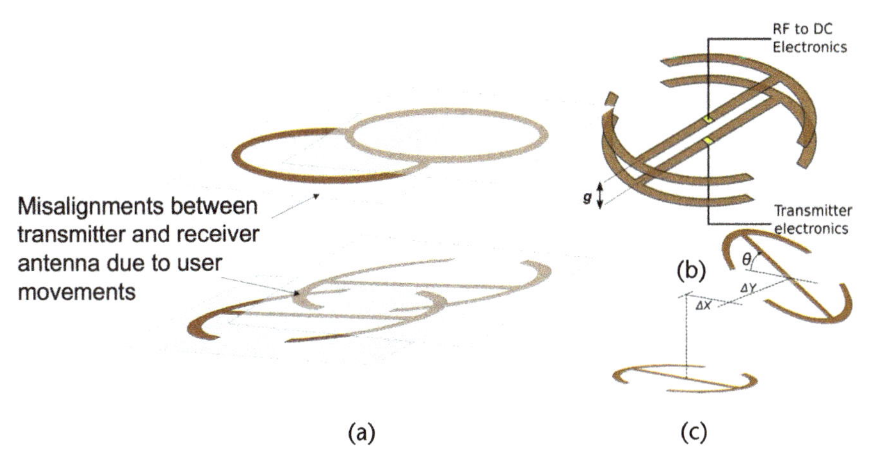

Figure 6.11 (a) Misalignment between the transmitter and the receiver antenna for near-zone power transfer, (b) geometry and configuration of the proposed misalignment resilient power transfer system, and (c) a typical misaligned case where angular misalignment of θ and lateral misalignment of x and y are shown. (©2020 IEEE [42].)

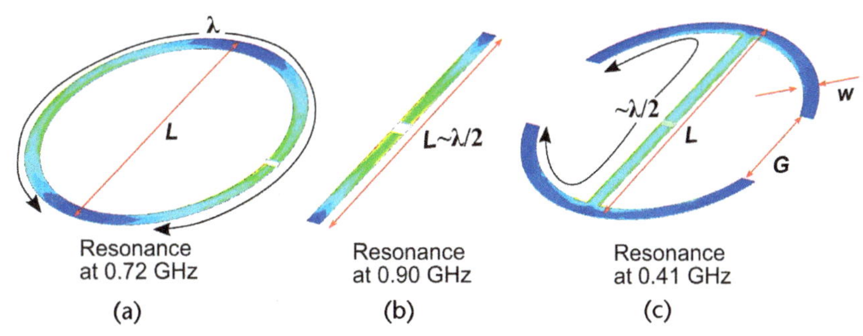

Figure 6.12 Qualitative comparison of the resonant frequency and corresponding current distributions (not to scale, light color represents maxima, and dark color represents minima) for the related resonant antenna configurations (a) single loop, (b) dipole, and (c) anchor-shaped. The length L = 15 cm is chosen in this example, the width of the antennas was w = 0.75 cm, G = 5 cm. (©2020 IEEE [42].)

distribution. Remarkably, the properties of these two resonators are put together to generate those of the anchor-shaped antenna. It can be observed that approximately a half-wavelength-long current distribution is found along the geometry of the anchor-shaped antenna. More details will be discussed next.

The half-wavelength resonance leads to the miniaturization (decreased resonance frequency longer wavelength) and an anchor-shape allows increased coupled surface (aperture) and a strong fringing electric field in the vicinity of the resonator through the fringe-enabling cavities on either side of the structure. As a result, the proposed structure adds an electric coupling mode on top of magnetic coupling already present in loop antennas. These effects lead to improved couplings under misalignments, as further discussed in the next sections. This is a result of the miniaturization property of the anchor-shaped topology, where lower resonant frequency is attained for the same footprint. The miniaturization of the anchor-shaped antenna can be seen from the input impedance and reflection coefficient properties.

A comparison of these properties for the anchor and loop is shown in Figure 6.13(a) and (b). As is known, the resonance is associated with $\chi(f) = 0$ and $\chi'(f) > 0$,

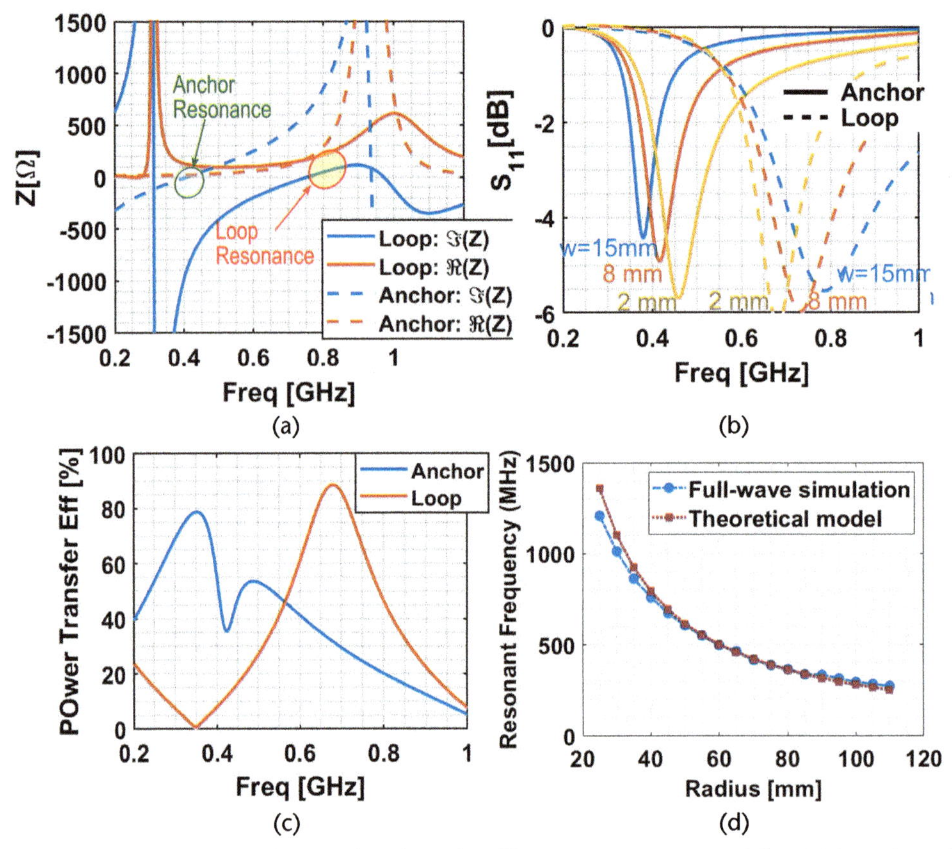

Figure 6.13 Comparison of the anchor-shaped and loop antenna in terms of their (a) input impedance, (b) S11 matching with changing strip width, (c) power transfer efficiency (%) in the presence of a similar receiver antenna positioned at 1 cm distance, and (d) comparison of the theoretical and numerical model of the resonance frequency of the anchor-shaped antenna. In these simulations, we use the dimensions of G = 5 cm, w = 0.75 cm, and L = 15 cm. (©2020 IEEE [42].)

where χ is the reactance profile and χ^t is the first derivative of the reactance profile [47]. Considering this in Figure 6.13(a), the resonant frequency changes from 0.8 GHz for the loop to about 0.4 GHz for the anchor. This verifies the miniaturization property of the antenna. This is further evident in Figure 6.13(b), where the resonant frequency moves down along the frequency axis for the anchor antenna. In Figure 6.13(c), the PTE is calculated using

$$\eta = \frac{|S_{21}|^2}{\left(1-|S_{11}|^2\right)\left(1-|S_{22}|^2\right)} \tag{6.5}$$

between the transmitting and receiving antennas and is shown when the antennas of same type are considered for the transmitter and receiver (as shown in Figure 6.11(b)). We observe the frequency bifurcation effect caused by two PTE peaks in the proposed antenna. This is typical for coupled structures that simultaneously have electric and magnetic coupling [45]. A gap g was chosen to be 1 cm for this comparison. The power coupling for the loop is slightly better in this aligned case. However, a more in-depth study considering different degrees of misalignments revealed that the PTE for the anchor-shaped antenna is better than that of its loop counterpart.

The above discussion points to the conclusion that the anchor antenna is a miniaturized form of the loop antenna combined with dipole to realize the enhancement of the PTE explained via two mechanisms such as (1) the extension of the wave length and (2) the enhancement of the fringing fields. The RF properties of the anchor-shaped antenna is ultimately deduced from the loop antenna due to the fact that the loop is used as the genesis for the anchor. For a more effective PTE, it is reported that the magneto-inductive effects are used in conjunction with resonance to compensate for the leakage inductance [48]. In this work, the resonant frequency of the loop is the frequency at which the quality factor, Q [49] (which represents the ratio of the energy stored via inductive effects and the energy dissipated per oscillation) is the highest. Thus, the resonance frequency is written as

$$f_{loop} = \frac{c^{8/7}\mu_o^{1/7}\rho^{1/7}}{4\times 15^{2/7}\pi^{11/7}r_c^{2/7}a^{6/7}} \tag{6.6}$$

where c is the speed of light, μ_o is the permeability of free space, ρ is the resistivity of the conductive material, r_c is the radius of the conductor, and a is the outer radius of the antenna, is empirically deduced [49]. For the anchor-shaped antenna, the only unknown is set to be the outer radius, a, as all the other parameters such as the radius of the conductor and the dimensions of the fringe-enabling cavities ($G = 2/3$ a) are indeed a function of the outer radius. The resonant frequency of the anchor was thus empirically deduced to be

$$f_{theoretical-anchor} = \frac{10^{9/7}c^{8/7}\mu_o^{1/7}\rho^{1/7}}{4\times 15^{2/7}\pi^{11/7}a^{8/7}} \tag{6.7}$$

In (6.7), the cavities have a dimension of two thirds of the outer radius of the anchor and the width (w) of the conductive traces, one tenth of the outer radius. These values were chosen to assure the scalability of the anchor shaped antenna at high frequencies (i.e., at smaller footprints). The performance of the anchor-shaped antenna as a function of these parameters was simulated and is shown in Figure 6.14. The theoretical analysis from (6.7) was verified by a full-wave simulation from an Ansys HFSS. The comparison between the two models is presented in Figure 6.13(d). As can be observed, the full-wave simulation and theoretical values of the resonant frequency are in good agreement.

The corresponding inductance of the anchor-shaped antenna was inspired from [49] and can be written as

$$L_{anchor} = \frac{7\mu_o a}{2\pi}\times\left[\ln\left(\frac{28a}{\sqrt{\pi w s}}\right)-2\right] \tag{6.8}$$

where w and s are the dimensions of the conductive traces. This self-inductance value is found to be 363 nH, which is higher than 319 nH found for the single loop

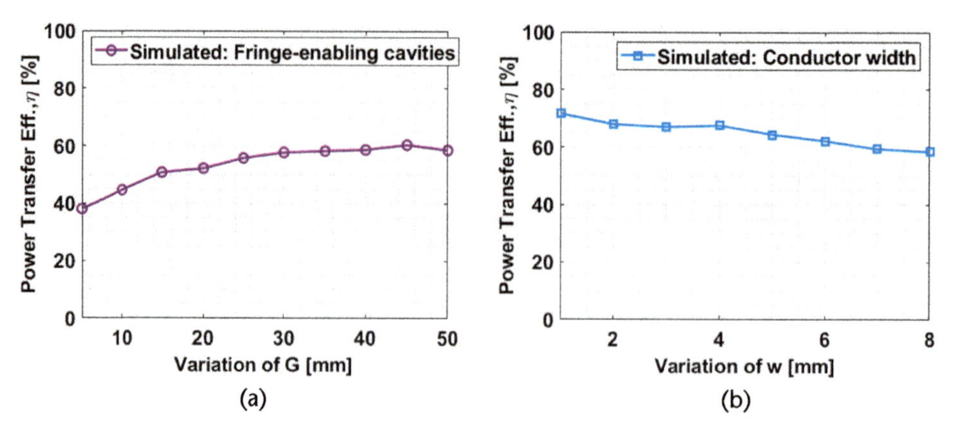

Figure 6.14 Power transfer efficiency of the anchor antenna as a function of (a) the fringe-enabling cavity, and (b) the width of the stripline. In these simulations, G = 5 cm when w varies, w = 0.75 cm when G varies, and L = 15 cm. (©2020 IEEE [42].)

as reported in [49]. The latter was expected as the length of the conductor in anchor ($P \approx 7a$) is more than that of the loop ($P = 2\pi a$). This is one of the indicators of the reduction of the resonant frequency given by $f_r = 1/(2\pi\sqrt{LC})$. Another factor contributing to the reduction of the frequency are the fringing fields that enhance the capacitance of the anchor, again decreasing the resonant frequency.

From the resonant frequency and the inductance, we deduce the self-capacitance, which is:

$$C_{anchor} = \frac{50 \times a}{7\pi\mu_o c^2}\left[\ln\left(\frac{28a}{\sqrt{\pi w s}}\right) - 2\right]^{-1} \tag{6.9}$$

where $C_{anchor} = C_{self} + C_{fringe}$. The capacitance associated to the fringing fields can be taken from [50] and can be written as:

$$C_{fringe} = \varepsilon_0 \left(\left(\frac{\varepsilon_r w}{h}\right)^n + \left((\varepsilon_r + 1)\pi\right)^n \left(\frac{1}{\ln\left(\frac{8h}{w+1}\right)} - \frac{w}{8h} \right)^n \right)^{\frac{1}{n}} \tag{6.10}$$

In this case, $n = 1.08$, w is the width of the strip line, h is the distance separating the two conductive surfaces to generate the fringing fields, and E_r is the dielectric of the transmitting medium. This theoretical model was demonstrated and tested in [50] for dielectric constants from 1 to 24 and for values of h extending from 0 to ∞. For anchor shaped antennas, h can be the fringe-enabling cavity, G or the fixed distance separating the antennas, g. It should be noted that, in this model, the parallel plates are assumed to be perfectly aligned with each other. In the case of misalignments, the new value of the separation becomes $h = \sqrt{d2 + \Delta 2}$ (for lateral misalignments) and $\bar{h} = \sqrt{2a2(1 - cos\vartheta)}$ (for angular misalignments). Each anchor antenna has a surface area of

$$A_{anchor} = \pi\left[\alpha^2 - (a-w)^2\right] - 2Gw + 2(a-w)w = \frac{57\pi a^2 + 14a^2}{300} \tag{6.11}$$

The expression of (6.11) is written based on the value of $G = 2a/3$ and $w = a/10$ and the mutual capacitance deduced from it will be:

$$C_{mutual} = \varepsilon_o \frac{A_{anchor}}{h} = \varepsilon_o \left(\frac{57\pi a^2 + 14a^2}{300h}\right) \tag{6.12}$$

Again, the value of h can be g, G, $h = \sqrt{d2 + \Delta 2}$, and $\bar{h} = \sqrt{2a2(1 - cos\vartheta)}$ where Δ is the varying distance in lateral misalignments. Overall, the capacitance of the whole system can be written as:

$$C_{total} = C_{self} + C_{fringe} + C_{mutual} \tag{6.13}$$

The total capacitance will then be:

$$C_{total} = C_{Anchor} + C_{mutual} = \frac{50 \times a}{7\pi\mu_o c^2}\left[\ln\left(\frac{28a}{\sqrt{\pi ws}}\right) - 2\right]^{-1} + \varepsilon_o\left(\frac{57\pi a^2 + 14a^2}{300h}\right) \tag{6.14}$$

The inequality (6.15) suggests that the electric coupling has a higher influence in enhancing the wavelength through the extent of the fringing fields. The introduction of the fringe-enabling cavities improves the PTE when transiting from a single loop to an anchor. This is a result of the increase of the spanning area for coupling between the transmitter and receiver. Therefore, we can shift the frequency based on the inequality below:

$$\frac{1}{2\pi\sqrt{L(C + C_m)}} < f_{broadside} < \frac{1}{2\pi\sqrt{(L - L_m)C}} \tag{6.15}$$

where C_m and L_m are the system's mutual capacitance and inductance, respectively. The latter is subject to change based on the misalignment due to the fact that each misalignment strengthens a particular set of fringing fields as explained in [51] When taking the dipole-like performance of the anchor-shaped antenna into account, a more simplistic model can be derived. From [52] we deduced that, for the anchor antenna, $Perimeter_{anchor} = 2\pi a + 2a - 2G = 2\pi a + 2a - 2(2a/3) \approx 7a = 0.7\lambda$. We then deduce the following expression for the frequency:

$$f_{simplistic-anchor} = \frac{c}{10 \times a} \tag{6.16}$$

where c is the speed of light and a is the outer radius of the anchor antenna. As can be seen in Figure 6.16, the simplistic model agrees with the simulated results

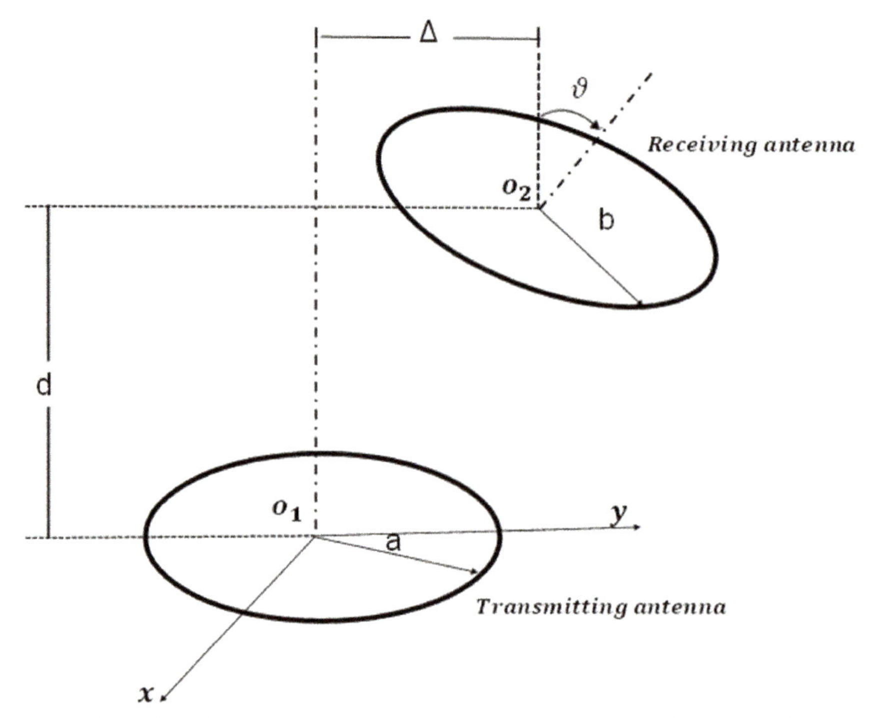

Figure 6.15 A detailed view of the wireless power transfer system with all lateral and angular misalignments, where "b"is the fixed distance between the transmitter and the receiver's initial position. It is represented by "g" in the previous sections. "Δ" is the lateral misalignment distance and "θ" is the angular misalignment angle. "a" and "b" are the Tx and Rx radii.

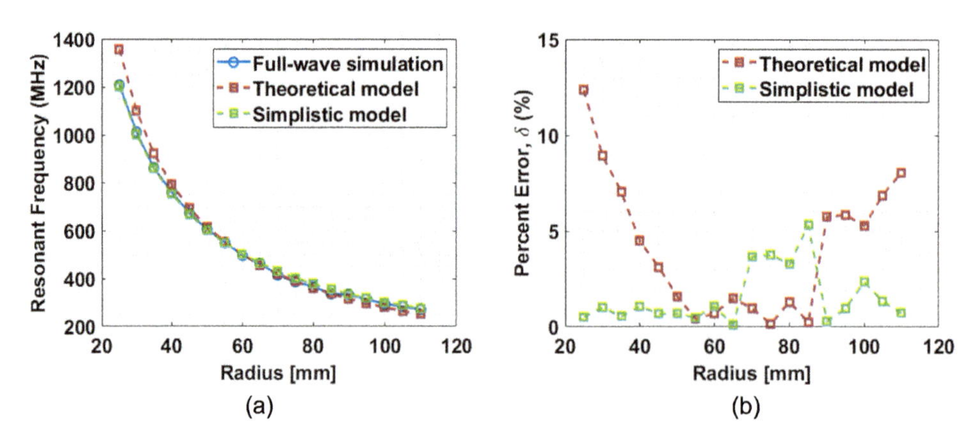

Figure 6.16 Representation of the theoretical and simplistic model of the anchor-shaped frequency modulations compared to their simulated counterpart: (a) theoretical and simplistic models compared to simulation, and (b) percent error of the aforementioned models with respect to their simulation counterpart.

perfectly. For anchor antennas of a radius of 3 cm or longer, the simplistic model from (6.16) differs from the simulation by only 5% while the model from (6.7) varies from the simulation model by up to 9%. The percent error was calculated using the equation

$$\delta = \frac{Theoretical/Simplistic - Simulated}{Simulated} \times 100\% \tag{6.17}$$

This result suggests that the anchor-shaped antenna has a dipole-like functionality. Indeed, the equations only take into account the outer radius, a, of the antenna. The mutual inductance, using the work published in [53], can be expressed as

$$L_{m(anchor)} = \frac{\mu_o \pi a^2 b^2}{2\left(a^2 + b^2 + z^2\right)^{3/2}}\left(1 + \frac{15}{32}\gamma^2 + \frac{315}{1024}\gamma^4\right) \tag{6.18}$$

where a and b are the radii of the transmitting and receiving antennas, z is the distance between the axes of the transmitter (Tx) and receiver (Rx), and $\gamma = \frac{2ab}{a^2 + b^2 + z^2}$. In this work $a = b$ and (6.18) will be

$$L_{m(anchor)} = \frac{\mu_o \pi a^4}{2\left(2a^2 + z^2\right)^{3/2}}\left(1 + \frac{15}{32}\gamma^2 + \frac{315}{1024}\gamma^4\right) \tag{6.19}$$

1. *Lateral misalignments:* The length z becomes z^t and (6.19) can be expressed as

$$L'_{m(anchor)} = \frac{\mu_o \pi a^4}{2\left(2a^2 + z'^2\right)^{3/2}}\left(1 + \frac{15}{32}\gamma'^2 + \frac{315}{1024}\gamma'^4\right) \tag{6.20}$$

$z' = \sqrt{D^2 + g^2}$ is the misaligned distance between the centers of the Rx and Tx, where g is the fixed distance from the receiving antenna initial center found in (6.11), and $\gamma' = \frac{2ab}{a^2 + b^2 + z'^2}$

2. *Angular misalignments:* The length z becomes z' and (6.19) can be expressed as

$$Z(inductive) < 377\Omega < Z(capacitive) \tag{6.21}$$

$z' = \sqrt{2a^2(1 - cos\theta)}$ is the misaligned distance between the centers of the Rx and Tx, where a is the outer radius of the receiver/transmitter found in (6.11), and $\gamma' = \frac{2ab}{a^2 + b^2 + z'^2}$. This is a reinforcement of the results found in Figure 6.12.

6.2.2.2 Near-Field Characteristics

We next focus on the near-field characteristics of the antenna, which have a bearing on the PTE. In Figure 6.17, we show the vector profile of electric and magnetic near-fields of the loop antenna in its principal plane. These fields are then compared with similar field profiles for the anchor-shaped antenna shown in Figure 6.18. For

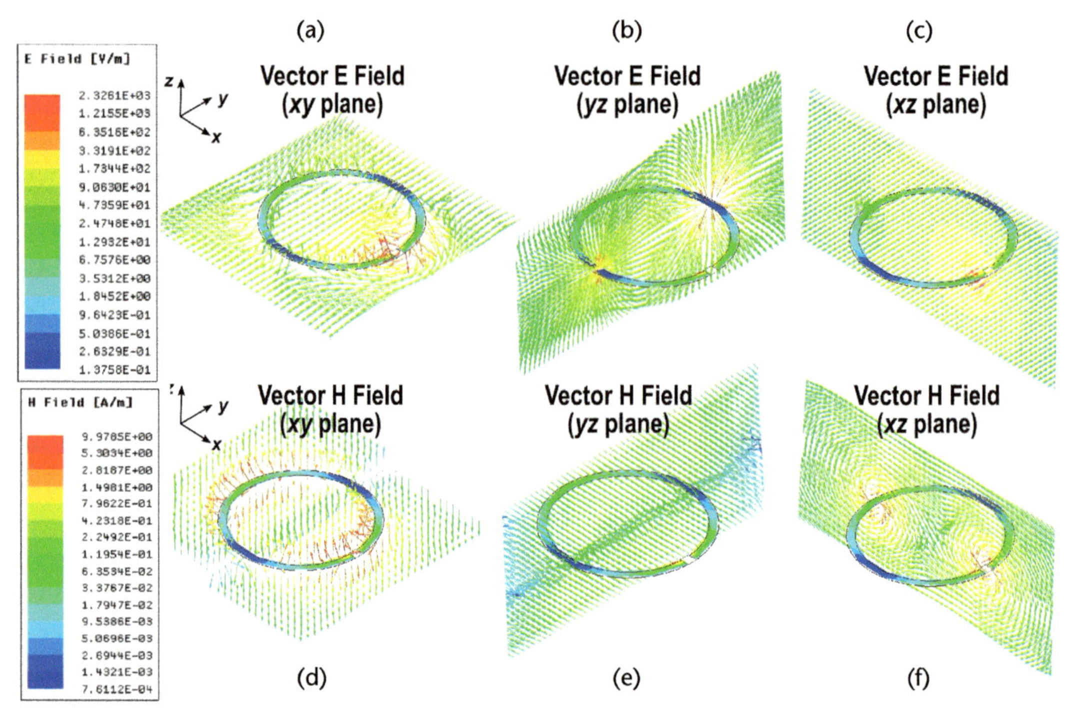

Figure 6.17 Vector E- and H fields at the resonant frequency of the resonant loop antenna in orthogonal planes through the geometry of the antenna. (a–c) Vector E field in xy, yz, and xz planes. (d–f) Vector H-fields in xy, yz, and xz planes. The same dimensions as in Figure 6.13 were used.

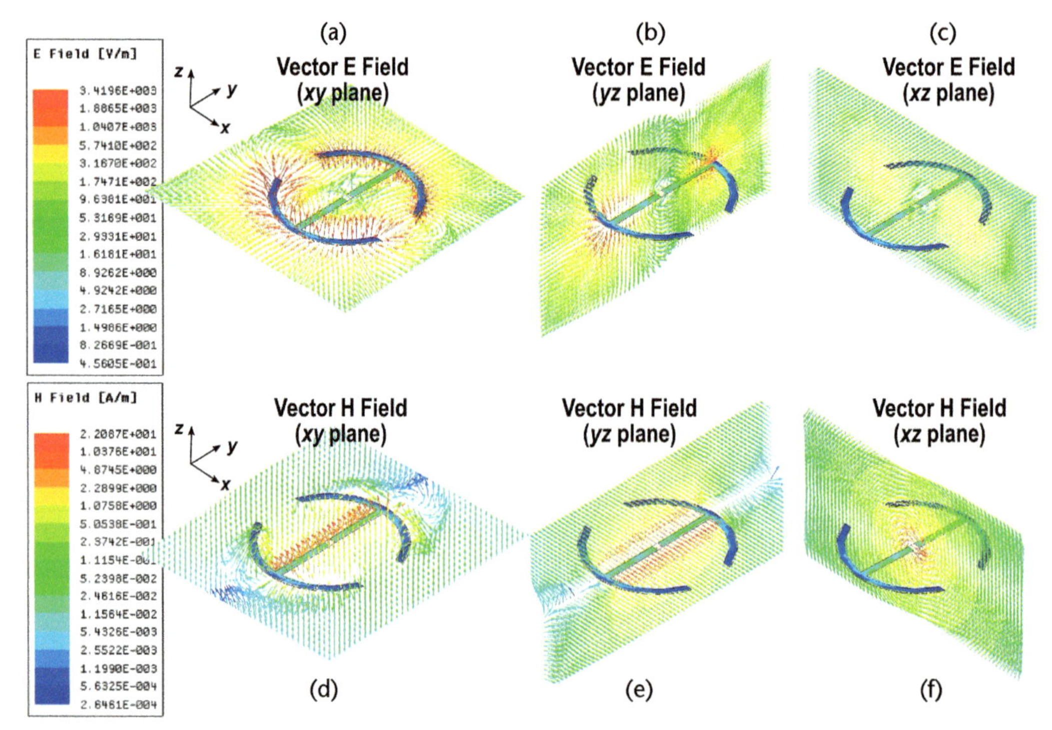

Figure 6.18 Vector E- and H fields at the resonant frequency of the anchor-shaped antenna in three orthogonal planes through the geometry of the antenna. (a–c) Vector E-field in xy, yz, and xz planes. (d–f) Vector H-fields in xy, yz, and xz planes. (©2020 IEEE [42].)

the loop antenna, E-field lines exist across the diametrically opposite points of the loop. These points are associated with the current minimum in the loop. Likewise, the nature of H-field lines can be derived from the vector plots in Figures 6.17(d), (e), and (f), suggesting that these fields exist around the conductor near the current maximum points. The input power was chosen to be the default value of 1W provided by HFSS. A summary of field lines is shown in Figure 6.19(a). The vector field profile of the anchor-shaped structure is shown in Figure 6.18. From Figure 6.18(a–c), it can be inferred that the E-field vector is aligned along the central strip (shank) of the anchor. Furthermore, electric fields also extend beyond the coupled surface (aperture) due to the fringing fields associated to the two open ends of the anchor (see Figure 6.19(b)). It is noted that the extended E-fields are also due to the large wavelength, as compared to loop, where the total loop radius is $\lambda/2\pi$ as compared to anchor, where the structure is similar to a miniaturized 0.7–λ dipole. These extended fields allow coverage between the transmitter and receiver even under misalignment cases.

6.2.2.3 Power Transfer Characteristics under Misalignment between Transmitter and Receiver

To validate the effect of the fringing field and increased wavelength, the power transfer efficiency under different misalignment scenarios is shown in Figures 6.21 and 6.22. Specifically, we consider four degrees of freedom where the receiver and transmitter antennas are misaligned. These include the two lateral misalignment movements (represented by Δx, Δy) and the two angular misalignment movements (represented by ϕ, θ). These improvements are associated with (a) longer wavelength (at resonance frequency) allowing a greater extent of electric field lines, and (b) the presence of fringing fields owing to the open ends of the anchor. The chosen geometrical dimensions of the antennas are the same as shown in Figure 6.12. The misalignment resilience of the PTE for the lateral misalignment cases are shown in Figure 6.22(a). It is noted that for the aligned case ($\Delta x = 0$ or $\Delta y = 0$), the efficiency is smaller for the anchor antenna than for the loop antenna. We continued the evaluation of the PTE in the broadside direction and found that the single loop antenna performed slightly better than its anchor counterpart (see Figure 6.20). The

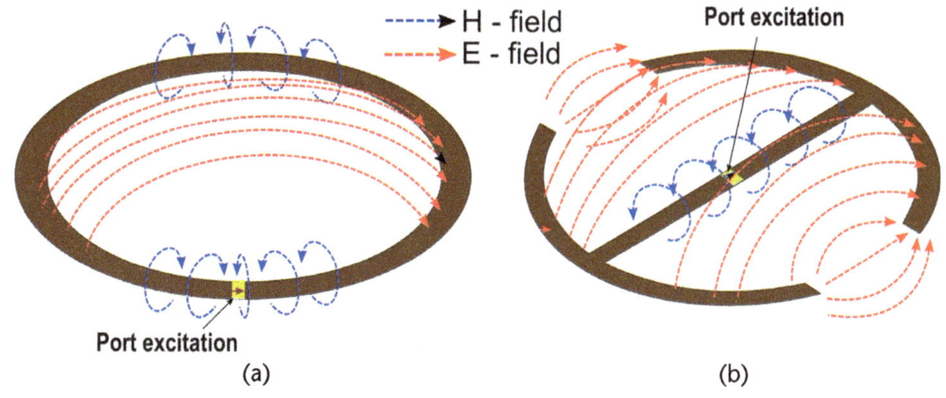

Figure 6.19 Summary of electric and magnetic field lines in the (a) PEC loop, and (b) PEC anchor-shaped resonators. In these simulations, G = 5 cm, w = 0.75 cm, and L = 15 cm. (©2020 IEEE [42].)

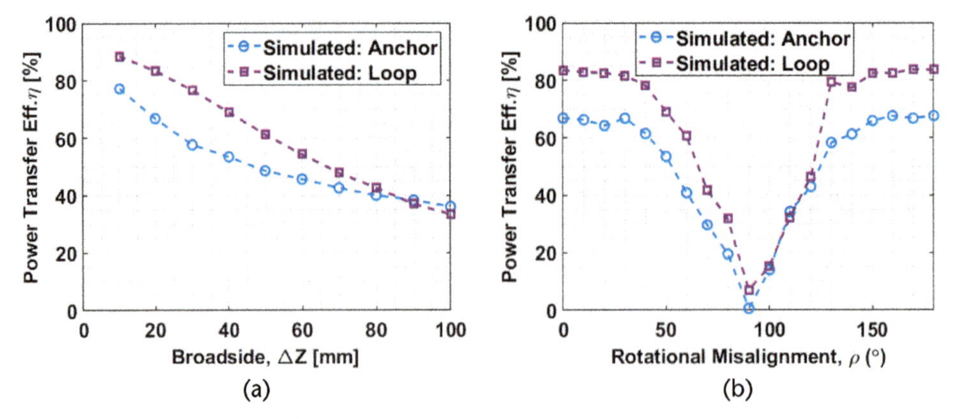

Figure 6.20 Power transfer efficiency of the anchor antenna compared to the loop, when (a) the antennas are separated away from one another along the orthogonal axis, and (b) rotating the receiving antenna around the orthogonal axis while keeping the distance constant with respect to one another. We chose G = 5 cm, w = 0.75 cm, and L = 15 cm for these simulations. (©2020 IEEE [42].)

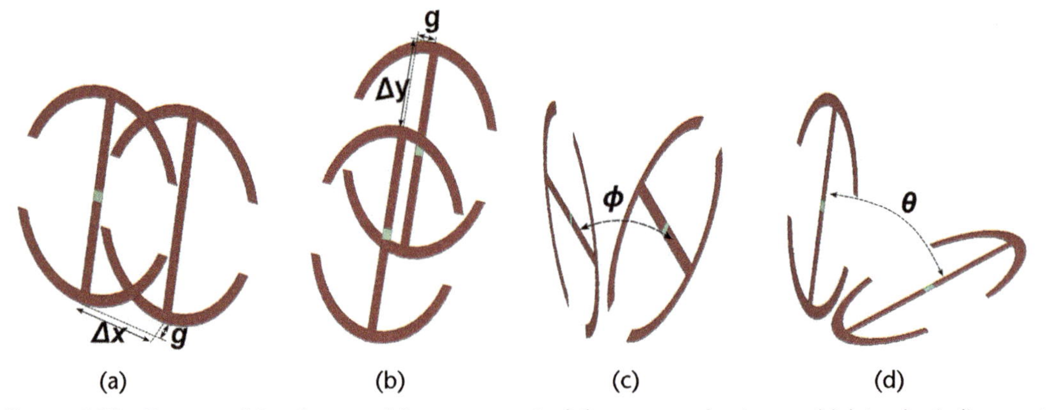

Figure 6.21 Degrees of freedom used for assessment of the proposed antenna: (a) lateral misalignment along openings, (b) lateral misalignment along the shank, (c) azimuthal rotational misalignment around the shank, and (d) elevational rotational misalignment along the crown. (©2020 IEEE [42].)

frequency of operation was found to be between 340 and 380 MHz for the anchor and 678 and 680 MHz for the loop. But the PTE performance becomes better for the anchor-shaped antenna in cases when $\Delta x > 5$ cm or $\Delta y > 5$ cm lateral misalignment. This represents misalignment of 33% of the aperture size.

The resonance frequency for the loop antenna is around 0.72 GHz and for the anchor-shaped antenna it is 0.41 GHz. Since we postulate that the improvement in PTE is due to both the frequency reduction and the fringing fields, we would like to isolate the role of fringing fields for the improvement in the PTE. Therefore, we plot the PTE as a function of electrical distances (normalized by the wavelength of operation) as shown in Figure 6.22(b). We do see improvement in the PTE for 0.4λ and $\Delta y > 0.6\lambda$ in this case, which suggests that the fringing fields are indeed responsible for an improvement in PTE under misalignment cases. Furthermore, fringing fields also affect the azimuthal and elevation misalignments, as shown in Figures 6.22(c) and (d). Once again, the proposed anchor shaped structure shows improved PTE for $\theta > 30°$ and $\phi > 30°$, owing to the decreased resonant frequency and the extent of the fringing fields. The PTE was also evaluated while changing

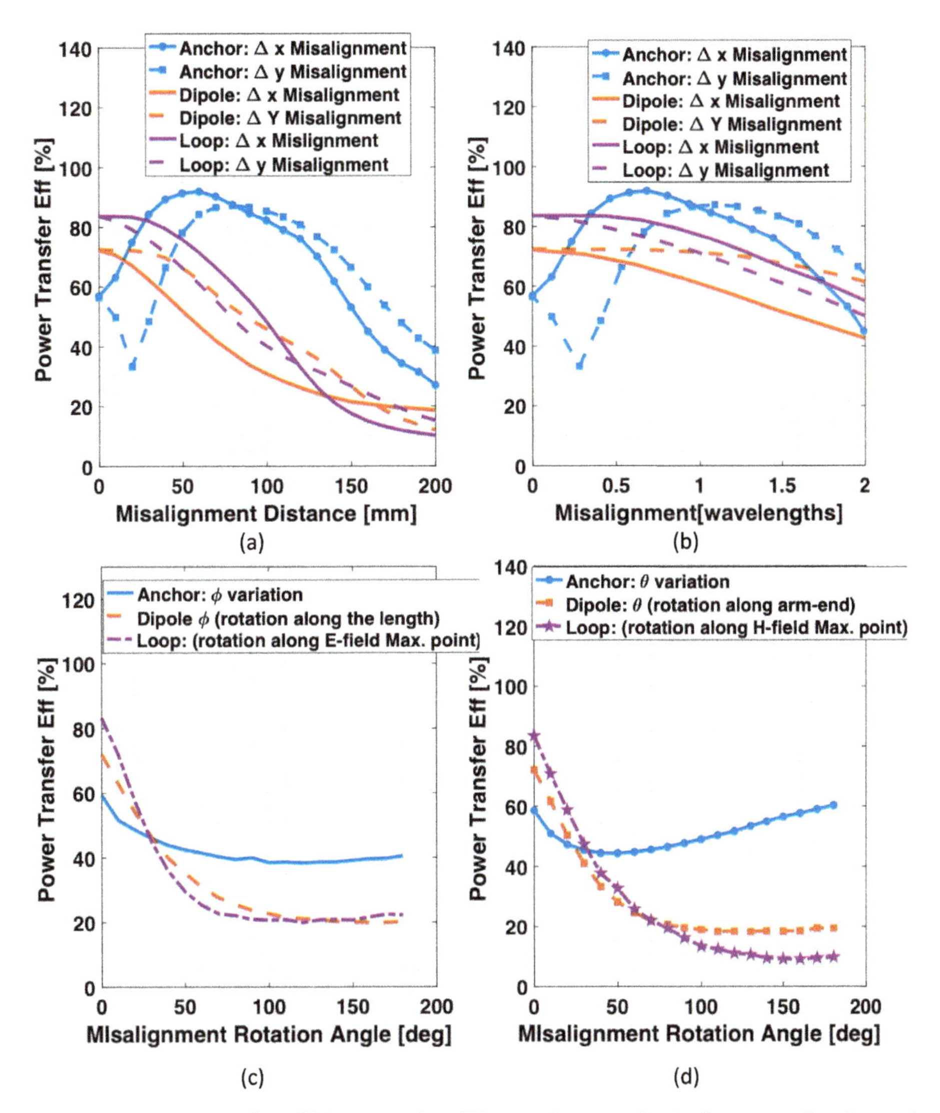

Figure 6.22 Power transfer efficiency under different degrees of misalignment, for the anchor-shaped structure, compared with the traditional resonant loop antenna. g is 2 cm, G = 5 cm, w = 0.75 cm, and L = 15 cm for these simulations. (©2020 IEEE [42].)

the orientation of the transmitter relative to the receiver (Figure 6.20(b)). We notice that when the shanks of the receiver and transmitter are orthogonal, the power transfer efficiency shows a null. These results will be further supplemented by measurements in Section 6.2.2.5.

6.2.2.4 Magnetic versus Electric Coupling in the Lateral Misalignment

For the proposed antenna, the modality of power transfer changes from inductive to capacitive under different lateral misalignment scenarios. An artifact of this is also the slightly reduced PTE for the proposed antenna when compared to the loop antenna under small lateral misalignment cases (Figure 6.22(a)). To understand these features, we propose a new method for analyzing the type of coupling from transmitter to receiver. Specifically, it is well-known that the EM fields in electric (capacitive) coupling are associated to a wave-impedance greater than 377 Ω and the EM fields in magnetic (inductive) coupling are associated to wave impedance smaller than 377 Ω [54]. Therefore, the inequality

$$L'_{m(anchor)} = \frac{\mu_o \pi a^4}{2\left(2a^2 + z'^2\right)^{3/2}}\left(1 + \frac{15}{32}\gamma'^2 + \frac{315}{1024}\gamma'^4\right) \tag{6.22}$$

lays out the conditions for these cases. We note that for a strong inductive coupling Z should be much smaller than 377 Ω, and by the same token for strong capacitive coupling Z should be much greater than 377 Ω. When Z is close to 377 Ω, the impedance of the coupled structure is matched with the free-space impedance, which is the radiation condition. This refers to radiation loss condition and is associated to loss of efficiency. As stated in [45], in the vicinity of the cavities, the electric field is stronger. By the same token, away from the cavities (in the vicinity of the central bar, which is the shank of the anchor), the magnetic field is stronger. To test these scenarios, a specific apparatus is considered for each:

1. The transmitter and receiver are kept at a fixed distance of 20 mm and the receiver is moved laterally along the shank of the transmitter (Figure 6.23(c), along the shank). In this experiment, we anticipate magnetic coupling.
2. The transmitter and receiver are kept at a fixed distance of 20 mm and the receiver is moved laterally across the cavity of the transmitter (Figure 6.23(c), along the cavity). In this experiment, we anticipate electric coupling.

The extent of couplings (magnetic or electric) can be compared using the maximum E-field or H-field values in the respective cases.

A full-wave simulation was carried out for the above cases. The conductive traces of the antennas were assigned PEC boundary conditions, an airbox was used to model free space radiation, and a 50-Ω lumped port was assigned to each antenna's excitation port. Medium-sized mesh grids were used for the setup. While displacing the receiving antenna by up to 10 cm, the peak electric and magnetic fields were recorded for each case. The recorded values are plotted in Figure 6.23(a) and (b). A qualitative parameter to determine the type of coupling (i.e., the ratio of the peak electric and magnetic fields) is plotted in Figure 6.23(d). The comparison impedance through inequality (6.22), knowing the free-space wave impedance (Z_o = 377 Ω), is also shown. Figure 6.23(d) shows a monotonic increase in the wave-impedance as the antenna is moved along the shank, and also shows an increase and then decrease for the case along the cavity. That is, for the either of the cases, the system loses energy in radiation for small misalignment distances due to Z close to 377 Ω. Furthermore, the peak in the "Along Shank" curve is associated to the dip in the PTE for the ΔY misalignment in Figure 6.22(a), since the peak represents an impedance closer to 377 Ω.

This analysis further serves to verify that the use of fringing fields invokes capacitive coupling, aiding PTE for the misalignment cases. Through this, the system

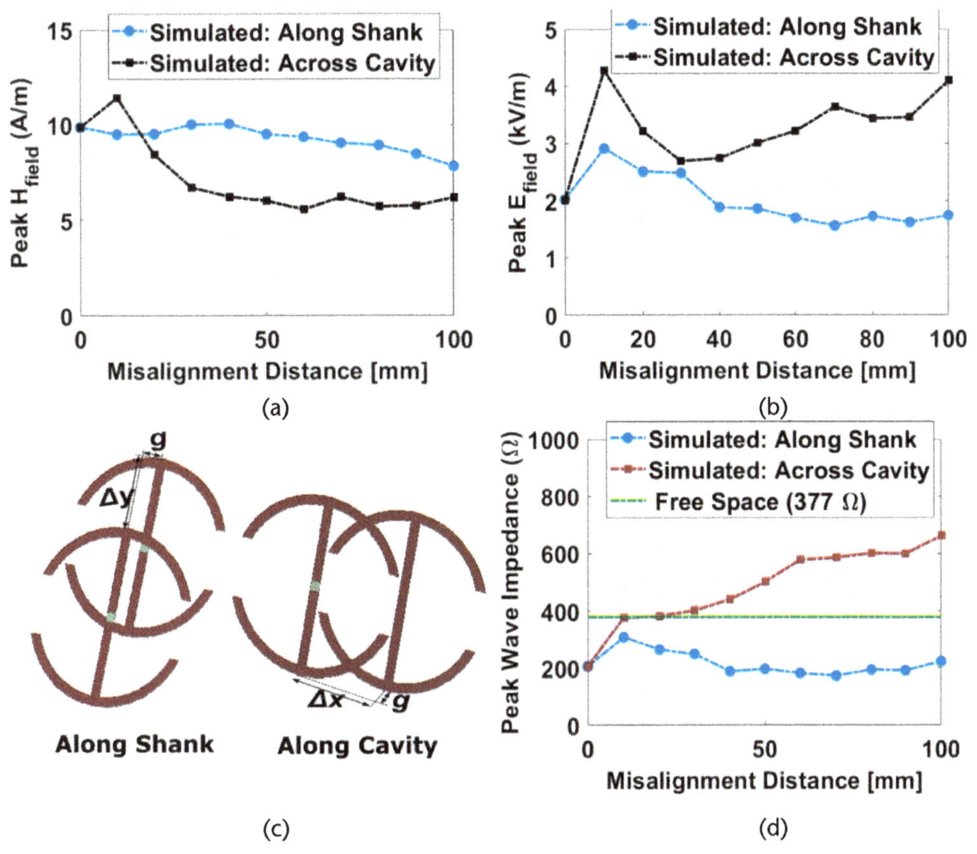

Figure 6.23 Near-field and wave-impedance characteristics of the proposed anchor antenna (G = 5 cm, w = 0.75 cm, and L = 15 cm): (a) the peak magnetic field exhibited by the transmitter as the receiver slides along the shank and across the crowns of the transmitter, (b) the peak magnetic field exhibited by the transmitter as the receiver slides along the shank and across the crowns of the transmitter, (c) the misalignment setup for E-field and H-field testing, and (d) peak wave impedance of the air gap between the transmitter and receiver obtained from the corresponding electric and magnetic fields compared to impedance of plane waves in free space. (©2020 IEEE [42].)

exploits mixed couplings by virtue of its unique shape. The proposed principle also holds for different types of the polygons used in lieu of the circular periphery. As can be seen in Figure 6.24, regardless of the nature of the polygon forming the crown of the anchor, the performance of the power transfer efficiency remains the same.

6.2.2.5 Measurements

The validation of the proposed anchor-shaped antenna is realized by measuring the PTE of the antenna and comparing it with the PTE of the loop antenna. Both loop and anchor antennas were prototyped using FR4 substrate of dielectric constant E_r = 4.4, loss tangent *tanδ* = 0.017, and thickness 1.54 mm. The measurement setup and measured results are shown in Figures 6.25 and 6.26, respectively.

First, a comparison of the PTE performance of the loop antenna and the anchor antenna is conducted when the antennas had the same footprint of L = 15 cm. The other dimensions of the fabricated anchor antenna were G = 5 cm and w

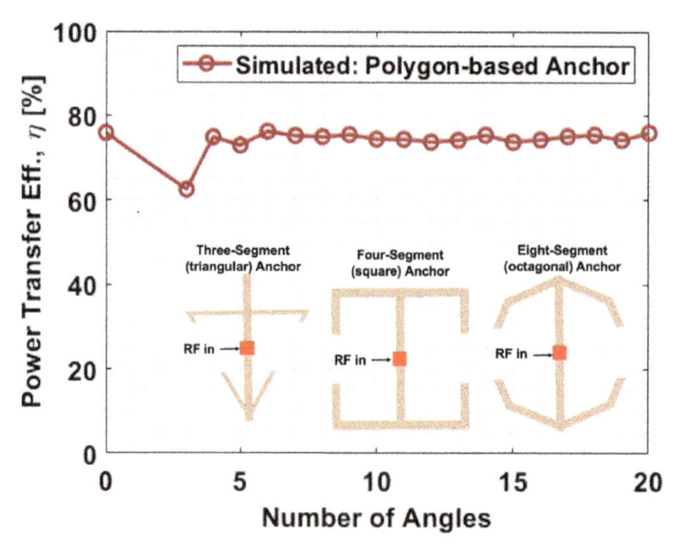

Figure 6.24 PTE performance of different polygonal configurations (various segments or angles) used to generate the crowns of the anchor. Examples of three different anchors generated from different polygonal peripheries are shown in the inset. (©2020 IEEE [42].)

= 0.75 cm. These results are shown in Figure 6.26 (a–c). For these cases, the resonant frequency of the anchor-shaped antenna was found to be approximately half of the loop antenna. To isolate the effect of only the shape, in the second case, we compare the performances of the anchor and loop designed for the same operating frequency. For this case, the anchor antenna with footprint $L = 9$ cm is characterized and compared with the loop antenna (Figure 6.26 (d–h)). Within this case, the radii of the loop and anchor are different, and therefore, two choices exist for the axis of rotation. These choices are associated with two different results as shown in the inset of Figure 6.26. Corresponding measurement cases are considered for a fair and rigorous comparison.

6.2.2.6 PTE Improvements for Lateral Misalignment

For lateral misalignment along the x-axis as shown in Figure 6.26(a), the anchor antenna maintains a PTE almost constant at around 85% for varying misalignment cases. This was achieved when the receiver and transmitter are positioned with

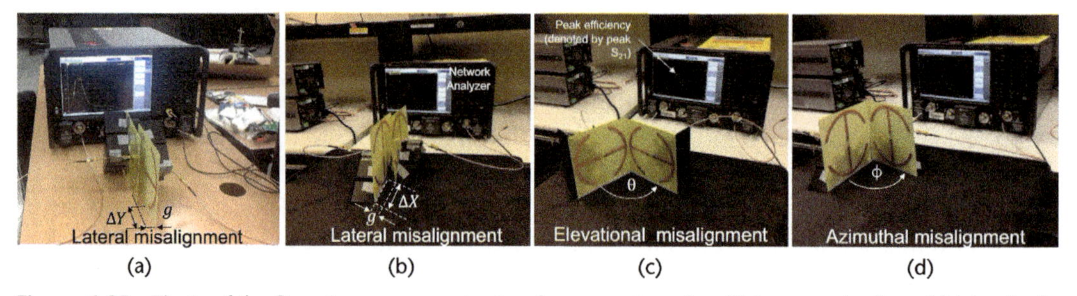

Figure 6.25 Photo of the 2-port measurement setup for power transfer efficiency evaluation: (a) lateral misalignment (ΔY) with g = 5 cm, (b) lateral misalignment (ΔX) with g = 5 cm, (c) angular misalignment due to elevation, and (d) angular misalignment due to azimuth. (©2020 IEEE [42].)

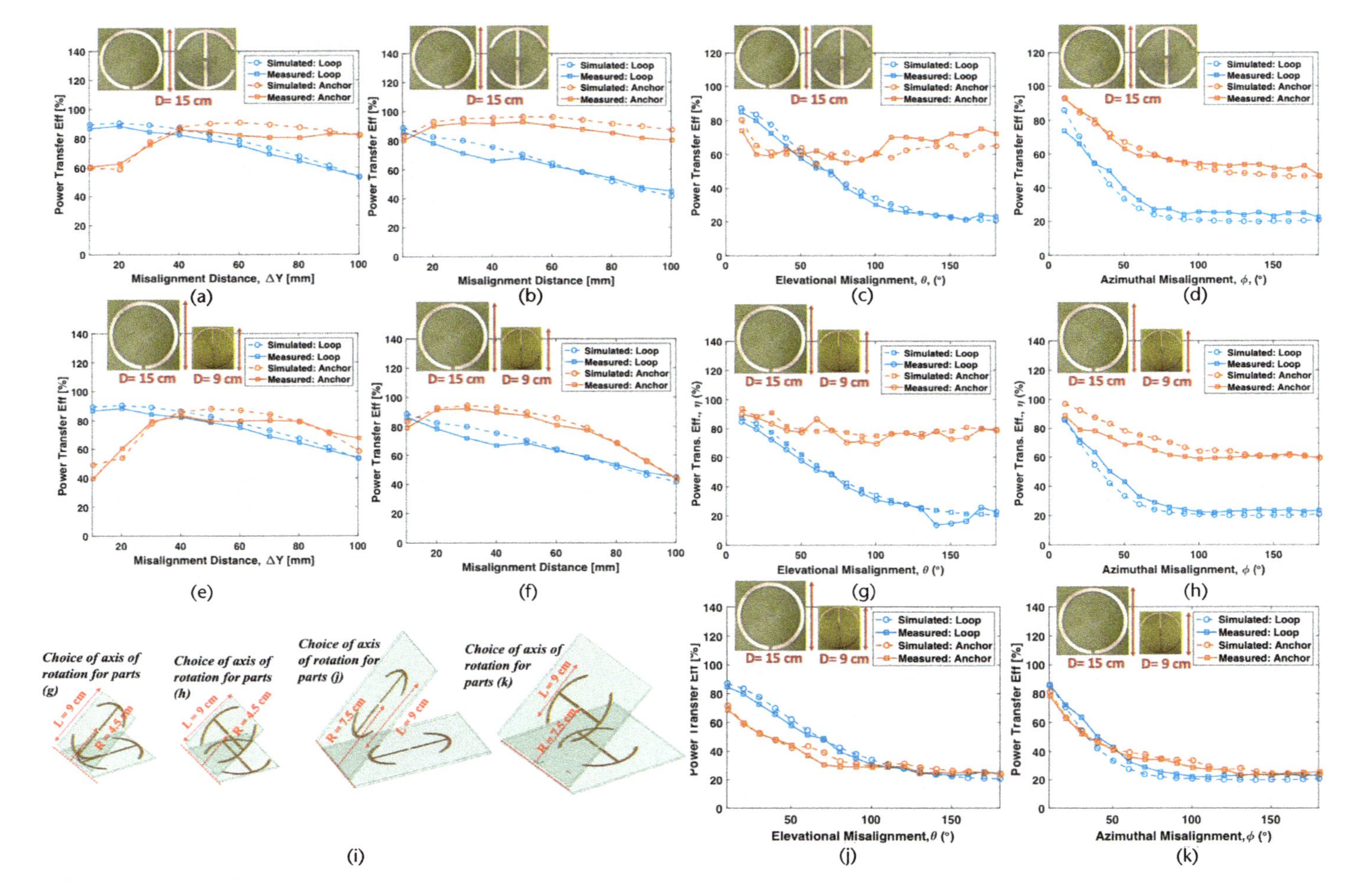

Figure 6.26 Misalignment tests for the large anchor as compared to the loop antenna for the same sizes of the two types of antennas shown in (a–d) and same operational frequency for the two types of the two antennas (e–h). (i) The distinction in the measurement configuration for angular misalignment cases (g), (h), (j), and (k). The dimensions of the fabricated anchor-shaped antenna are G = 5 cm, w = 0.75 cm, and L = 15 cm. For the smaller anchor antenna shown in (j) and (k), the dimensions were chosen to be G = 3 cm, w = 0.2 cm, and L = 9 cm. The dimensions for the loop antenna were w = 0.75 cm, and L = 15 cm. (©2020 IEEE [42].)

initial gap $g = 1$ cm and the receiver is moved sideways Δx from 1 cm to 10 cm. The loop antenna, however, experiences a decaying PTE from 90% to 40%. Further, Figure 6.26(a) shows a good agreement between simulated and measured power transfer efficiency.

In Figure 6.26(d), the results are compared for the small-anchor and loop antenna. Once again, even with 60% miniaturized size, we observe better PTE performance of the anchor antenna over the considered range of misalignment. In Figure 6.26(a) and (e), a trough is observed in the power transfer efficiency for small misalignment cases, which is consistent with the analysis presented in Section 6.2.2.4.

6.2.2.7 Angular Misalignment in the Elevation Plane

In this case, the rotational misalignment is introduced such that the central bars of the receiver and the transmitter create an angle θ, which is varied from 10° to 180°. As shown in Figure 6.26(b), when considering the anchor of the same size as the loop antenna, an average PTE of 65% is shown by the anchor antenna and the PTE is almost uniform throughout the whole angular range. For the loop antenna, however, the PTE decreased from 85% to 20% for this angular range. Similar comparisons are made between the small-anchor ($G = 3$ cm, $w = 0.2$ cm, and $L = 9$ cm) and the loop ($G = 5$ cm, $w = 0.75$ cm, and $L = 15$ cm) as shown in Figure 6.26(e). The anchor antenna exhibited an average PTE of 80%, while the loop experienced a rapid decay. Another case that we considered was when the small anchor was placed at the center of square substrate whose side is of the same dimension as the loop antenna (15 cm). This is to create angular misalignment with the same radius of rotation as the larger loop antenna with dimension of $L = 15$ cm (Figure 6.26(g)). Due to the miniaturized size and extended radius of rotation, the performance of the anchor antenna was found to be degraded. Even so, the proposed antenna holds the advantage of miniaturization of 60%.

6.2.2.8 Angular Misalignment in Azimuthal Plane

The relative angular misalignment along the open ends of the anchor shape is referred to as azimuthal rotation (Figure 6.25, right). For the same size case (Figure 6.26 (c)), for $\phi \geq 50°$, the average PTE exhibited by the anchor antenna was found to be 55% while that of the loop antenna was 20%. For the same frequency operation or the small-anchor case (Figure 6.26(f)), the average PTE was found to be 60%. For the small-anchor case with a large rotation radius (Figure 6.26(h)), no improvement in PTE over the larger loop was noted due to the elongated radius of rotation. As shown previously, the proposed geometry can take advantage of miniaturization for this case. In conclusion, the anchor-shaped antenna performs better than the loop antenna either due to increased power-transfer efficiency or due to miniaturized size. For all four misalignment degrees of freedom, the value of S_{11} was found to be between –6 and –44 dB for the anchor antenna and –8 to –37 dB for the loop antenna. The 3-dB fractional bandwidth ranged from 26% to 72% for the anchor and between 16% and 29% for the loop.

6.2.2.9 Potential Applications

Due to the ability of the proposed power transfer system to sustain higher RF performance when subjected to misalignments, it can be incorporated into wearable and other IoT-type devices. It is an excellent solution to smart and adaptive wireless charging, where some level of movement is anticipated by the wearer. Because of the low profile and ease of fabrication, anchor-shaped antennas present themselves as a strong candidate for charging platforms for on-clothing implementation of power transfer and harvesting and inductive charging for mobile phones and medical devices like an insulin pump. In the next section (Section 6.3.1), we present a full textile version of the anchor-shaped antenna system for power transfer and harvesting for wearable applications (see Figure 6.10). This system is comprised of a chair whose upholstery is used to implement the transmitter and a dress where the receiver (anchor-shaped antenna + rectifier) is embrodered on its back. The design, fabrication, and measurements related to the whole system are presented next.

6.2.3 Textile-Integration of an Anchor-Shaped Antenna and Its Ergonomic Applications

6.2.3.1 Misalignment-Resilient Antenna Geometries for Ergonomic Textile Integration

In the previous sections, an anchor-shaped antenna geometry was proposed that has shown high WPT efficiency under misalignments [51]. Specifically, this geometry showed approximately 65% improvement in efficiency when compared to a loop antenna of the same size and under linear and angular misalignment cases.

Notably, these advantages arise from the two-pronged advantage of the anchor shape (see Figure 6.28). First, due to an open arm structure, the fringing electrical fields are exhibited near the open ends of the arms, and these fields extend beyond the area of the aperture. This allowed for wireless power transfer when the receiver and transmitters were misaligned. Second, due to a decreased frequency of operation, the anchor-shaped structure was miniaturized as well. This allowed for extended near-zone fields due to an increased wavelength and therefore an increase in the power transfer efficiency under misalignments. The same measurements were carried out for antennas of the same geometry and dimensions that were made of conductive textiles embroidered onto denim fabric. The antennas performed as well as their PCB-based ancestors and the results are shown in Figure 6.27.

In this section, we will demonstrate the wireless charging system that uses the previously proposed antennas. To test their performances for wearable applications, we conducted tests of misalignment resiliency for the antenna when wrapped around a cylindrical surface. We further conducted simulations to understand the SARs of the skin tissue when using these antennas.

6.2.3.2 The Effects of Mechanical Deformation and Misalignment on PTE

Figure 6.29 shows the efficiency of wireless power transfer when anchor-shaped antennas are used for transmitters and receivers. Specifically, the tests consisted of a receiver antenna, which was wrapped around a cylindrical surface to mimic the wrapping of a flexible antenna over an arm or a leg ($R_{curvature}$ = 37 mm). We

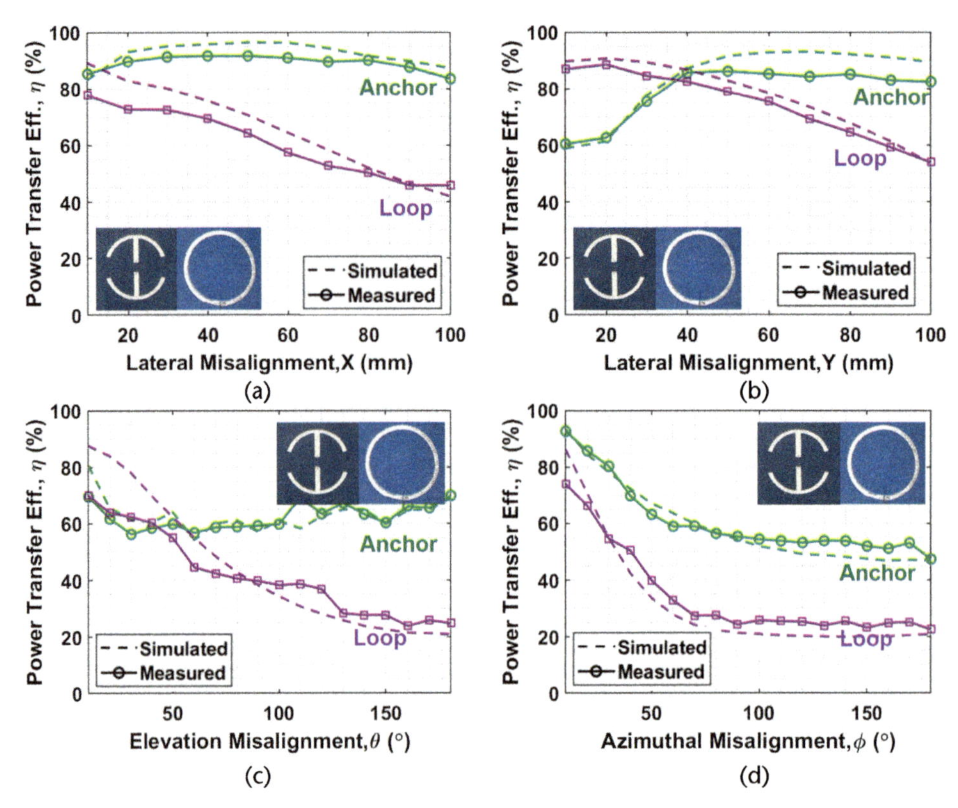

Figure 6.27 PTE performance of the textile-based anchor-shaped antenna compared to its single loop counterpart: (a) PTE performance when the Tx and Rx underwent lateral misalignments across the fringe-enabling cavities, (b) PTE performance when the Tx and Rx underwent lateral misalignments along the shank of the anchors, (c) PTE performance when the Tx and Rx underwent elevational misalignments, and (d) PTE performance when the Tx and Rx underwent azimuthal misalignments. (©2020 IEEE [55].)

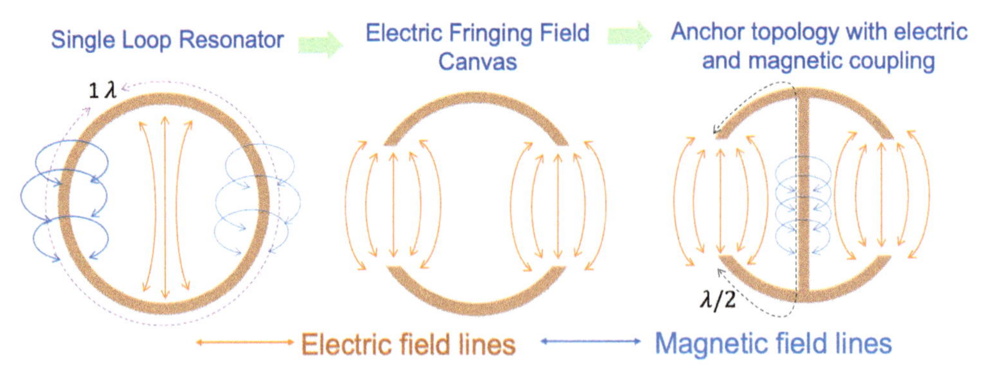

Figure 6.28 Evolution of the anchor topology from its single loop antenna. The single-loop resonator represents the starting point. Electric fringe-enabling cavities are introduced into the single loop. The resulting anchor topology is attained by using a central bar for ringing fields and increased resonance wavelength within the same area. (©2021 IEEE [43].)

also considered the misalignment between the transmitter and receiver antennas to determine the effect of minor movements on the PTE. For this study, we chose an anchor-shaped antenna with a diameter of 15 cm and strip width of 0.75 cm. The results are supported by the distribution of electric and magnetic fields that can

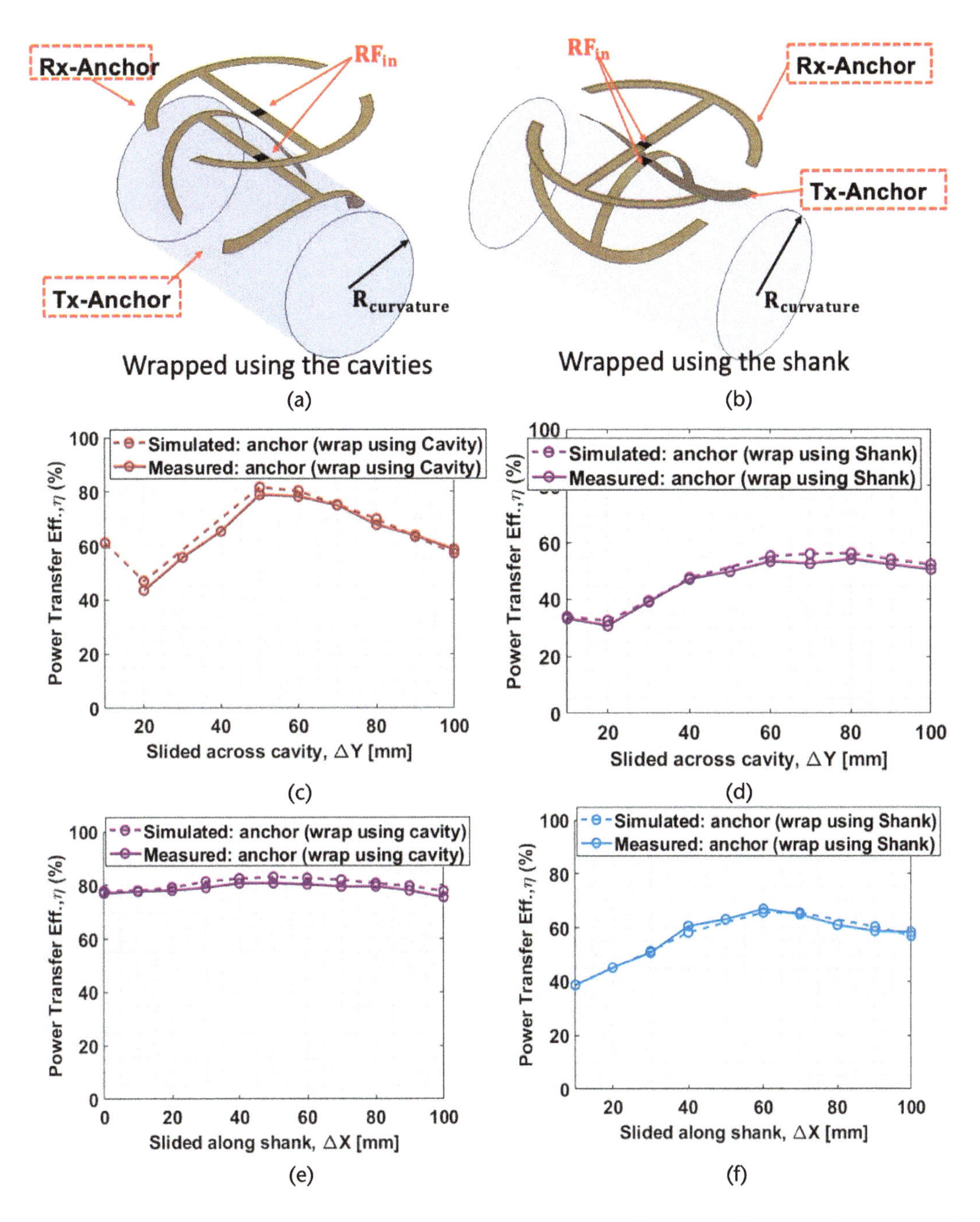

Figure 6.29 The influence of wrapping over cylindrical surfaces and lateral misalignments (Rcurvature = 37 mm) emulating wearable antennas wrapped over arms and legs: (a) wrapping along the cavity, (b) wrapping along the shank, (c) power transfer efficiency for case (a) and movement along the shank, (d) power transfer efficiency for case (b) and movement along the shank, (e) power transfer efficiency for case (a) and movement along the cavities (a) power transfer efficiency for case (b) and movement along the cavities. (©2021 IEEE [43].)

be seen in Figure 6.30. The coupled antennas exhibited a resonant frequency of 360 MHz. This frequency was chosen considering certain defense and commercial bands such as 300 and 433 MHz being commonly used. The size of the antenna

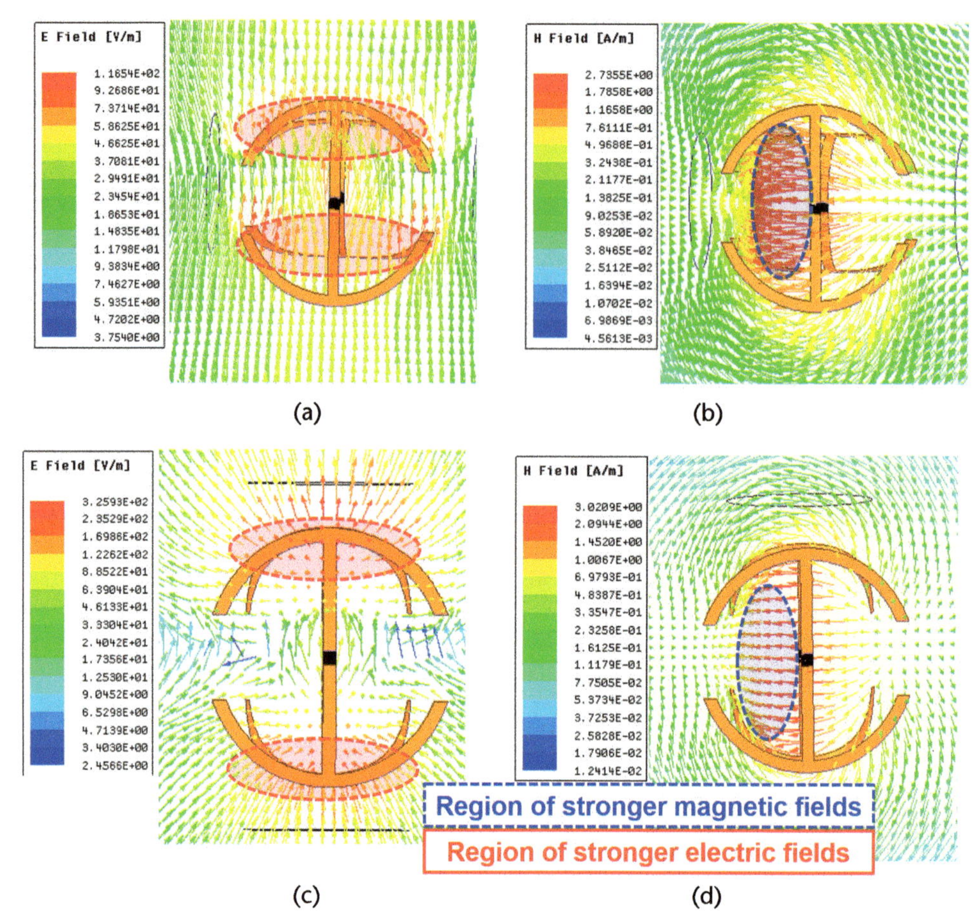

Figure 6.30 Field distribution resulted from wrapping the antenna over cylindrical surfaces and lateral misalignments (Rcurvature = 37 mm) emulating wearable antennas wrapped over arms and legs. (a) E-fields resulted from wrapping using the crown of the anchor, (b) H-fields resulted from wrapping using the crown of the anchor, (c) E-fields resulted from wrapping using the fringe-enabling cavities of the anchor, and (d) E-fields resulted from wrapping using the fringe-enabling cavities of the anchor.

can be scaled to align it exactly with these bands. More details on the RF properties of the antenna are shown in [51]. The efficiency (η) was obtained using scattering parameters evaluated from full-wave simulations. The simulation results were further verified by measurement tests using fabric substrate antennas fabricated using embroidery of conductive threads on denim + stabilizer textile (see Figure 6.33(c)). The fabrication methods were adopted from [1]. We also noted a dielectric constant of $E_r = 1.67$ and a loss tangent $\tan\delta = 0.07$ for the textile surface from this work. The conductive thread, Elektrisola-7, was used for embroidery to prototype the anchor-shaped antennas. SMA connectors with 50-Ω impedance were soldered at the feeding ports for measurements. Simulation and measurement results show the PTE of the wrapped receiver antenna under misalignment (see Figure 6.29).

The radius of curvature of the cylinder was chosen to be 37 mm, which represents the radius of an average wrist. The normal distance between the transmitter and receiver antennas was chosen to be 1 cm, which represents a typical close distance power transfer while sitting or lying down (see Figure 6.10). Depending on the direction of misalignment and two possible wrapping configurations (see Figure 6.29(a,b)), four potential cases arise for this investigation. First, we consider a case of misalignment along the cavities. For this case, when the receiver antenna is wrapped across the cavities, the PTE was found to be 70% on an average upon misalignment (Figure 6.29(c)), while an average PTE of 50% was found when the receiver was wrapped along the shank (Figure 6.29(d)). The second case involves the misalignment movement along the shank where an average of 80% of PTE was exhibited when the receiver antenna was wrapped across the cavities (Figure 6.29(e)) and 55% when it is wrapped along the shank (Figure 6.29(f)). A maximum of 10-cm misalignment is shown, which represents a movement of 66% of the total size of the antenna. For these scenarios, the value of S_{11} was found to be around –19 dB at the resonant frequency. The measured and simulated results agree quite well and further show that the PTE remains better than 38% and is found to be as high as 80%.

Specifically, the anchor-shaped structure allows increased reach of fringing fields, and the PTE values remain high under wrapping and under misalignment cases. These results validate the misalignment resilience of the anchor-shaped antenna, which was previously reported for nonconformal or planar cases [51].

6.2.3.3 SAR Considerations of the Proposed Wireless Power Transfer

The influence of electromagnetic fields in the near zone of a receiver antenna on human tissue was characterized using full-wave simulation of the transmitter and receiver antenna. The receiver antenna was backed by a ferrite sheet to reduce human exposure to radiation. We noted only a nominal effect of the ferrite sheet on the S_{11} of the antennas and on the S_{21} between the antennas. The receiver antenna is placed near a human torso to evaluate the SAR (see Figure 6.31). A built-in torso model from HFSS was used for this simulation. The input power at the transmitter was chosen to be 1W transferred over a distance of at least 6 inches to a surface area of 675 cm^2 (representing the receiver front). The maximum received power is close to 15 mW, which is well within the ICNIRP [56] limit of 1000 μW/cm^2 from 30 to 400 MHz. Figure 6.31(a) depicts the simulation setup and results are shown in Figure 6.31(b). The evaluation was done for two cases, where the transmitting and receiving devices were laterally misaligned along the direction of (1) the shank and (2) the end cavities (see Figure 6.31(b)). As can be seen, when the antennas are laterally misaligned up to 10 cm, the maximum average SAR value of 0.3 W/kg was noted and is found to be within the regulated limit of 0.4 W/kg [56].

6.2.4 RF-to DC Rectifier Design and Optimization

The textile integration of rectifier requires a minimalistic approach while using only embroidered transmission lines and a small number of RF components. An RF-to-DC rectification circuit that uses transmission line resonances and a single rectifier diode was designed and fabricated. The goal was to minimize the number of RF components in the circuit for an easy textile integration, while also to maximize the RF-to-DC power conversion efficiency.

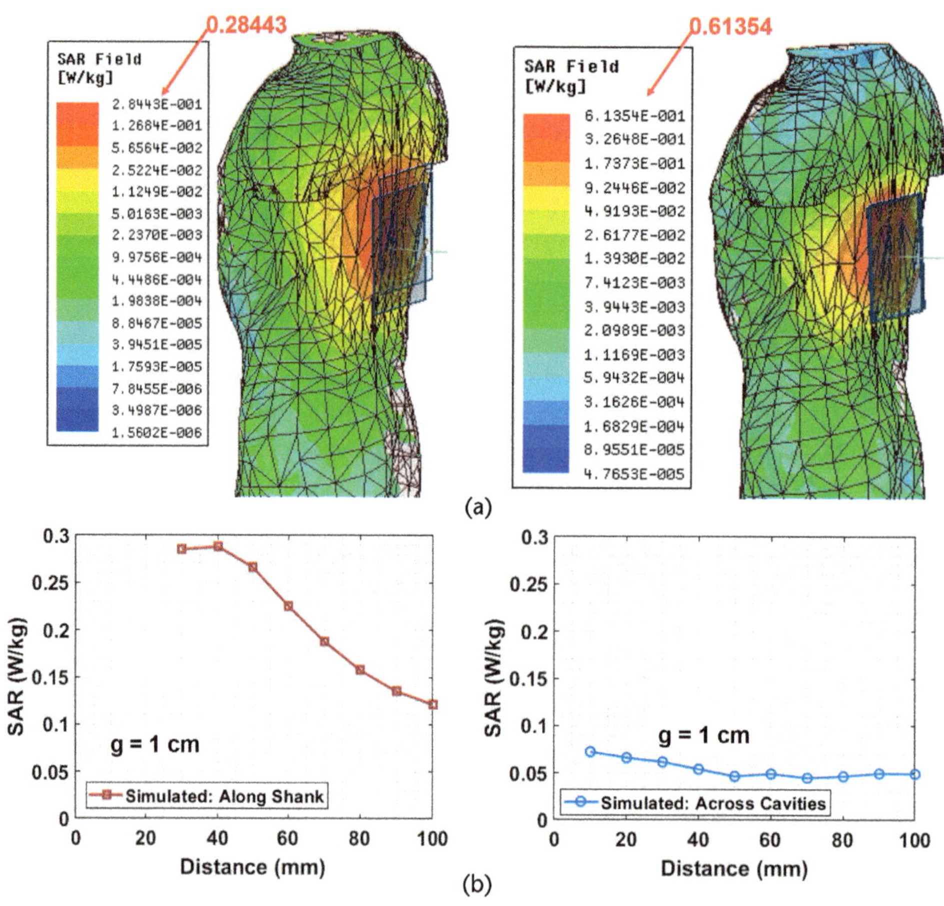

Figure 6.31 SAR evaluation of the system when it is subject to both lateral misalignments along the shank and across the cavity: (a) photo of the human model with extracted SAR value, and (b) SAR values extracted from human model for various misalignment distances (input RF power = 1W). (©2021 IEEE [43].)

6.2.4.1 Working Principle of the $\lambda g/8$ Shorted Stub Rectifier

The circuit-representation of the rectifier is illustrated in Figure 6.32(a). According to [57], the Schottky diode is a parallel circuit of an internal capacitor C_e and resistor R_e. The impedance of the diode is expressed as:

$$Z_{Diode} = \frac{R_e}{1+\left(R_e C_e \omega\right)^2} - j\frac{R_e^2 C_e \omega}{1+\left(R_e C_e \omega\right)^2} \tag{6.23}$$

Placing a small capacitor in parallel with the diode will make the impedance of the circuit look like:

$$Z_{Diode+shunt} = \frac{R_e}{1+\left(R_e C_{total} \omega\right)^2} - j\frac{R_e^2 C_{total} \omega}{1+\left(R_e C_{total} \omega\right)^2} \tag{6.24}$$

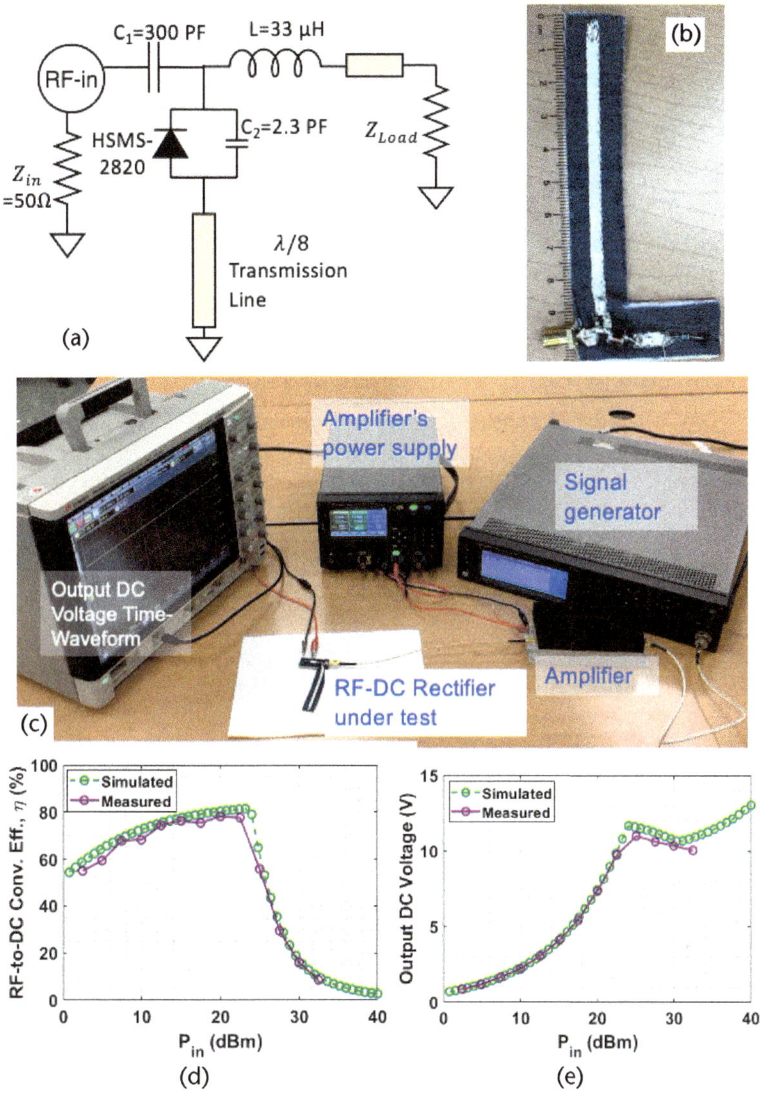

Figure 6.32 Fabric-based RF-to-DC conversion rectifying circuit made of conductive thread embroidered onto fabric substrate: (a) circuit diagram of single diode rectifier using transmission lines (the circuit is powered by a 50-Ω RF source), (b) fabricated prototype using denim substrate, (c) measurement setup used for rectifier efficiency characterizations, (d) simulated and measured efficiency as a function of RF input power, and (e) simulated and measured collected DC voltage as a function of RF input power. (©2021 IEEE [43].)

where $C_{total} = C_e + C_2$. To maximize the RF-to-DC conversion efficiency, the imaginary part of the diode impedance $\Im m(Z_{Diode+shunt}) = \frac{R_e^2 C_{total}\omega}{1+(R_e C_{total}\omega)^2}$ should be matched with the impedance of a shorted stub. Therefore, the series $\lambda/8$ shorted stub is used to achieve the conjugate matching. The impedance of the shorted stub will depend on the frequencies that are multiples of the resonant frequency (f_r). The multiples of the resonant frequency are the harmonics and their orders can be labeled as $k = f/fr$ where f represents the harmonics. The impedance of the shorted stub will be

$$Z_{\lambda/8-stub} = jZ_l \tan(\beta l) = jZ_l \tan\left(\frac{\pi}{4}k\right) = \begin{cases} 0, & if\ k = 0(DC\ signal) \\ jZ_l, & if\ k = 1(Fundamental\ signal) \\ \infty, & if\ k = 2(Second\ harmonic) \\ -jZ_l, & if\ k = 3(Third\ harmonic) \end{cases} \quad (6.25)$$

Equation (6.25) suggests that the second harmonic will be terminated due to the high impedance represented by the shorted stub. Therefore, the currents related to the second harmonics that will be flowing through the diode will be exponentially decayed. The third harmonics will contribute to making the diode more capacitive and any related current that will flow through the diode will be filtered away by the 33-nH DC pass filter (inductive counterpart). The expression of the impedance of the rectifying circuit will be:

$$Z_{rect.circuit} = \frac{R_e}{1+(R_e C_{total}\omega)^2} - j\frac{R_e^2 C_{total}\omega}{1+(R_e C_{total}\omega)^2} + jZ_l \quad (6.26)$$

The conjugate matching will be achieved with $Z_l = \frac{R_e^2 C_{total}\omega}{1+(R_e C_{total}\omega)^2}$. At the resonant frequency, the value of C2 will be tuned to make:

$$Z_{rect.circuit} = \frac{R_e}{1+(R_e C_{total}\omega)^2} \quad (6.27)$$

which should be significantly small with respect to the load resistor, Z_{load} to achieve high RF-to-DC conversion efficiencies.

$$Z_{rect.circuit} = \frac{R_e}{1+(R_e C_{total}\omega)^2} \ll Z_{load} \quad (6.28)$$

The fact that the topology considered in this work is a single diode, when $Z_{rect.circuit} = Z_{load}$, the maximum power transfer will be capped at 50% as the RF-to-DC conversion efficiency depends on the ratio $\frac{Z_{load}}{Z_{rect.circuit} + Z_{load}}$. Therefore, a very significant decay in $Z_{rect.circuit}$ will help achieve an RF-to-DC conversion efficiency up to 95% for lossless circuits. The efficiency will slightly decrease based on the loss factor of the circuit. As shown in prior works [1], these features are realizable by using a resonant shorted transmission line section and by positioning a capacitor connected in parallel with the rectifying diode (see Figure 6.32(a)). The length of the transmission line was optimized to be $\lambda_g/8$ to achieve this effect. A DC block capacitor C_1 is added at the input, while L, and C_2 were optimized for a 50-Ω input impedance along with high RF-to-DC power conversion efficiency at 360 MHz. The choke inductor L was used to remove the ripples from the output DC waveform. For illustration, the design and measurements were conducted with a resistive load of Z_{load}= 683 Ω: however, exact load conditions will be subject to the output devices, such as IoT/IoHT, wearable sensors, or rechargeable supercapacitors, which are connected at the output of the rectifier. The values of these RF components are summarized in Table 6.6. After optimizing the components and lengths of transmission lines in ADS circuit simulation software, the rectifiers were implemented on a denim textile substrate (see Figure 6.32(b)). The conductive regions were realized using embroidery of conductive threads on the substrate. A conductive thread, Elektrisola-7, consisting of silver-coated copper strands (Cu/Ag50 amalgam), was used for the embroidery process. Details of the embroidery steps can be found in [1, 58, 59]. The measurements were conducted by providing an RF signal of frequency 360 MHz with power level changing between 0 and 32 dBm (Figure 6.32(d, e)). The simulated and measured results show a good agreement. This agreement is better than some of the prior comparisons to a nonlinear model for the diode in ADS, which was available from the vendor. The required input signal was achieved using a Keysight MXG signal generator (N5183B) and power amplifier (Mini-circuits ZHL-20W-13+). The output of the rectifier, the DC voltage waveform, was recorded through a mixed signal oscilloscope (MSOS254A). The waveforms showed almost flat output voltage with no ripples, which was then used to calculate the output DC power from the rectifier using V^2/Z_{load}.

The power conversion performance of the rectifier is shown in Figure 6.32(d) and as shown, a peak of 77.27% of RF-to-DC conversion efficiency was achieved at around 22.5 dBm input. Beyond input power of 22 dBm, the voltage level was found to be at least 10V (Figure 6.32 (b)). The comparison with state-of-the-art is done in Table 6.3. To the authors' knowledge, this is the highest RF-to-DC efficiency reported for a textile-based rectifying circuit. Moreover, we note a 50%, 60%, and 70% of RF-to-DC conversion efficiency at the input power range values of 26, 22, and 14 dB, respectively. These results are comparable or better than the current state of the art reported in prior works (see Table 6.3).

Table 6.6 Optimized Parameters of the Textile-Based Rectifying Circuit Based on a $\lambda_g/8$ Short-Circuited Stub

C_1	300 pF (DC Block)
Operating Frequency	360 MHz
C_2	2.3 pF (tuned for diode impedance matching)
Z_{Load}	683 Ω (optimized for high power collection)
L	33 μH (RF choke)
Diode	HSMS2820 (for medium power rectification)
Substrate	denim (E_r = 1.67 and tanδ = 0.07)
Conductive Surfaces	Elektrisola-7 ($\sigma = 2.44\times10^6$ S/m [1])
Length of TL	$\lambda_g/8$* (for 360 MHz)
Skin Depth, δ	17μm (for 360 MHz)

*$\lambda_g/8 = \lambda/(8\sqrt{\varepsilon_r}) = \lambda/(8\sqrt{1.67}) = 8.06$ cm.

6.2.5 System Design and Tests Using RF Rectifier and Anchor-Shaped Antenna

6.2.5.1 Illustration of Ergonomic Wireless Power Transfer

Having optimized the misalignment resilient antenna and corresponding rectifier for RF-to-DC rectification, we consider the performance of the textile-integrated rectenna system using these components. Specifically, first the ergonomic nature of the setup is illustrated for a user seated in a chair with a rectenna receiver integrated in their clothing (see Figure 6.33). Notably, due to textile nature of the antenna substrate, and due to a conductive fiber-based antenna and circuit fabrication techniques, such an integration is now possible. Likewise, the transmitter antenna is embroidered in the upholstery of chair. The transmitter antenna and receiver rectenna are shown in Figures 6.33(c) and (d), respectively.

To illustrate the wireless power transfer, three LEDs were connected in parallel to the output DC line of the rectifier [55]. As shown, the LEDs were lit when the receiver was in the vicinity of the transmitter and for the cases when the receiver underwent misalignments. Total transmit power of about 1W was used for this experiment. The misalignments were done by rotating and laterally translating the receiver as shown in Figure 6.33(b). This experiment serves to demonstrate the

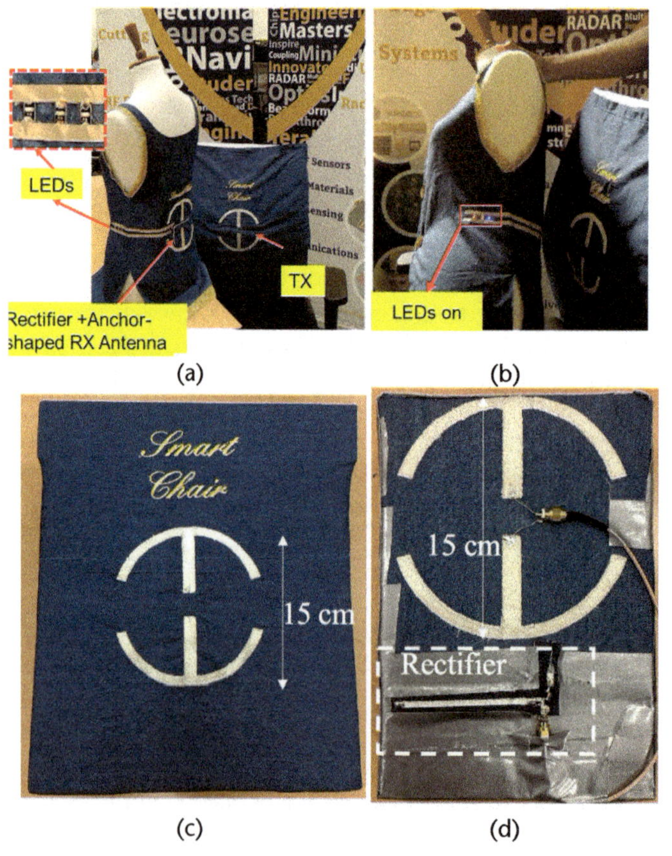

Figure 6.33 Illustration of practical use of the ergonomic wireless charging platform: (a) fabric embroidered antenna and rectifier circuits in dressing and chair upholstery, (b) LEDs turned on under wireless powering and under misalignments, (c) close-up of the transmitting antenna, and (d) close-up of the receiver antenna and rectifier circuit. (©2021 IEEE [43].)

usefulness of the proposed ergonomic wireless charging system and successful operation of the receiver antenna and rectifier circuits.

6.2.5.2 Quantitative Power Collection Setup

To quantitatively characterize the performance of the system, we conducted comprehensive experiments to measure the wireless power transfer in the inductive near-field and radiative near-field regions of the transmitter antenna. The performance of the receiver in a very close proximity (1 to 10 cm) of the transmitter antenna has been covered in the prior works [51, 55]. In this work, we wanted to learn if a user could walk around in the vicinity of the transmitter antenna while allowing wireless power transfer through the wearable surfaces. These determinations are made by collecting power under exaggerated misalignment between the transmitter and receiver antennas, ranging up to 4 ft. from the transmitter and 2 ft. of lateral shifts. The setup adopted for this measurement is shown in Figure 6.34. The power transmitter (Tx) was positioned at the edge of the grid and was stationary. The transmitter consisted of a Keysight MXG signal generator (N5183B) and power amplifier (Mini-circuits ZHL-20W-13+), which provided a total output power of about 1W. The receiver antenna was placed at translating locations, each 6 inches apart, covering an area of 4 ft. × 4 ft. (see Figure 6.34(b)). We note that

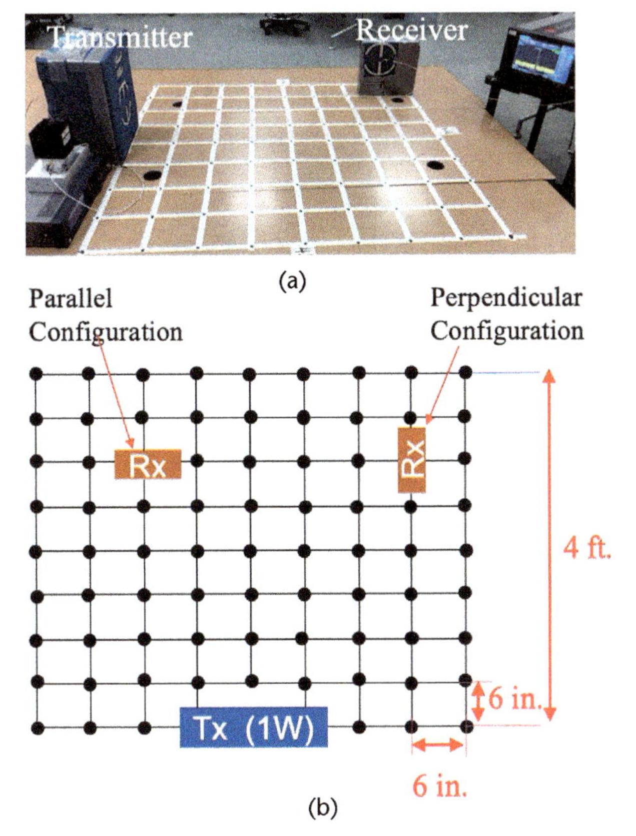

Figure 6.34 (a) Photo of the measurement setup with 2-D grid where the receiver was translated for individual measurements, and (b) schematic of the 2-D grid, perpendicular/parallel configuration, and dimensions of the grid points used for the measurements. (©2021 IEEE [43].)

in a practical setting (as shown in Figure 6.33), the receiver antenna's plane may rotate with respect to the transmitter antenna's plane. Therefore, we consider two cases where the receiver antenna placed in parallel with and perpendicular to the transmitter antenna. Received power or voltage information for an arbitrary angle can be interpreted by knowing these two orthogonal cases.

6.2.5.3 Measurement Results

The tests characterized the RF power received by the receiver antenna at each of the grid points. To test the efficacy of the rectifier in providing a DC output, measurements were also conducted for rectifier's output of DC voltage and DC power. These measured values, as a function of 2-dimensional space, are provided in Figure 6.35. RF power measurements are shown in Figure 6.35(a), while rectified DC voltage and DC power measurements are shown in Figure 6.35(b) and (c), respectively. First, we consider Figure 6.35(a). As shown, above 10 mW of power is recorded by the receiver within a distance of 1.5 ft. normal distance from the transmitter antenna. Remarkably, the power levels remain >10 mW across 4 ft. of lateral region at 1.5 ft. normal distance from transmitter. Further, at a normal distance of 2 ft., 12-mW power is received by the antenna. The power decrease to about 2 mW when

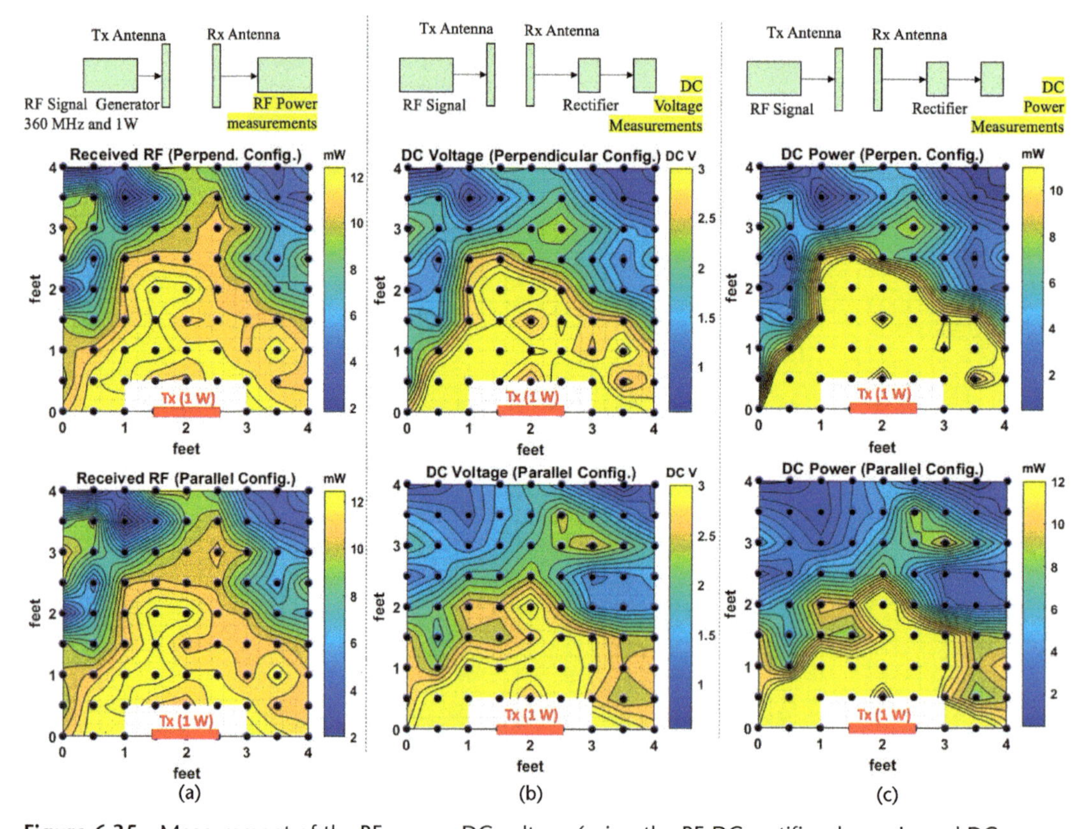

Figure 6.35 Measurement of the RF power, DC voltage (using the RF-DC rectifier shown in and DC power across the 2-D region in the front of the transmitter antenna. The black marker points represent the measurement points, while the colors in the 2-D region represent the interpolated values. (a) RF power recorded using a spectrum analyzer. (b) DC voltage recorded using a multimeter. (c) DC power measured calculated by V 2/R, with R = 683 Ω. (©2021 IEEE [43].)

we move away to a normal distance of 4 ft. from the transmitter antenna. We also note that the orientation of the receiver antenna has a rather small impact on the collected power, as we compare the top and bottom 2-D plots in Figure 6.35(a).

Next in Figure 6.35(b) and (c), we present the output of the antenna + rectifier circuit of Figure 6.32. The output voltages were recorded and found to be up to 3V in the vicinity of the transmitter, which is relevant for the state-of-the-art IoT sensors and charging devices. Note that the output voltages can be increased by using multiple diode-based rectifiers for voltage multiplication [60]. The corresponding DC power was calculated using $P = V^2/R$ with $R = 683\ \Omega$ as shown in Figure 6.35(c). Furthermore, we note that the rectified DC power is in the same order as the RF power collected by the antenna. Notably, between 0 and 2 ft., the incident RF power is around 10 to 12 mW (or 10 to 11 dBm). At this input power level, the rectifier efficiency is close to 60% (see Figure 6.32), allowing around 6 to 10 mW of DC power availability. This verifies the power conversion efficiency of the rectifier. Overall, this verifies that for a 1W transmitter, the choice of diode and components (which determines the efficiency for a given input power) is well optimized for wireless power collection at up to approximately 2 ft. from the transmitter. We note that received power levels are of interest for low-power wearable sensors and IoT devices [61, 62]. Power consumption of these devices is in pW to μW range and therefore the proposed wireless powering through daily life activities is appropriate for wireless powering or charging these devices. Specifically, low-power devices, such as temperature sensors with power consumption between 113 pW and 1.4 μW [33, 63–66], current and voltage sensors, as well as biosensors with power consumption of 9.3 nW and 436 μW [34, 67–69] are already reported in prior works. Furthermore, the collected DC power level is enough to power ECG monitoring systems, pulse oximeters, or neural recording systems with typical power requirements between 5 to 12 mW [70–72].

6.3 Conclusion

In this chapter, two power transfer and harvesting systems were proposed. The first system consisted of a harvesting jacket that embodies the clothing integration of antennas and power-harvesting circuits for wearable applications. An array of low-profile antennas (patches) is combined with single-diode rectifiers where each antenna has its own rectifier and the outputs of all the rectennas (antenna + rectifier) are combined for high and robust DC power collection. Using boosted Wi-Fi signals, we were able to achieve a DC power collection of 0.6 mW, which is enough to drive a low-power sensor. This jacket can be used to wirelessly power low-power biosensors for medical applications. The second system was made of a new class of misalignment-resilient antenna topology referred to as anchor-shaped combined with a single-diode rectifying circuit for DC power collection. The anchor-shaped antenna allows for both lateral and angular misalignments within the ranges of 1 to 10 cm as well as 10° and 180°, respectively. With respect to its single-loop counterpart, the PTE performance of the proposed anchor-shaped antenna is better and that of the single-loop antenna experiences a rapid decrease. The excellent performance of the anchor-shaped antenna is due to the fringe-enabling cavities that are introduced in the structure, which boosts both electric and magnetic coupling modes that are responsible for high and stable PTEs. This design is well-suited for contactless and misalignment-free wireless charging applications.

Antenna topology was used to develop a wireless charging platform that is embedded into clothing to be used by patients with charging needs for the devices, that enables medical professionals to charge their wearable devices while focusing on saving lives, and that allows consumers to charge their everyday wearable/portable devices. The transmitters used to illuminate the receivers are set to transmit a maximum of 1W (per FCC/FDA regulations). This power level is enough to collect up to 10 mW of DC power, which is enough to drive a wide range of IoT devices.

References

[1] Vital, D., S. Bhardwaj, and J. L. Volakis, "Textile Based Large Area RF-Power Harvesting System for Wearable Applications," *IEEE Transactions on Antennas and Propagation*, Vol. 68, No. 3, 2019, pp. 2323–2331.

[2] Zhang, H. S., S.- L. Chai, K. Xiao, and L. F. Ye, "Numerical and Experimental Analysis of Wideband E-Shape Patch Textile Antenna," *Progress in Electromagnetics Research*, Vol. 45, 2013, pp. 163–178.

[3] Moro, R., S. Agneessens, H. Rogier, and M. Bozzi, "Wearable Textile Antenna in substrate integrated waveguide technology," *Electronics Letters*, Vol. 48, August 2012, pp. 985–987.

[4] Lemey, S., F. Declercq, and H. Rogier, "Textile Antennas as Hybrid Energy-Harvesting Platforms," *Proceedings of the IEEE*, Vol. 102, No. 11, 2014, pp. 1833–1857.

[5] Song, L., A. C. Myers, J. J. Adams, and Y. Zhu, "Stretchable and Reversibly Deformable Radio Frequency Antennas Based on Silver Nanowires," *ACS Applied Materials & Interfaces*, Vol. 6, No. 6, 2014, pp. 4248–4253.

[6] Cook, B. S., and A. Shamim, "Inkjet Printing of Novel Wideband and High Gain Antennas on Low-Cost Paper Substrate," *IEEE Transactions on Antennas and Propagation*, Vol. 60, No. 9, 2012, pp. 4148–4156.

[7] Cheng, S., Z. Wu, P. Hallbjorner, K. Hjort, and A. Rydberg, "Foldable and Stretchable Liquid Metal Planar Inverted Cone Antenna," *IEEE Transactions on Antennas and Propagation*, Vol. 57, No. 12, 2009, pp. 3765–3771.

[8] Zhong, J., A. Kiourti, T. Sebastian, Y. Bayram, and J. L. Volakis, "Conformal Load-Bearing Spiral Antenna on Conductive Textile Threads," *IEEE Antennas and Wireless Propagation Letters*, Vol. 16, 2017, pp. 230–233.

[9] Wang, Z., L. Zhang, Y. Bay Ram, et al., "Embroidered Conductive Fibers on Polymer Composite for Conformal Antennas," *IEEE Transactions on Antennas and Propagation*, Vol. 60, No. 9, 2012, p. 4141.

[10] Kiourti, A., C. Lee, and J. L. Volakis, "Fabrication of Textile Antennas and Circuits with 0.1 mm Precision," *IEEE Antennas and Wireless Propagation Letters*, Vol. 15, 2015, pp. 151–153.

[11] Zhong, J., A. Kiourti, J. L. Volakis, T. Sebastian, and Y. Bayram, "Mechanical and Thermal Tests of Textile Antennas for Load Bearing Applications," in *2016 IEEE International Symposium on Antennas and Propagation (APSURSI)*, IEEE, 2016, pp. 1943–1944.

[12] Corchia, L., G. Monti, and L. Tarricone, "Wearable Antennas: Nontextile Versus Fully Textile Solutions," *IEEE Antennas and Propagation Magazine*, Vol. 61, No. 2, 2019, pp. 71–83.

[13] Shawalil, S., K. N. Abdul Rani, and H. A. Rahim, "2.45 GHz Wearable Rectenna Array Design for Microwave Energy Harvesting," *Indonesian Journal of Electrical Engineering and Computer Science*, Vol. 14, May 2019, pp. 677–687.

[14] Vital, D., J. L. Volakis, and S. Bhardwaj, "Loss-Characterization and Guidelines for Embroidery of Conductive Textiles," in *2018 IEEE Antennas and Propagation Society International Symposium (APSURSI)*, July 2018.

[15] Kiourti, A., C. Lee, and J. L. Volakis, "Fabrication of Textile Antennas and Circuits with 0.1 mm Precision," *IEEE Antennas and Wireless Propagation Letters*, Vol. 15, 2016, pp. 151–153.

[16] Balanis, C. A., *Antenna Theory: Analysis and Design,* Fourth Edition, Hoboken, NJ: John Wiley and Sons, 2015.

[17] Wang, Z., L. Zhang, Y. Bayram, and J. L. Volakis, "Embroidered Conductive Fibers on Polymer Composite for Conformal Antennas," *IEEE Transactions on Antennas and Propagation*, Vol. 60, No. 9, 2012, pp. 4141–4147.

[18] Vallejo, M., J. Recas, P. del Valle, and J. Ayala, "Accurate Human Tissue Characterization for Energy-Efficient Wireless On-Body Communications," *Sensors*, Vol. 13, No. 6, 2013, pp. 7546–7569.

[19] IEEE C95.1-2005, IEEE Standard for Safety Levels with Respect to Human Exposure to Radio Frequency Electromagnetic Fields, 3 kHz to 300 GHz.

[20] Sun, H., Y.- X. Guo, M. He, and Z. Zhong, "Design of a High-Efficiency 2.45-GHz Rectenna for Low-Input-Power Energy Harvesting," *IEEE Antennas and Wireless Propagation Letters*, Vol. 11, 2012, pp. 929–932.

[21] Valenta, C. R., and G. D. Durgin, "Harvesting Wireless Power: Survey of Energy-Harvester Conversion Efficiency in Far-Field, Wireless Power Transfer Systems," *IEEE Microwave Magazine*, Vol. 15, June 2014, pp. 108–120.

[22] Mohamed, M. M., G. A. Fahmy, A. B. Abdel-Rahman, et al., "High-Efficiency CMOS RF-to-DC Rectifier Based on Dynamic Threshold Reduction Technique for Wireless Charging Applications," *IEEE Access*, Vol. 6, 2018, pp. 46826–46832.

[23] Hemour, S., Y. Zhao, C. H. P. Lorenz, et al., "Towards Low-Power High-Efficiency RF and Microwave Energy Harvesting," *IEEE Transactions on Microwave Theory and Techniques*, Vol. 62, April 2014, pp. 965–976.

[24] Paing, T., E. Falkenstein, R. Zane, and Z. Popovic, "Custom IC for Ultra-Low Power RF Energy Harvesting," *2009 Twenty-Fourth Annual IEEE Applied Power Electronics Conference and Exposition*, IEEE, 2009, pp. 1239–1245.

[25] Masotti, D., A. Costanzo, M. D. Prete, and V. Rizzoli, "Genetic-Based Design of a Tetra-Band High-Efficiency Radio-Frequency Energy Harvesting System," *IET Microwaves, Antennas Propagation*, Vol. 7, December 2013, pp. 1254–1263.

[26] Vera, G. A., A. Georgiadis, A. Collado, and S. Via, "Design of a 2.45 GHz Rectenna for Electromagnetic (EM) Energy Scavenging," *2010 IEEE Radio and Wireless Symposium (RWS)*, January 2010, pp. 61–64.

[27] Olgun, U., C. Chen, and J. L. Volakis, "Wireless Power Harvesting with Planar Rectennas for 2.45 GHz RFIDs," *2010 URSI International Symposium on Electromagnetic Theory*, August 2010, pp. 329–331.

[28] Ensworth, J. F., S. J. Thomas, S. Y. Shin, and M. S. Reynolds, "Waveform-Aware Ambient RF Energy Harvesting," *2014 IEEE International Conference on RFID (IEEE RFID)*, April 2014, pp. 67–73.

[29] DeLong, B., A. Kiourti, and J. L. Volakis, "A 2.4-GHz Wireless Sensor Network Using Single Diode Rectennas," *2016 IEEE International Symposium on Antennas and Propagation (APSURSI)*, June 2016, pp. 403–404.

[30] FCC Rules and Regulations, AIR802—2.4 & 5 GHz Bands.

[31] DeLong, B. J., *Integration of Radio Frequency Harvesting with Low Power Sensors*, PhD thesis, The Ohio State University, 2018.

[32] Talla, V., B. Kellogg, B. Ransford, S. Naderiparizi, J. R. Smith, and S. Gollakota, "Powering the Next Billion Devices with Wi-Fi," *Commun. ACM*, Vol. 60, 2015, pp. 83–91.

[33] Wang, H., and P. P. Mercier, "Near-Zero-Power Temperature Sensing Via Tunneling Currents through Complementary Metal-Oxide-Semiconductor Transistors," *Scientific Reports*, Vol. 7, No. 1, 2017, p. 4427.

[34] Hu, J., Y.- B. Kim, and J. Ayers, "A Low Power 100 MΩ CMOS Front-End Transimpedance Amplifier for Biosensing Applications," *2010 53rd IEEE International Midwest Symposium on Circuits and Systems (MWSCAS)*, IEEE, 2010, pp. 541–544.

[35] Adami, S.- E., P. Proynov, G. S. Hilton, et al., "A Flexible 2.45-GHz Power Harvesting Wristband with Net System Output from- 24.3 dBm of Rf power," *IEEE Transactions on Microwave Theory and Techniques*, Vol. 66, No. 1, 2017, pp. 380–395.

[36] Besnoff, J. S., T. Deyle, R. R. Harrison, and M. S. Reynolds, "Battery-Free Multichannel Digital ECG Biotelemetry Using UHF RFID techniques," in *2013 IEEE International Conference on RFID (RFID)*, IEEE, 2013, pp. 16–22.

[37] El Khosht, R. M., M. A. El Feshawy, M. A. El Shorbagy, et al., "A Foldable Textile-Based Broadband Archimedean Spiral Rectenna for RF Energy Harvesting," *2016 16th Mediterranean Microwave Symposium (MMS)*, IEEE, 2016, pp. 1–4.

[38] Naresh, B., V. K. Singh, V. Bhargavi, A. Garg, and A. K. Bhoi, "Dual-Band Wearable Rectenna for Low-Power RF Energy Harvesting," in *Advances in Power Systems and Energy Management*, S. N. Singh, F. Wen, and M. Jain (eds.), Singapore: Springer, pp. 13–21, 2018.

[39] Costanzo, A., D. Masotti, and M. Del Prete, "Wireless Power Supplying Flexible and Wearable Systems," in *2013 7th European Conference on Antennas and Propagation (EuCAP)*, IEEE, 2013, pp. 2843–2846.

[40] Masotti, D., A. Costanzo, and S. Adami, "Design and Realization of a Wearable Multifrequency RF Energy Harvesting System," *Proceedings of the 5th European Conference on Antennas and Propagation (EUCAP)*, IEEE, 2011, pp. 517–520.

[41] Dini, M., M. Filippi, A. Costanzo, et al., "A Fully-Autonomous Integrated RF Energy Harvesting System for Wearable Applications," *2013 European Microwave Conference*, IEEE, 2013, pp. 987–990.

[42] Vital, D., and S. Bhardwaj, "Misalignment Resilient Anchor-Shaped Antennas in Near-Field Wireless Power Transfer Using Electric and Magnetic Coupling Modes," *IEEE Transactions on Antennas and Propagation,* Vol. 69, No. 5, 2020, pp. 2513–2521.

[43] Vital, D., P. Gaire, S. Bhardwaj, and J. L. Volakis, "An Ergonomic Wireless Charging System for Integration with Daily Life Activities," *IEEE Transactions on Microwave Theory and Techniques*, Vol. 69, No. 1, 2020, pp. 947–954.

[44] Owen, N., P. B. Sparling, G. N. Healy, D. W. Dunstan, and C. E. Matthews, "Sedentary Behavior: Emerging Evidence for a New Health Risk," *Mayo Clinic Proceedings*, Vol. 85, No. 12, 2010, pp. 1138–1141.

[45] Hong, J.- S., and M. J. Lancaster, "Couplings of Microstrip Square Open-Loop Resonators for Cross-Coupled Planar Microwave Filters," *IEEE Transactions on Microwave Theory and Techniques*, Vol. 44, No. 11, 1996, pp. 2099–2109.

[46] Wang, C.- S., G. A. Covic, and O. H. Stielau, "Power Transfer Capability and Bifurcation Phenomena of Loosely Coupled Inductive Power Transfer Systems," *IEEE Transactions on Industrial Electronics*, Vol. 51, February 2004, pp. 148–157.

[47] Jin, P., and R. W. Ziolkowski, "Low-Q, Electrically Small, Efficient Near-Field Resonant Parasitic Antennas," *IEEE Transactions on Antennas and Propagation*, Vol. 57, No. 9, 2009, pp. 2548–2563.

[48] Hui, S. Y. R., W. Zhong, and C. K. Lee, "A Critical Review of Recent Progress in Mid-Range Wireless Power Transfer," *IEEE Transactions on Power Electronics*, Vol. 29, No. 9, 2013, pp. 4500–4511.

[49] Jonah, O., S. V. Georgakopoulos, and M. M. Tentzeris, "Optimal Design Parameters for Wireless Power Transfer by Resonance Magnetic," *IEEE Antennas and Wireless Propagation Letters*, Vol. 11, 2012, pp. 1390–1393.

[50] Chow, Y., K. Wan, T. Sarkar, and B. Kolundzija, "Microstrip Line on Ground Plane with Closely Spaced Perforations—Fringe Fields and Formulas by Synthetic Asymptote," *Microwave and Optical Technology Letters*, Vol. 32, No. 3, 2002, pp. 201–204.

[51] Vital, D., and S. Bhardwaj, "Misalignment Resilient Anchor-Shaped Antennas in Near-Field Wireless Power Transfer Using Electric and Magnetic Coupling Modes (Minor Revisions)," *IEEE Transactions on Antennas and Propagation*, Vol. 1, No. 1, pp. 1–1, 2020.

[52] Hamid, M., and R. Hamid, "Equivalent Circuit of Dipole Antenna of Arbitrary Length," *IEEE Transactions on Antennas and Propagation*, Vol. 45, No. 11, 1997, pp. 1695–1696.

[53] Raju, S., R. Wu, M. Chan, and C. P. Yue, "Modeling of Mutual Coupling between Planar Inductors in Wireless Power Applications," *IEEE Transactions on Power Electronics*, Vol. 29, No. 1, 2013, pp. 481–490.

[54] Ott, H. W., *Electromagnetic Compatibility Engineering*, Hoboken, NJ: John Wiley & Sons, 2009.

[55] Vital, D., J. L. Volakis, and S. Bhardwaj, "A Wireless Power Transfer System (WPTS) Using Misalignment Resilient, On-Fabric Resonators for Wearable Applications," in *2020 IEEE MTT-S International Microwave Symposium (IMS)*, IEEE, 2020, pp. 1184–1187.

[56] ICNIRP, "ICNIRP Guidelines for Limiting Exposure to Electromagnetic Fields (100 KHz to 300 GHz)," *Health Physics*, Vol. 118, No. 5, 2020, pp. 483–524.

[57] P. Wu, S. Y. Huang, W. Zhou, Z. H. Ren, Z. Liu, K. Huang, and C. Liu, "High-efficient rectifier with extended input power range based on self-tuning impedance matching," *IEEE Microwave and Wireless Components Letters*, Vol. 28, No. 12, pp. 1116–1118, 2018.

[58] Finkenzeller, K., *RFID Handbook: Radio-Frequency Identification Fundamentals and Applications*, Chichester, United Kingdom:John Wiley & Sons, 1999.

[59] Vital, D., J. Zhong, S. Bhardwaj, and J. L. Volakis, "Loss-Characterization and Guidelines for Embroidery of Conductive Textiles," in *2018 IEEE International Symposium on Antennas and Propagation & USNC/URSI National Radio Science Meeting*, IEEE, 2018, pp. 1301–1302.

[60] Dickson, J. F., "On-Chip High-Voltage Generation in MNOS Integrated Circuits Using an Improved Voltage Multiplier Technique," *IEEE Journal of Solid-State Circuits*, Vol. 11, No. 3, 1976, pp. 374–378.

[61] Olgun, U., C.- C. Chen, and J. L. Volakis, "Design of an Efficient Ambient WiFi Energy Harvesting System," *IET Microwaves, Antennas & Propagation*, Vol. 6, No. 11, 2012, pp. 1200–1206.

[62] Gilbert, J. M., and F. Balouchi, "Comparison of Energy Harvesting Systems for Wireless Sensor Networks," *International Journal of Automation and Computing*, Vol. 5, No. 4, 2008, pp. 334–347.

[63] F. Deng, Y. He, B. Li, L. Zhang, X. Wu, Z. Fu, and L. Zuo, "Design of an Embedded CMOS Temperature Sensor for Passive RFID Tag Chips," *Sensors*, Vol. 15, No. 5, 2015, pp. 11442–11453.

[64] Lee, J., and S. Cho, "A 1.4-μW 24.9-ppm/deg C Current Reference with Process-Insensitive Temperature Compensation in 0.18-μm CMOS," *IEEE Journal of Solid-State Circuits*, Vol. 47, No. 10, 2012, pp. 2527–2533.

[65] Souri, K., Y. Chae, F. Thus, and K. Makinwa, "12.7 A 0.85 V 600 nW All-CMOS Temperature Sensor with an Inaccuracy of ±0.4°C (3σ) from -40 to 125° C," in *Solid-State Circuits Conference Digest of Technical Papers (ISSCC), 2014 IEEE International*, IEEE, 2014, pp. 222–223.

[66] Law, M. K., and A. Bermak, "A 405-nW CMOS Temperature Sensor Based on Linear MOS Operation," *IEEE Transactions on Circuits and Systems II: Express Briefs*, Vol. 56, No. 12, 2009, pp. 891–895.

[67] Ji, Y., C. Jeon, H. Son, B. Kim, H.-J. Park, and J.-Y. Sim, "A 9.3 nW All-in- One Bandgap Voltage and Current Reference Circuit Using Leakage-Based PTAT Generation and DIBL Characteristic," in *Proceedings of the 23rd Asia and South Pacific Design Automation Conference*, IEEE Press, 2018, pp. 309–310.

[68] Ueno, K., T. Hirose, T. Asai, and Y. Amemiya, "A 300 nW, 15 ppm/C, 20 ppm/V CMOS Voltage Reference Circuit Consisting of Subthreshold MOS- FETs," *IEEE Journal of Solid-State Circuits*, Vol. 44, No. 7, 2009, pp. 2047–2054.

[69] Salvia, J., P. Lajevardi, M. Hekmat, and B. Murmann, "A 56MΩ CMOS TIA for MEMS Applications," *Custom Integrated Circuits Conference, 2009. CICC'09. IEEE*, IEEE, 2009, pp. 199–202.

[70] Spanò, E., S. Di Pascoli, and G. Iannaccone, "Low-Power Wearable ECG Monitoring System for Multiple-Patient Remote Monitoring," *IEEE Sensors Journal*, Vol. 16, No. 13, 2016, pp. 5452–5462.

[71] Tavakoli, M., L. Turicchia, and R. Sarpeshkar, "An Ultra-Low-Power Pulse Oximeter Implemented with an Energy-Efficient Transimpedance Amplifier," *IEEE Transactions on Biomedical Circuits and Systems*, Vol. 4, No. 1, 2009, pp. 27–38.

[72] Harrison, R. R., R. J. Kier, C. A. Chestek, et al., "Wireless Neural Recording with Single Low-Power Integrated Circuit," *IEEE Transactions on Neural Systems and Rehabilitation Engineering*, Vol. 17, No. 4, 2009, pp. 322–329.

CHAPTER 7

Radiofrequency Finger Augmentation Devices for the Tactile Internet

G. M. Bianco and G. Marrocco

7.1 Introduction

The tactile internet (TI) is an emerging technology consisting of extremely low-latency systems with tactile inputs and audio/video feedback [1]. A maximum round-trip delay of about 1 ms is required not to have the users suffer from cybersickness [2, 3], and enable applications ranging from exoskeleton controlling to remote driving. Figure 7.1(a) shows the evolution from the mobile internet to the TI and their main features. The ultimate goal of the TI is the remote delivery of physical senses through the internet, creating the internet of skills [4] that merges the TI with artificial intelligence to achieve the perception of zero-delay networks. Although the 5G network can enable new body-centric communication networks, the TI is still in its infancy. The implementation of TI systems still needs research concerning the physical layer, resource management and networking issues, as well as the combined use of different protocols [5].

A 5G hand-worn system capable of capturing the human tactile experience has yet to be developed, although progress can come from finger augmentation devices (FADs), which are defined as "finger-worn devices with an additional augmentation other than their form, that provide a supplemental capability for one or more fingers using the finger itself as a central element" [6]. Two examples of FADs are an on-nail display [7] and a touch-sensitive pad [8]. The main limitation of FADs is their powering since wires hinder any free-hand gestures, resulting in discomfort for the wearer. Instead, wireless FADs usually are paired with an off-body companion device, and the resulting system is not wearable as a whole [6]. More comfortable fingertip devices can be fabricated by resorting to epidermal electronics, which are ultrathin, soft electronics adhering directly onto the human skin and mechanically imperceptible for the user [9]. The on-body powering issue was overcome by combining FADs with the RFID paradigm, creating a new class of systems named radiofrequency finger augmentation devices (R-FADs).

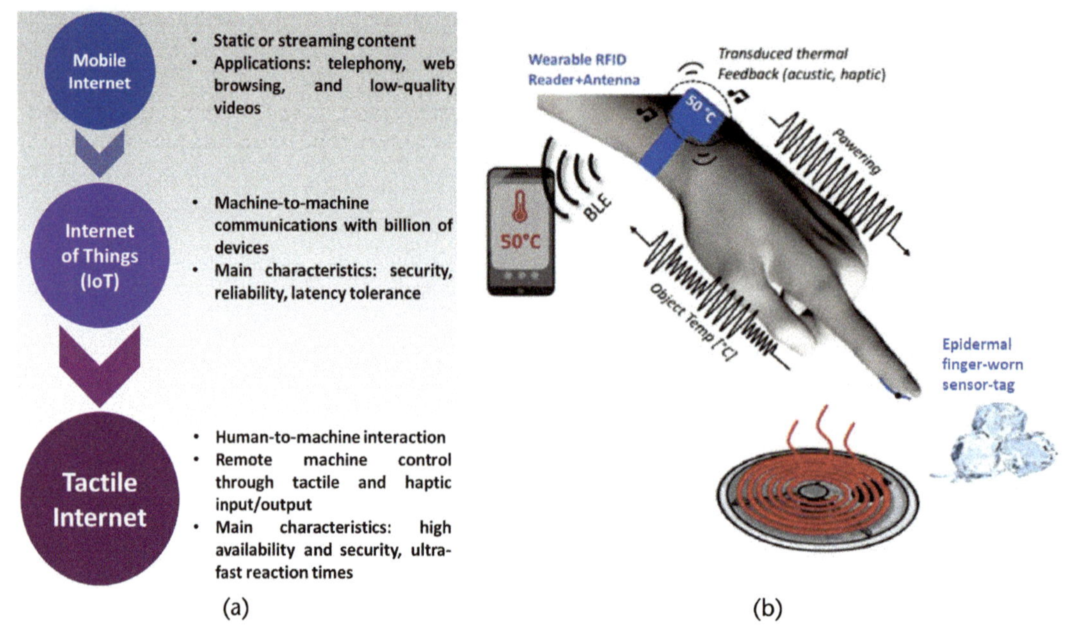

Figure 7.1 (a) Leaps of the tactile internet. (Adapted from [5].) (b) Sketch of R-FADs for temperature-sensing. (©2019 IEEE [16].)

Although many RFID-based FADs were proposed (e.g., [10–13]), the term R-FAD here specifically refers to a system composed of finger-worn sensors provided with sensing capabilities that communicate with a reader antenna placed on the wrist. The system is powered by a portable reader source that utilizes a processing unit to give feedback to the user (Figure 7.1(b)). The EM wave generated by the reader's antenna energizes the fingertip sensor, which, in turn, modulates and backscatters the wave toward the reader. In this way, there is no wire connected to the fingertips, the system is wholly wearable, and natural free-hand gestures are preserved. Current R-FADs communicate through UHF (860–960 MHz) links, and their latency is higher than 1 ms as required by TI applications. Nonetheless, the ongoing research on 5G RFID systems [14] allows envisaging 5G R-FADs in the near future. While the 5G technology becomes more mature, current UHF R-FADs can be exploited in the short term to test and study the wrist-finger links through backscattering modulation.

R-FADs have recently been tested to recover thermal sensitivity, which can be lost because of life-saving medical treatments and is vital to perform everyday activities, like cooking, without suffering injuries. R-FAD systems can also provide the wearer with sensorial ultrability. R. Shilkrot in [15] first used this term to describe new types of perception not inherent to the natural human capabilities and obtained through an artificial interface. As a practical example, the dielectric-sensing R-FAD can provide the wearer with the ultrability of sensing the electric permittivity of the material being touched.

This ultrability is obtained through a recently introduced class of self-tuning integrated circuits (ICs) that can automatically modify their internal impedance to make the matching to the hosting antenna nearly insensitive to changes in local boundary conditions.

This chapter reviews the state of the art on R-FADs concerning the following topics:

1. Communication models for the fingertip-wrist backscattering link and its variability;
2. Design method for fingertip RFID dielectric sensors;
3. Epidermal flexible and stretchable devices for transforming our fingertips into touch scanners;
4. Applications of R-FADs to aid sensorially impaired people;
5. Digital therapy for studying the cognitive remapping applied to the lost thermal feeling.

7.2 Communication Models for the Fingertip-Wrist Backscattering Link and Its Variability

The wrist-worn reader communicates via backscattering with the fingertip sensors in the R-FAD architecture. Thus, the EM challenge consists of establishing a robust fingertip-wrist link in the presence of the lossy human body and during natural gestures. Reference [17] introduces and analyzes the wrist-finger links. There are several challenges in establishing such links.

1. Since a wearable battery-fed reader is needed for a fully wearable R-FAD, the available power is reduced with respect to a fixed interrogation station;
2. The direct contact between the antennas and the wearer's skin generates high power losses;
3. The wrist-finger channel is significantly time-variant due to the free-hand gestures, which continuously change the mutual arrangement of antennas;
4. The EM interactions occur near both the antennas and the human body;
5. Multiple sensors can be used simultaneously; hence the intersensor coupling could further complicate the interaction.

Since the antennas' arrangements are dependent on the morphology of the user's hand and on the gesture performed, the link is characterized by intersubject (user-dependent) and intrasubject (no user-dependent) variabilities.

The complex wrist-finger communication operates in the midfield and can be analyzed through a two-port model (Figure 7.2(a)) that accounts for everything between the reader's antenna feeding port and the IC. The lossy two-port network is characterized by its 2×2 impedance (Z), admittance (Y), or scattering (S) matrix. The antennas are represented through their Thevenin (Z-, S-matrix) or Norton (Y-matrix) equivalent circuits. The metric to characterize the links is the transducer power gain (G_T), defined as the ratio between the power $P_{R \to T}$ delivered by the reader to the chip in the specific impedance matching condition and the available power emitted by the reader generator $P_{av,R}$. By assuming a two-port network characterized through the impedance Z-matrix, the G_T can be evaluated as [19]

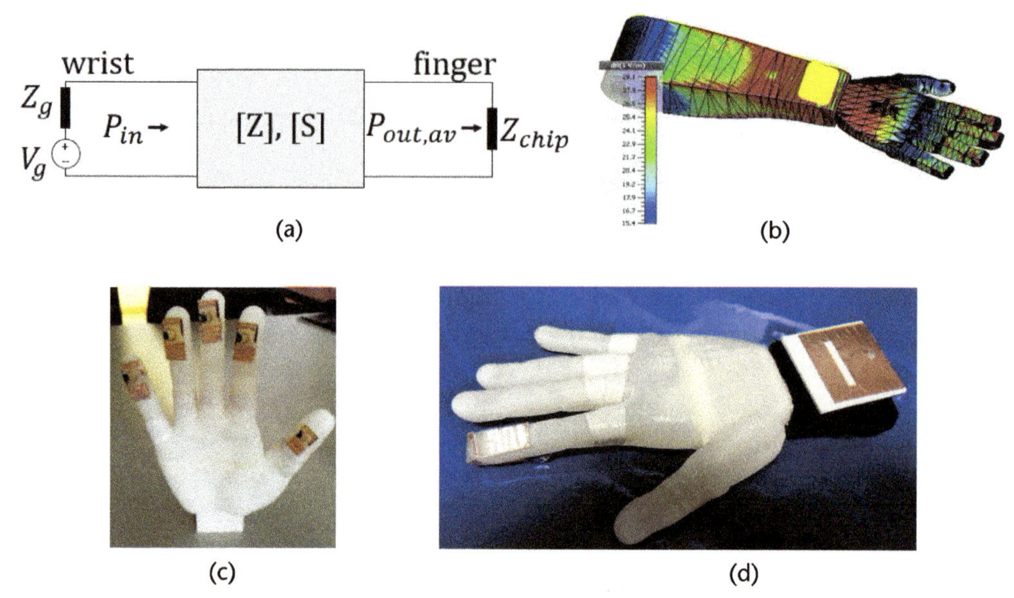

Figure 7.2 (a) Two-port model of the wrist-finger link. (©2019 IEEE [17].) (b) Homogeneous numerical hand phantom and E-field when a wrist patch (yellow rectangle) is active and fed with 1W source power. (©2017 IEEE [18].) A 3-D-printed hand phantom equipped with (c) five finger sensors and (d) an R-FAD prototype with both the antennas of the sensor and the reader. (©2017 IEEE [18].)

$$G_T = \frac{P_{R\to T}}{P_{av,R}} = \frac{4R_{IC}R_G\left|Z_{21}\right|^2}{\left|\left(Z_{11}+Z_G\right)\left(Z_{22}+Z_{IC}\right)-Z_{12}Z_{21}\right|^2} \tag{7.1}$$

where $Z_{IC} = R_{IC} + jX_{IC}$ is the microchip impedance, and $Z_G = R_G + jX_G$ is the generator impedance. Similarly, the G_T can be evaluated by the admittance Y-matrix [20]

$$G_T = \frac{4G_{IC}G_G\left|Y_{21}\right|^2}{\left|\left(Y_{11}+Y_G\right)\left(Y_{22}+Y_{IC}\right)-Y_{12}Y_{21}\right|^2} \tag{7.2}$$

where $Y_{IC} = Y_{IC} + jB_{IC}$ is the chip admittance, and $Y_G = G_G + jB_G$ is the generator admittance. In the ideal case of perfect impedance matching at both ports, G_T is maximum and it is called system power gain (g), which can be evaluated from the scattering S-matrix [21]

$$g = \frac{\left|S_{21}\right|^2}{\left(1-\left|S_{11}\right|^2\right)\left(1-\left|S_{22}\right|^2\right)} \tag{7.3}$$

The system power gain is usually employed to study the phenomenology of fingertip-wrist communications through numerical simulations[1] and to evaluate the optimum achievable link performance.

1. The numerical phantoms and the EM simulations in this chapter were obtained by CST® Microwave Studio 2018.

All the numerical analyses hereafter reviewed consider a reference R-FAD and a homogeneous hand phantom (Figure 7.2(b)). First, the feasibility of energizing R-FAD devices through a portable reader was numerically evaluated, accounting for the power losses caused mainly by the hand [17]. Let P_{IC} be the IC sensitivity (i.e., the minimum power needed by the IC to activate and interact with the reader) in dBm, and p_S an arbitrary safety margin accounting for the uncertain knowledge of the IC and the limited control on the actual power emitted by the reader. The analysis of the link also considers a set of M known hand gestures. The minimum input power P_{in} required to activate the communication in the overall set of considered gestures is, then,

$$P_{in,min} = \frac{p_{IC} + p_S}{\min\limits_{1 \le m \le M} \{g_m\}} \tag{7.4}$$

Body-worn R-FADs have been extensively numerically analyzed in [17] concerning bending, SAR, link quality when performing different gestures, and intersensor coupling. First, the fingertip sensor impedance match is not significantly affected by bending or hand gestures. Similarly, when considering a wrist-mounted patch as the reader's antenna, the H-plane bendings are negligible [22]. Regarding the SAR, the R-FAD causes minimal power absorption. When sourcing an R-FAD reader by 1W of input power, the wrist-mounted patch generates a SAR of 9 mW/kg, which is significantly lower than the EU regulation of 4 W/kg.

Concerning the intrasubject variability due to different gestures, differences in the g of about 15 dB are observed between the best-case and the worst-case gestures. This difference also depends on the sensor antenna layout. Finally, if multiple fingertip sensors are worn simultaneously, the strong power absorption by the human tissues causes the on-finger antennas to be negligibly EM coupled. Accordingly, the six-port equivalent model of the multichannel R-FAD can be reduced to five equivalent two-port networks, each representing a reader-tag link.

In [18], the first R-FAD prototype was tested on a 3-D-printed hand model, filled with a liquid reproducing the human body electric properties (Figure 7.2(c,d)). When evaluating the same link multiple times by removing and repositioning the sensor, the measured G_T experiences some fluctuations due to the different adherence to the hand phantom. Thus, sensor prototypes ensuring a reproducible and stable adherence are required to obtain a highly reliable link.

Based on the mathematical formulation in (7.4), simulations revealed that a value of $P_{in,min} \le 27$ dBm allows establishing a robust link over three common gestures (hand open, index touching, and prehension) [17]. Accordingly, a commercially available RFID portable reader can successfully power R-FAD systems. The first R-FAD prototype tested on real users is shown in Figure 7.3(a). The different sensor-reader links for each finger were measured to assess the intrasubject variability of the R-FAD communications due to the specific channels. A maximum difference of 5 dB in the measured G_T for a given gesture and user was observed [17]. Statistical analysis was performed on 10 different users (the hand form-factors in Figure 7.3(b)) wearing the sensor on the index finger, and each performing five given gestures (Figure 7.3(c)). The evaluated maximum difference was 13 dB (Fig-

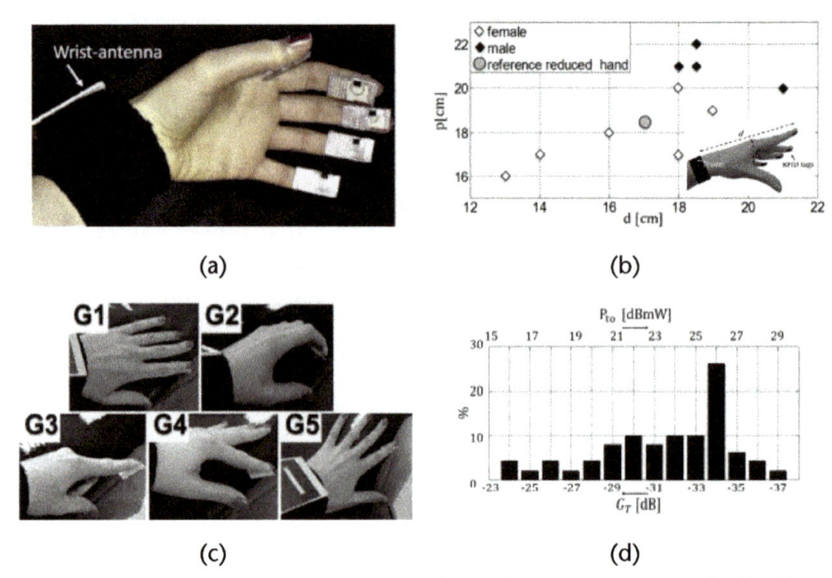

Figure 7.3 (a) Multisensor R-FAD prototype worn by a volunteer. (©2019 IEEE [17].) (b) Hand sizes of the test population. (©2019 IEEE [17].) (c) Five gestures performed by each volunteer of the test population. (©2019 IEEE [17].) (d) Normalized distribution of Gt at 870 MHz of the R-FAD for all the considered volunteers and hand gestures and corresponding Pto in the case of Pic = –5 dBm. (©2019 IEEE [17].)

ure 7.3(d)), but a portable reader emitting 27 dBm can still energize the single-worn sensor in 95% of the considered combinations anyway [17].

7.3 Constrained Design of R-FADs

When wearing an R-FAD during the day, the user may have to interact with multiple objects and materials. During the interactions, the environment where the sensor operates changes, and the consequent impedance mismatch between the fingertip antenna and the IC can compromise the wrist-finger link and prevent any communication. Accordingly, the design of the R-FADs will be constrained to make the on-hand link reliable independent of the interactions performed by the wearer. This reliability is possible by resorting to temperature-sensing ICs capable of autotuning (also known as self-tuning; an example of such microchips is the Magnus S3 by Axzon).

Autotuning ICs are capable of modifying their internal impedance to compensate for changes in the boundary conditions. They can be modeled as a resistance in parallel to a variable capacitance, which in turn can be considered as a switchable ladder of equal capacitors of capacitance C_0 so that the IC overall capacitance is $Cic(n) = Cmin + nC_0$ [23]. With reference to Figure 7.4(a), the number n of equal equivalent capacitors to compensate for the admittance $Y_T(\psi)$,which depends on the boundary conditions ψ seen by the IC, is determined by the self-tuning equation [23]:

$$\left|B_{IC}(n)+B_T(\psi)\right|=0 \tag{7.5}$$

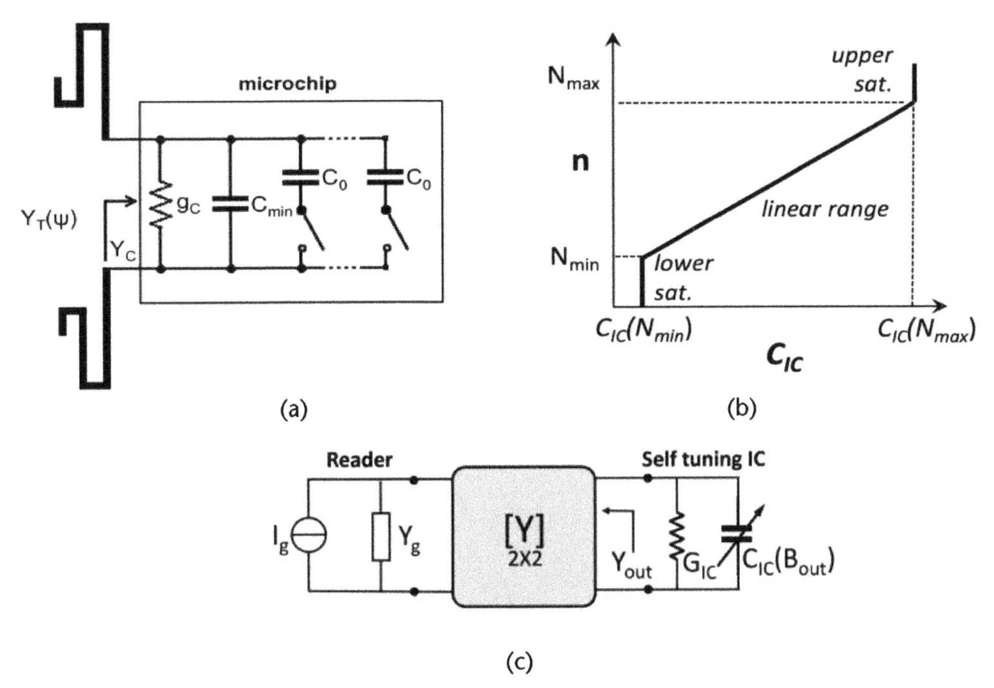

Figure 7.4 (a) Equivalent circuit of an autotuning IC. (©2018 IEEE [23].) (b) Lower and upper saturation zones of an autotuning IC, and linear range wherein the self-tuning equation holds. (©2020 IEEE [24].) (c) Two-port model of a wrist-finger link including an autotuning IC. (©2020 IEEE [24].)

where $B_T(\psi) = \text{Im}[Y_T(\psi)]$, $B_{IC}(n) = 2\pi f C_{IC}(n)$, and f is the frequency. The self-tuning equation holds if n is comprised between N_{min} and N_{max}, whose values depend on the actual IC considered. If $n \leq N_{min}$ or $n \geq N_{max}$, saturation occurs (Figure 7.4(b)) [24]. The tuning parameter n is also returned by the IC to the reader through a metric called sensor code (SC). The SC can be exploited for sensing, too, as shown next in Section 7.5.2.

When considering a self-tuning fingertip sensor, the two-port model of the wrist-finger link must account for the variable IC capacitance (Figure 7.4(c)) The model can be employed to design finger sensors, optimizing the communication link with the reader. Let us assume that all the J expected interactions are fully characterized by a set of values $\psi = \{\psi_1, ..., \psi_J\}$ of a physical parameter ψ. As a practical example, $\psi = \varepsilon = Re(\varepsilon_C)$, where ε_C is the complex permittivity of the touched material. Then, the vector $\psi = \varepsilon = \{\varepsilon_1, ..., \varepsilon_J\}$ represents the boundary conditions wherein the fingertip sensor must work. The optimal design of the sensor must ensure a reliable wrist-finger link in the above boundary conditions and, to optimally utilize the self-tuning feature of the IC, the antenna susceptance $B_T(\psi)$ has to be exactly mapped in the linear range of the IC (Figure 7.4 (b)). With reference to the loop-matched [25] dipole layout in Figure 7.5(a), the set of geometrical parameters to be optimized is a = $\{a_1, a_2\}$. The optimization problem is formalized as the minimization of the penalty function U(**a**):

$$U(\mathbf{a}) = \sum_{i=1}^{3} w_i u_i(\mathbf{a}) \tag{7.6}$$

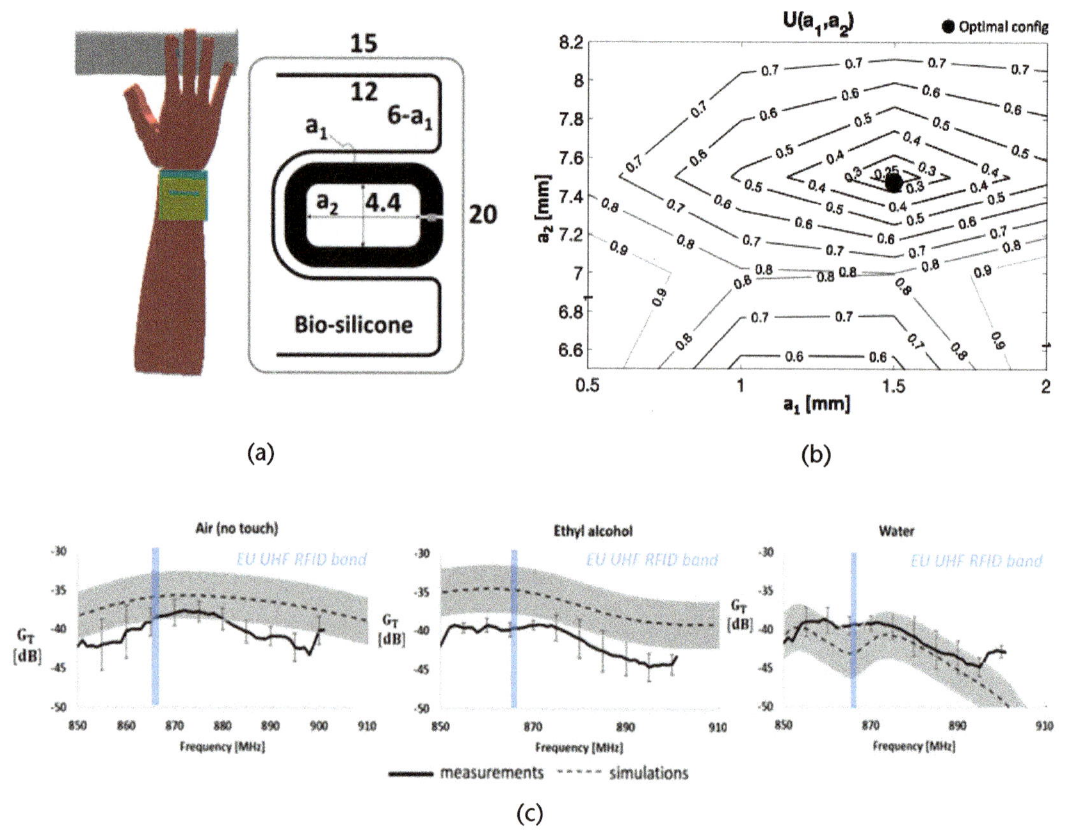

Figure 7.5 (a) Layout of a loop-matched fingertip antenna to be optimized through constrained design (left) and numerical model of the hand contacting a cylindrical container (right). (©2020 IEEE [24].) (b) Penalty function U(a) with the indication of the optimal couplet {a1, a2}, with regard to the antenna layout in Figure 7.5(a). (©2020 IEEE [24].) (c) Transducer power gain of the constrained-designed fingertip sensor (from Figure 7.5(a,b)) when contacting PET bottles filled with three different materials. The ±3-dB shadowed region around the simulated values account for the expected intrasubject variabilities, and the EU UHF RFID band is highlighted in blue. (©2020 IEEE [24].)

The w_i terms are weights, whereas the three addends $u_i(\mathbf{a})$ enforce the minimization of (1) the activation power, (2) the difference $\min_{1 \le j \le J}(n_j) - N_{min}$, and (3) the difference $N_{max} - \max_{1 \le j \le J}(n_j)$. In this way, the communication performance is optimized and the autotuning linear range is fully exploited, thus avoiding saturation [24]. When numerically evaluating (7.6), the minimum $U(\mathbf{a})$ identifies the optimal design of the fingetip antenna (Figure 7.5(b)) that achieves stable performances in the considered variable boundary conditions.

An example with three boundary conditions (air, ethyl alcohol, and water) in a vast range of permittivities (ε = {1, 17, 78}, respectively) is shown in Figure 7.5(c). The average standard deviation in the 865.6–867.6-MHz band [26] is about just 1 dB over the three materials. Overall, the achieved performance can be considered rather insensitive to the touched item.

7.4 R-FAD Manufacturing

State-of-the-art wrist antennas in R-FADs are the folded patches, as shown in Figure 7.2(d). These antennas are worn on the wrist using cloth bracelets directly contacting the human skin (Figure 7.3(a)) and hosting the reader module as well (Figure 7.6(a)). Their substrate is a slightly stretchable low-permittivity closed-cell polyvinyl-chloride foam ($\varepsilon = 1.55$, $\sigma = 6 \cdot 10^{-4}$ S/m) [17], which can moderately bend to conform to the user's wrist morphology. The most appropriate placement of the reader's antenna when considering both the communication link and the user's comfort is the topside of the wrist [17].

The fingertip sensors have to be soft and possibly stretchable because of their positioning; otherwise, the interactions with objects will be hindered. Different kinds of finger sensors were designed, such as (Figure 7.7): (1) on-silicone sensors, (2) plaster sensors, and (3) in-silicone sensors. The first type of finger sensors is made of adhesive copper placed onto substrates of biocompatible silicone to adhere to the human skin. The copper is cut by carving machines (e.g., the Secabo S60II plotter). Examples of on-silicone sensors are the U-shaped meander-line dipole (U-tag; Figure 7.6(b)), and the flat-meandered dipole (F-tag; Figure 7.6(c)). The proximity to the human body is a critical factor when designing epidermal sensors, and millimetric differences in the antenna-body distance significantly impact the antenna's impedance parameters and, consequently, the antenna-IC impedance matching.

On-silicone sensors are moderately stretchable; they use silicone as support to be worn and can be removed and reused again. However, their limited stretchability and flexibility can be uncomfortable for the wearer, such as when grasping objects. The comfortability is greatly improved utilizing thin plaster tags [27]. The plaster tags are disposable sensors fixed onto the fingertip through self-adhering thin breathable films (Figure 7.8(a,b)). These sensors are highly flexible and noninvasive. Due to the high losses caused by the human body, a small square silicone spacer between the IC and the skin is still included to preserve the finger-wrist link. Since these epidermal sensors adhere directly to the human skin, they must follow every digit movement without breaking or degrading their performance. The adhesion is obtained by utilizing a medical tape, for example, 3M™ Tegaderm™, or Fixomull® by BSN Medical.

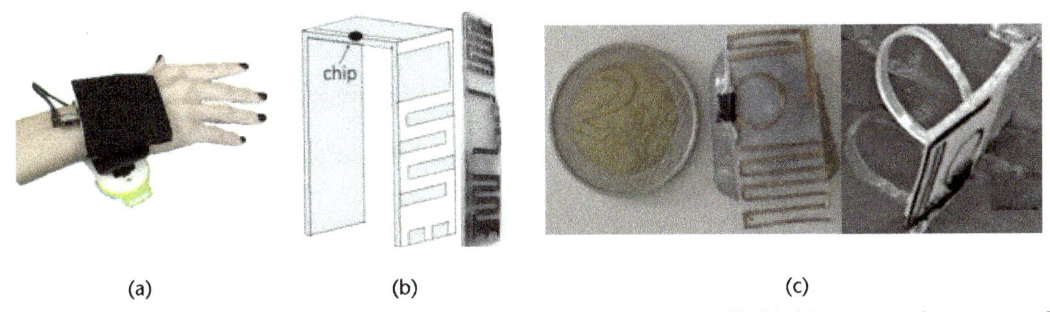

Figure 7.6 (a) Reader and antenna of the wrist-worn readers. (©2019 IEEE [16].) (b) Layout and prototype of the fingertip U-tag. (©2019 IEEE [17, 18]). (c) Two F-tag prototypes using different silicone supports. (©2019 IEEE [17].)

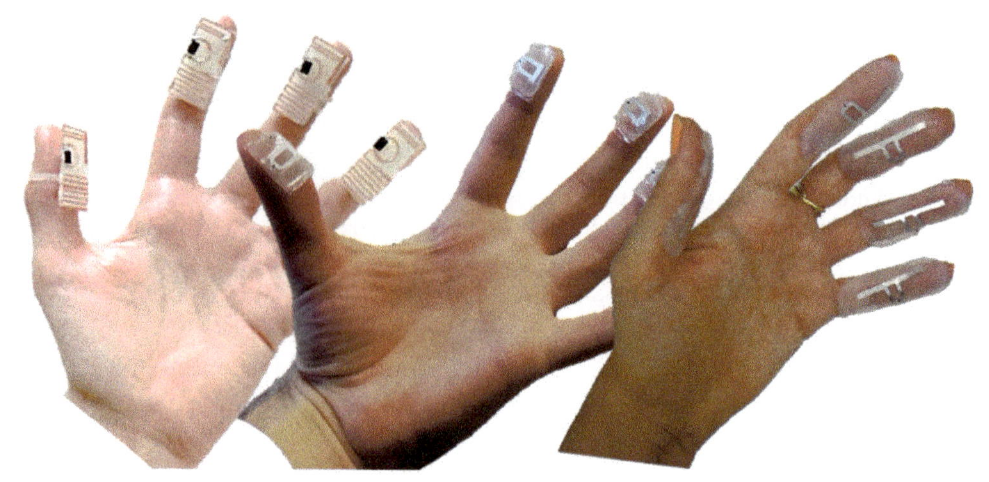

Figure 7.7 Multichannel R-FAD systems employing different types of finger sensors. From left to right: on-silicone sensors, plaster-sensors, and in-silicone sensors.

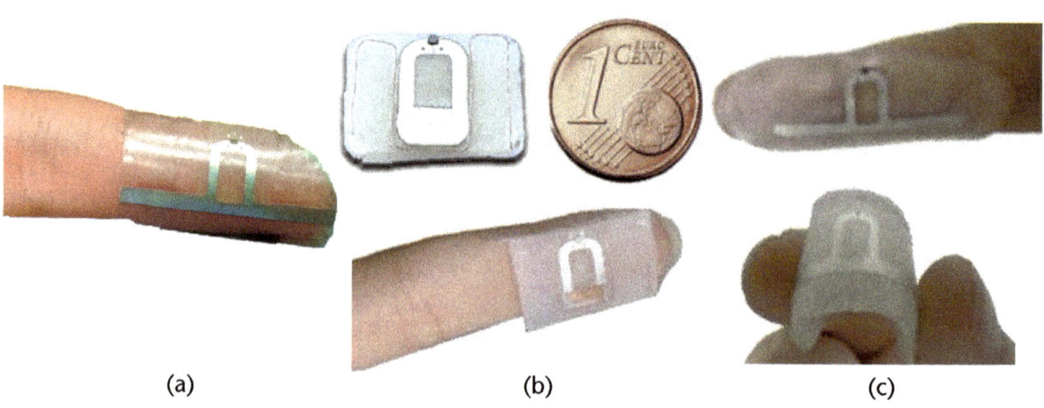

Figure 7.8 Plaster-finger sensors: (a) double-phalanx antenna. (Adapted from [28].) (b) Distal phalanx antenna, designed as in Figure 7.5. (©2020 IEEE [24].) (c) In-silicone double-phalanx finger sensors. (Tag from [29].)

In all cases, adherence is a critical issue that can affect the robustness of sensing and communication [24]. Consequently, the size of such sensors is more crucial than those of on-silicone ones since bending the fingers can cause the epidermal plaster sensors to lose adhesion to the skin. To minimize the antenna's size and mechanical stiffness, thin copper-wire antennas can be used, as the loop-matched dipole in Figure 7.8(b). Such thin-wire epidermal antennas suffer from degradation in their radiation performance if excessively stressed for longer than a day [30]. This is not expected to happen with disposable plaster sensors due to their limited use time. Because of hand movements, small sensors placed on a single phalanx (Figure 7.8(b)) are more effective than utilizing antennas covering both the intermediate and the distal phalanxes (Figure 7.8(a)), even if double-phalanx antennas can achieve better radiation performance thanks to the broader available surface. In any case, due to the submillimetric plaster and the thin copper dipole, this second type of sensor gets damaged when removed from the fingertip and cannot be reused.

Finally, in-silicone epidermal fingertip antennas manage to preserve both the reusability of the on-silicone sensors and the stretchability and comfort of the plaster sensors. The in-silicone sensors consist of an antenna embedded within ultrathin elastomeric layers that protect the antenna and simultaneously adheres to the human skin (Figure 7.8(c)). The antenna layout is the same as the plaster sensors in Figure 7.8(a), whereas the embedding silicone is composed of two different stretchable silicone components having different Young's moduli (Figure 7.9(a)). Two Ecoflex™ 00-30 (by Smooth On; thickness ~0.3 mm each) layers embed the antenna, and one layer of Silbione™ (by Elkem Silicones; thickness ~0.1 mm) fixes the sensor onto the finger's skin (Figure 7.9(b,c)). The Silbione™ layer allows the epidermal antenna to follow the finger movements even in the case of double-phalanx sensors, like those in Figure 7.8(c). In-silicone sensors are highly flexible, have a total thickness lower than 1 millimeter, and can easily be removed, cleaned, sanitized, and reapplied on the finger several times.

7.5 R-FAD Applications to Aid Sensorially Impaired People

The R-FAD systems were initially proposed to aid people suffering from hypoesthesia (i.e., a sensory impairment resulting in a reduction of sensitivity to specific kinds of stimuli). In particular, R-FADs can be used as soft tools to permanently overcome thermal and tactile hypoesthesia, as well as visual impairment. Thermal and tactile hypoesthesia are usually symptoms of peripheral neuropathy [31], a common disorder that is increasingly widespread due to obesity and aging and can incidentally be caused by many pathologies and medical treatments, like sickle cell disorder [32], diabetic neuropathy [33], or chemotherapy [34]. Sensorially impaired individuals incur several issues during daily activities, like getting severely burned without noticing. Regarding visual impairments, 285 million people were estimated

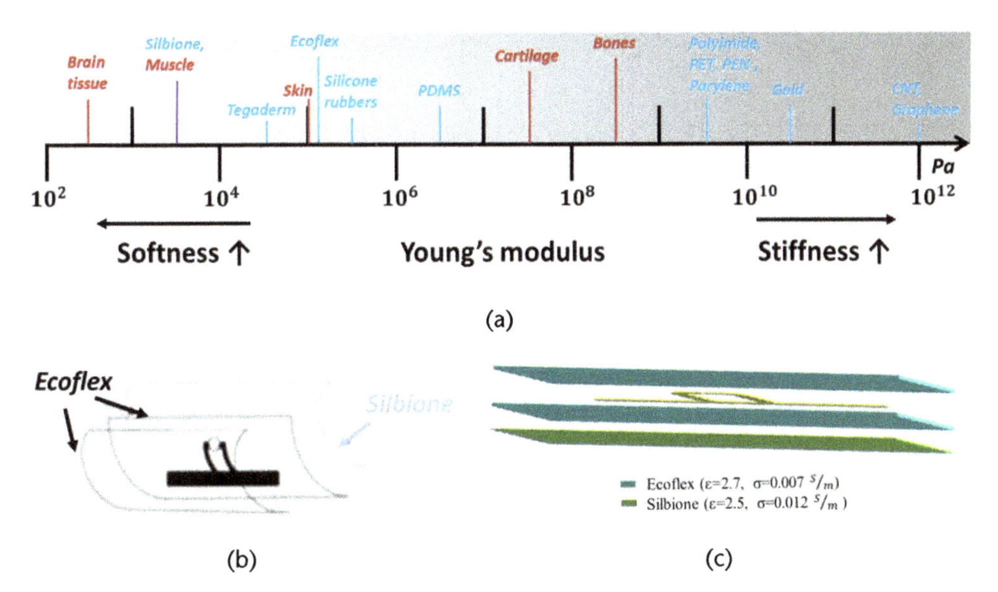

Figure 7.9 (a) Young's moduli of some human tissues (in red) and common materials used to fabricate sensors (in blue). (b) Layers of the multilayer in-silicone sensor in Figure 7.8(c). The dipole is 33 mm long. (Adapted from [29].) (c) Numerical model of the multilayer in-silicone sensor.

to be visually impaired globally in 2010 [35], and they find it challenging to recognize the objects they have to interact with.

R-FADs with temperature-sensing capabilities can provide an artificial sense for the temperature of objects to recover the thermal feeling indirectly. Moreover, self-tuning R-FADs can be designed to recognize the touched materials based on their dielectric permittivity, thus enabling the dielectric-sensing ultrability. This ultrability could be exploited for several everyday tasks, such as sensing the wetness of hair or discriminating between liquids (e.g., water and alcohol) or powders inside plastic bottles.

7.5.1 Sensing an Item's Temperature

Warm/cold sense can be restored by R-FADs embedding a temperature-sensing chip. When the sensor contacts the skin, the chip senses both the temperatures of the body and the external environment. Accordingly, a transient regime longer than 10 seconds can be experienced, depending on the object's temperature (Figure 7.10(a)). Temperature equilibrium is reached after 4 seconds (cold water at 8.8°C) to 5 seconds (hot wood at 50°C) up to 12 seconds (ice at 0°C). Consequently, depending on the difference between the starting temperature of the R-FAD sensor and the object's temperature, a variable touch duration may be needed. The thermal inertia of the actual IC embedded in the finger sensor influences the transitory duration, too. The estimation of the correct temperature can speed up by using an extrapolator. Usually, the temperature returned by the IC after the touch depends on the duration of the contact itself (Figure 7.10(b)). Denoting with Δt the contact time, and with T_∞ the actual object's temperature, then [18]

$$T_\infty = \frac{T(\Delta t) - e^{-\frac{\Delta t}{\tau}} \bullet T_0}{1 - e^{-\frac{\Delta t}{\tau}}} \tag{7.7}$$

where T_0 is the R-FAD temperature measured before the contact, τ is the time-constant depending on the thermal inertia of the considered IC. Figure 7.10(c) shows an example for some touch durations when employing the IC EM4235 by EM Microelectronic, having $\tau = 6.5$ s. When applying (7.7) to the data plotted in Figure 7.10(c), the hot water temperature was estimated. Experiments demonstrated that thanks to the extrapolator, a contact as short as a half-second long is sufficient to assess the object's temperature correctly with a ±1°C accuracy [18].

Recorded live demonstrations of temperature-sensing R-FADs exploiting (7.7) are available online regarding, in [36], continuous temperature sensing and, in [37], a real-time application. In the second video of [37] a multichannel temperature-sensing R-FAD is shown. Two sensor tags are simultaneously queried by the reader, the thumb sensor, and the index sensor. Each sensor estimates the temperature of the contacted object only based on its data.

A temperature-sensing R-FAD was tested on a sensorially impaired volunteer [16], a 45-year-old female suffering from radiotherapy-induced peripheral

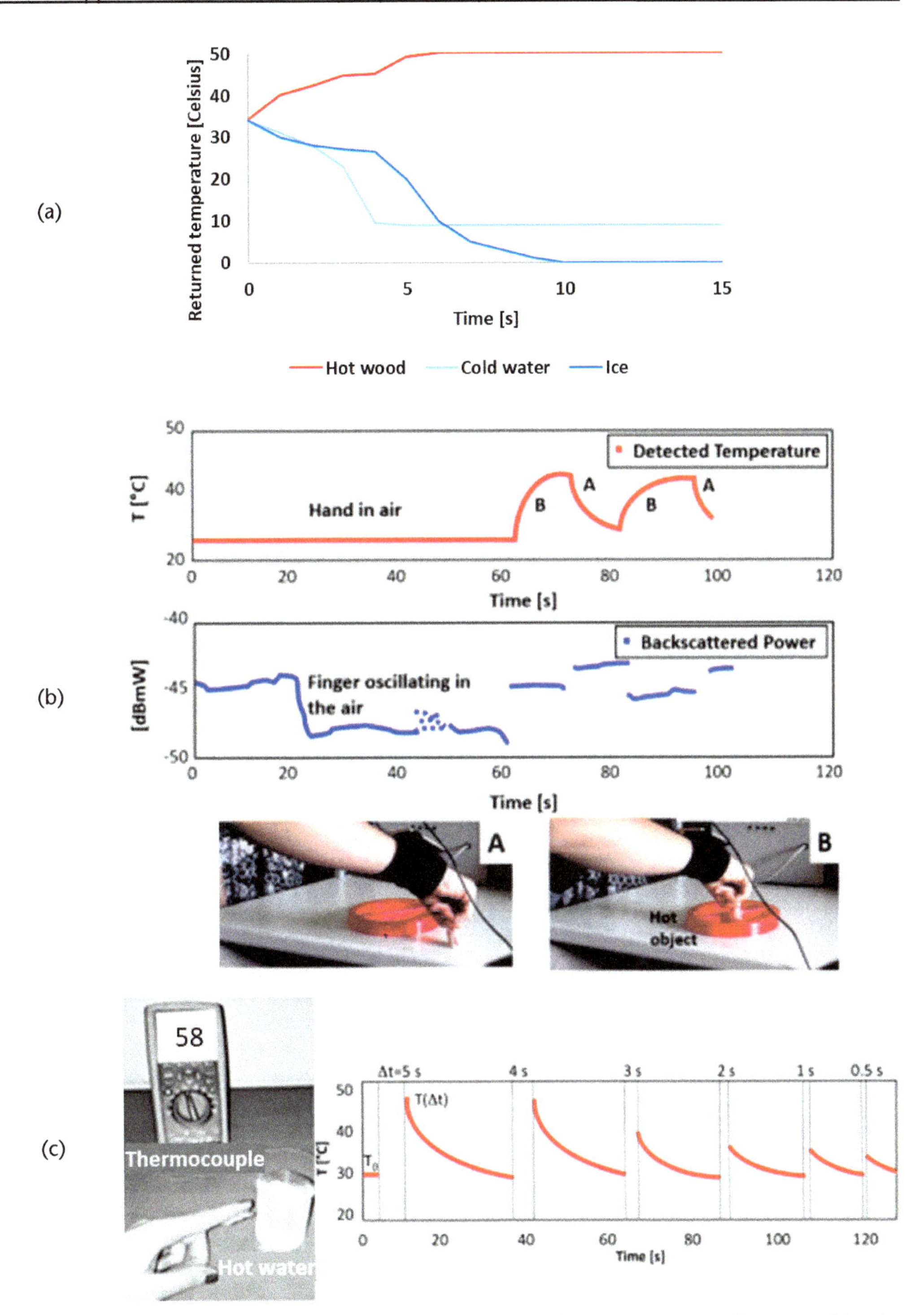

Figure 7.10 Temperature-sensing R-FADs. (a) Example of a temperature profile returned by the fingertip plaster sensor in Figure 7.8(a) embedding the Magnus-S3 IC when contacting objects. (b) Detected temperature and corresponding power backscattered by the U-tag in Figure 7.6(b) when the user cyclically touches the desk surface kept at environment temperature and a hot plastic object (approximative temperature of 40°C). (©2017 IEEE [18].) (c) Example of estimated temperature T_∞ of a hot glass starting from the temperature recorded by the finger sensor when contacting the target objects for different time intervals. (©2017 IEEE [18].)

neuropathy from more than 5 years before the experimentation. She lost thermal perception from both upper limbs, below the elbows, and over large parts of the trunk. The temperature-sensing R-FAD was a single-sensor system, including only one plaster-sensor on the index finger. It returned visual and acoustic feedback on the temperature of the touched object. The feedback produced by a tablet consisted of a color paired with a sound. Seven color–sound couplets were selected corresponding to seven levels (i.e., ranges) of temperature between –20°C and 70°C. After 15 consecutive days of at-home training with the R-FAD through 20-minute-long sessions, the patient was able to guess the contacted objects' temperatures with remarkable accuracy (Figure 7.11). Therefore, the R-FAD was successful in restoring thermal feeling.

7.5.2 Discrimination of Materials

Dielectric-sensing R-FADs were initially proposed to aid visually impaired people in [28]. They can estimate the permittivity of the touched object by exploiting the self-tuning mechanism detailed in Section 7.3. The SC metric returned by the autotuning IC is highly sensitive to the permittivity of the touched object throughout the mismatch of the fingertip antenna that the self-tuning IC tries to compensate. The raw SC was tested to recognize different materials inside PET bottles [24, 28], to find a light switch on the wall, and to sense the filling level of a bottle (Figure 7.12(a–c)). In another test, a volunteer touched his hair before and after wetting it, and the returned raw SC was 161 and 54, respectively. However, the raw SC is affected by very significant inter- and intrasubject variabilities such as the antenna layout, the sensor positioning, adherence to human skin, the gesture performed, and the specific IC used. Therefore, the raw SC has low reproducibility and has to be carefully managed.

User-specific baseline effects can be significantly reduced by resorting to the differential SC (ΔSC), defined as the difference between the raw SC and a calibration SC value. The ΔSC was proven to discriminate dielectric contrast even lower than a single unit [39]. By exploiting the ΔSC, the intersubject variability is significantly reduced [38], whereas the intrasubject variability due to different gestures performed is still not negligible (Figure 7.12(d,e)). A live demo involving dielectric-sensing R-FAD through the ΔSC is currently available online [40].

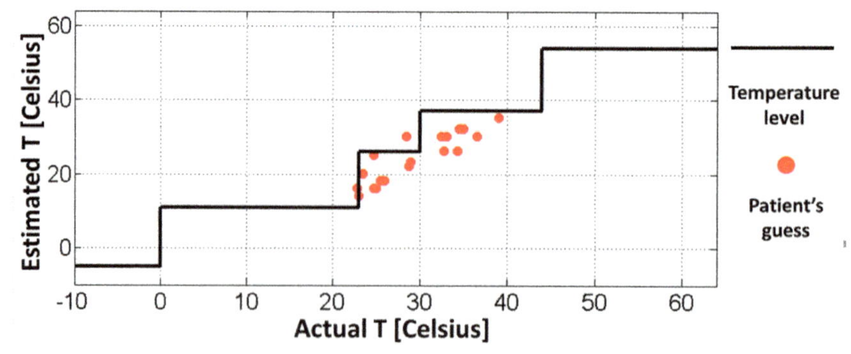

Figure 7.11 Object temperature vs. temperature estimated (red dots, one dot per touch) by a patient with thermal hypoesthesia through an R-FAD after 15 days of training. (©2019 IEEE [16].)

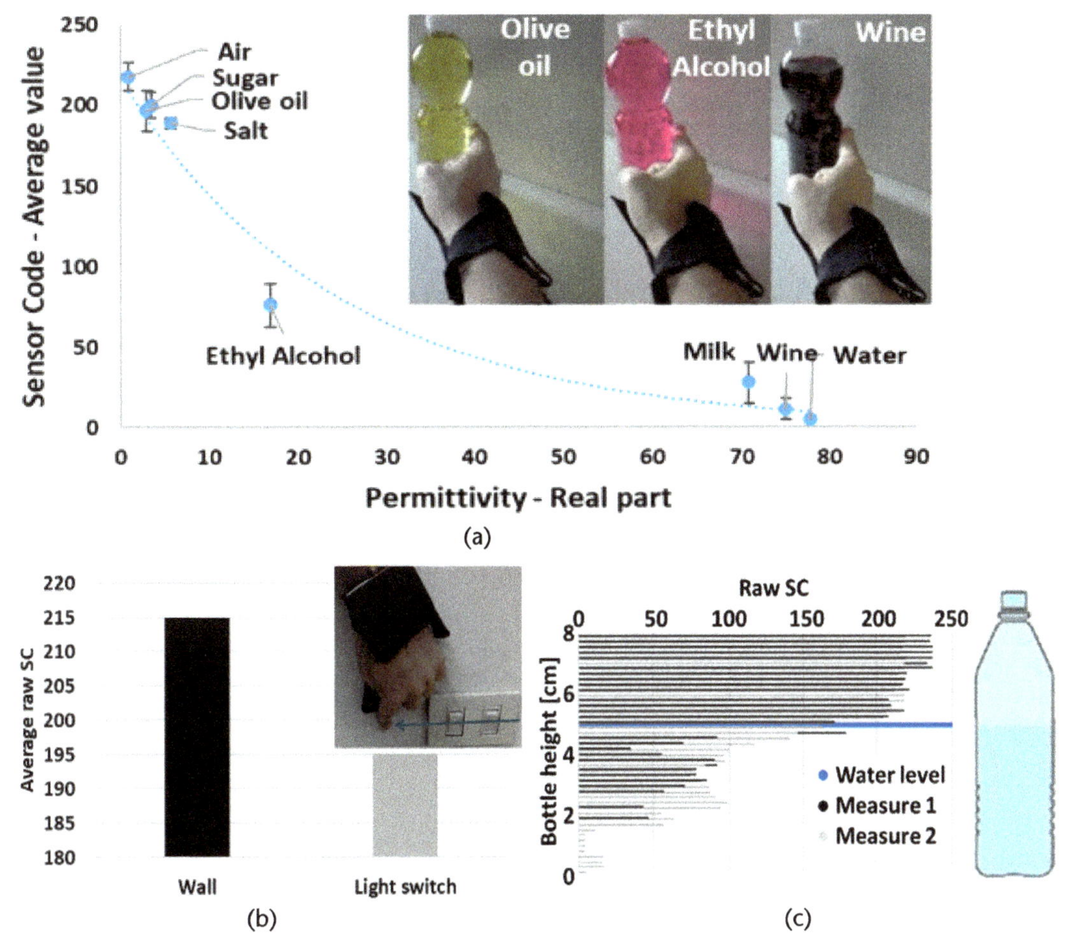

Figure 7.12 Dielectric-sensing R-FADs. (a) Mean SC value over five measurements when a single user contacts PET bottles filled with different powders or liquid. (Adapted from [28].) (b) Raw SC returned when contacting a wall and a light switch, and (c) when contacting a bottle half-filled with water twice from the top to the bottom. (d) Raw SC returned to four different users when touching nothing (air) or PET bottles filled with three liquids. (©2020 IEEE [38].) (e) Differential sensor code ΔSC corresponding to the raw SC in Figure 7.12(d) averaged on the users for two gestures. (©2020 IEEE [38].) (f) Digital fingerprints of a single user contacting three different materials, each identified by a color. (Adapted from [29].)

A further improvement on the robustness of sample data and the reliability of the identification of materials can be achieved by multichannel R-FADs, which are devices wherein all five fingers are provided with identical sensors [29]. In this case, the measurement metric is now a digital fingerprint, which is the vector of the ΔSC measured by each finger. Accordingly, when touching some materials, different digital fingerprints are observed (Figure 7.12(f)). The areas of digital fingerprints for four users and three touched objects having different permittivities are shown in Figure 7.13(a). Based on the digital fingerprints, the digital contrast can be defined as the gap of the $\overline{\Delta SC}$ (average of the returned ΔSC over all the five sensors) between two materials having different electric permittivities. The digital contrast in the case of a multichannel R-FAD achieves sensitivities up to two times those obtained by a single-sensor R-FAD using the same sensor (Figure 7.13(b)) [29].

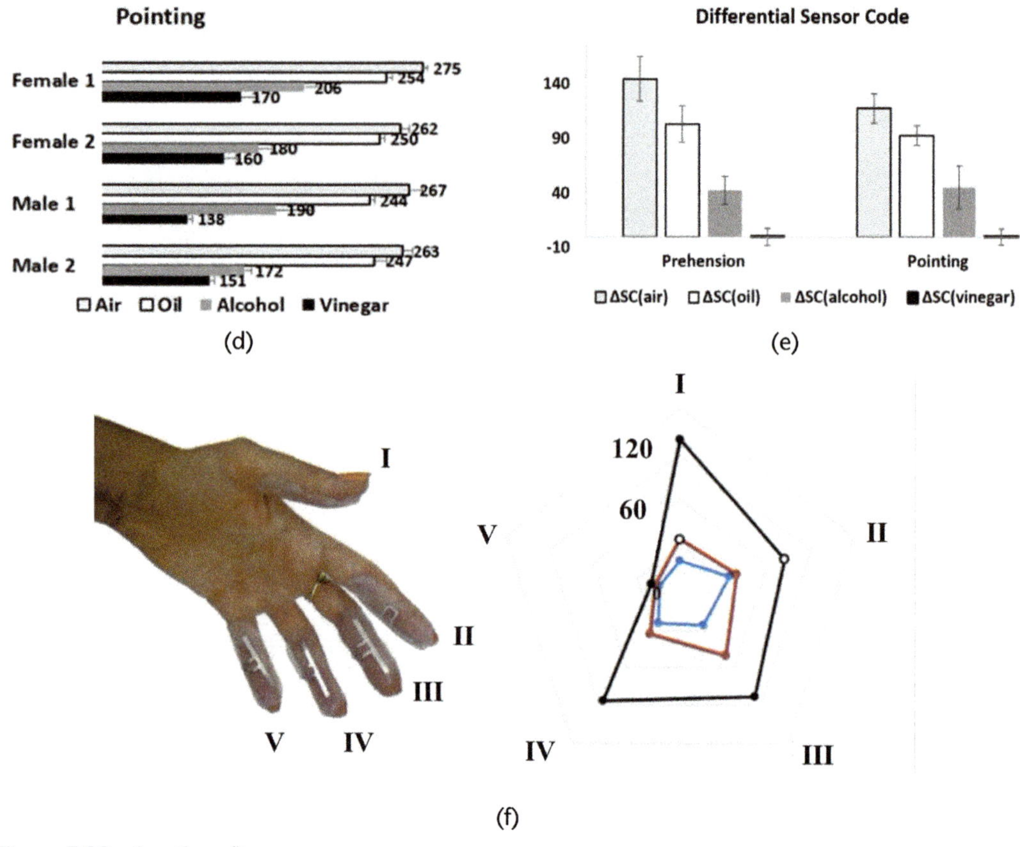

Figure 7.12 (continued)

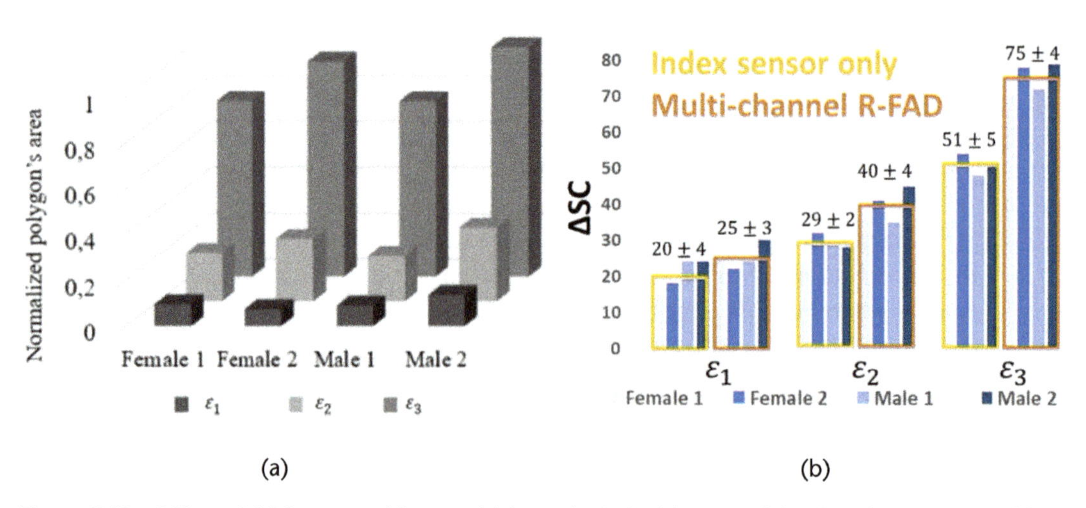

Figure 7.13 Differential SCs returned by a multichannel R-FAD. (a) Areas of the digital fingerprints of four volunteers contacting three materials having different permittivities $\varepsilon_1 < \varepsilon_2 << \varepsilon_3$. (Adapted from [29].) (b) Comparison between the ΔSC returned by a single sensor and the averaged differential $\overline{SC}$ measured by a multichannel R-FAD. (Adapted from [29].)

7.6 Application to Cognitive Remapping

According to the grounded theories of cognition, physical perception is intimately related to abstract representation [41]. For example, holding a warm or cold pack was proved to influence brain activation when performing a given task [42]. Consequently, a patient suffering from a prolonged sensorial degradation will experience a significant cognitive change, losing the emotional feeling of temperature and not perceiving either warm or cold as a pleasant concept when figuring out situations.

The temperature-sensing R-FAD was employed to study the possibility of cognitive remapping caused by the above problem by resorting to the same setup: the impaired patient as in Section 7.5.1 and a group of 20 neurologically healthy volunteers (age 30±9 years). The entity of the cognitive remapping was quantified through a Stroop-like task [43], where participants are requested to name the color of ink in which color names are printed (e.g., word "red" printed in blue). Cognitive dissonance is known to delay ink naming significantly [43]. Since a comparable effect occurs for words that are not color names but imply the concept of a color (color words such as "lemon," which implies the color "yellow"), a Stroop-like effect using color words implying temperature (such as "fire," related both to "red" and "hot") was also expected. The Stroop test was completed by the control group as well as by the hypoesthesia patient. The patient was asked to perform the test before and after the R-FAD training, as detailed in Section 7.5.1. A sketch of the experiment and the results are shown in Figure 7.14. More substantial cognitive dissonances cause longer response times (RTs); theoretically, a longer RT than the sensorially impaired patient's one before the R-FAD training is expected both after the R-FAD training and in the control group. Overall, the patient's RT delay before the R-FAD training was observed for color names but not for color words (Figure 7.14(c)). The patient's RT for color words was lower than those of 66% of the control group, suggesting that the prolonged loss of the physical thermal feeling probably weakened the cognitive dissonance related to the object's temperature. Instead, after the R-FAD training, the patient's RTs were more similar to those of the control group, with higher RTs returned for color words than before the training. Accordingly, a partial recovery of the concepts of warm and cold can be inferred.

7.7 Conclusion

This chapter reviewed the state of the art of R-FADs to understand how epidermal electromagnetics can help enable the tactile internet. Establishing a reliable fingertip-wrist link is still a critical issue. This link is varied for given antenna layouts mainly because of (1) the wearer's hand form-factor, (2) the performed gesture, (3) the sensorized finger, (4) the sensor's adherence to the skin, and (5) the touched material. Partial communication insensitivity to the type of contacted object can be achieved by a constrained design of the finger antenna involving a self-tuning IC. Among the experimented manufacturing options, the in-silicone sensors are the most comfortable and practical. Lastly, applications of R-FADs for aiding sensorially impaired people and studying cognitive neural remapping were reported.

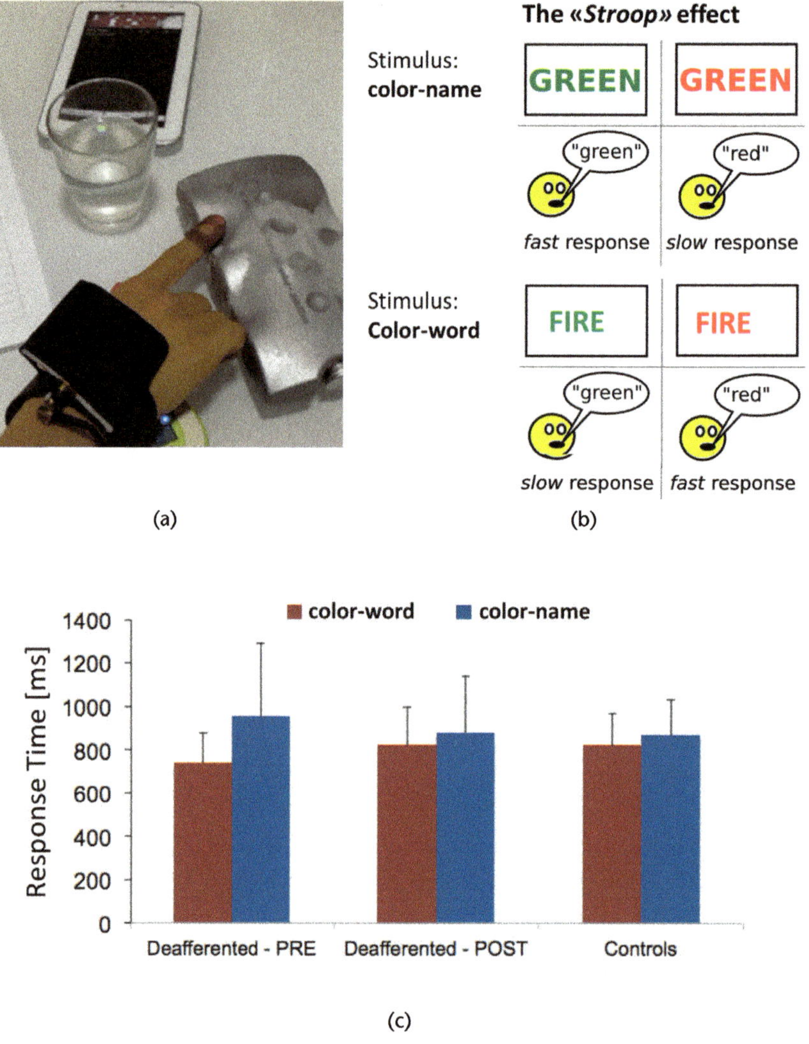

Figure 7.14 (a) Temperature-sensing R-FAD and experimental setup of R-FAD training sessions. The portable reader is connected via Bluetooth to the tablet, where an ad hoc app returns audio and visual feedback to the user. (©2019 IEEE [16].) (b) The Stroop effect and Stroop-like task. (©2019 IEEE [16].) (c) RTs of the deafferented patient before (pre) and after (post) R-FAD training are compared with those of the control group. (©2019 IEEE [16].)

R-FADs are an extremely promising technology to study the human cognitive sphere thanks to their ability to modify users' perceptions in a controlled way. According to our preliminary test, the temperature-sensing R-FAD was not only able to restore the physical thermal feeling, but also the abstract thermal feeling.

On-hand devices for enabling the TI are still little known, and future research can be envisaged. R-FADs provided with different sensing capabilities, as pressure or pH, can be included for applications such as telediagnosis; EM performance can be further enhanced by resorting to a directive wrist antenna made of the same soft-electronic technology as the fingertip sensors; and 5G R-FADs could use smart watches as readers. Moreover, cognitive remapping when conferring an ultrability to the user has never been studied and could also give helpful insight into human cognition.

7.8 Acknowledgments

The authors gratefully thank all the people in the Pervasive Electromagnetics Lab at the Tor Vergata University of Rome who worked on this topic over the years: Dr. Sara Amendola, Ms. Veronica Di Cecco, Ms. Valentina Greco, Ms. Silvia Guido, Ms. Federica Naccarata, and Ms. Cecilia Vivarelli.

References

[1] Fettweis, G. P., "The Tactile Internet: Applications and Challenges," *IEEE Vehicular Technology Magazine*, Vol. 9, No. 1, 2014, pp. 64–70.

[2] Burdea, G. C., and P. Coiffet, *Virtual Reality Technology*, Hoboken, NJ: John Wiley & Sons, 2003, "Cybersickness," pp. 269–274.

[3] Rebenitsch, L., and C. Owen, "Review on Cybersickness in Applications and Visual Displays," *Virtual Reality*, Vol. 20, No. 2, 2016, pp. 101–125.

[4] Dohler, M., et al., "Internet of Skills, Where Robotics Meets AI, 5G and the Tactile Internet," *European Conference on Networks and Communications*, Oulu, Finland, 2017.

[5] Maier, M., M. Chowdhury, B. Rimal, and D. Van, "The Tactile Internet: Vision, Recent Progress, and Open Challenges," *IEEE Communications Magazine*, Vol. 54, No. 5, 2016, pp. 138–145.

[6] Shilkrot, R. J., Huber, J. Steimle, S. Nanayakkara, and P. Maes, "Digital Digits: A Comprehensive Survey of Finger Augmentation Devices," *ACM Computing Surveys*, Vol. 40, No. 2, 2015, pp. 1–29.

[7] Su, C.- H., L. Chan, C.-T. Weng, R.-H. Liang, K.-Y. Cheng, and B.-Y. Chen, "NailDisplay: Bringing an Always-Available Visual Display to Fingertips," *Proceedings of Conference on Human Factors in Computing Systems*, Paris, France, 2013.

[8] Kao, H.- L., A. Dementyev, J. Paradiso, and C. Schmandt, "NailO: Fingernails as an Input Surface," *Proceedings of Conference on Human Factors in Computing Systems*, Seoul, Republic of Korea, 2015.

[9] Dahiya, R., "Epidermal Electronics–Flexible Electronics for Biomedical Applications," in *Handbook of Bioelectronics: Directly Interfacing Electronics and Biological Systems*, S. Carrara and K. Iniewski (eds.), Cambridge, United Kingdom: Cambridge University Press, 2015, pp. 245–255.

[10] Sedighi, P., H. Norouzi, and M. Delrobaei, "An RFID-Based Assistive Glove to Help the Visually Impaired," *IEEE Transactions on Instrumentation and Measurements*, Vol. 70, 2021, pp. 1–9.

[11] Taylor, P. S., and J. C. Batchelor, "Finger-Worn UHF Far-Field RFID Tag Antenna," *IEEE Antennas and Wireless Propagation Letters*, Vol. 18, No. 12, 2019, pp. 2513–2517.

[12] Liang, R.- H., S.-Y. Yang, and B.- Y. Chen, "InDexMo: Exploring Finger-Worn RFID Motion Tracking for Activity Recognition on Tagged Objects," *Proceedings of the 23rd International Symposium on Wearable Computers*, London, United Kingdom, 2019.

[13] Bainbridge, R. and J. A. Paradiso, "Wireless Hand Gesture Capture through Wearable Passive Tag Sensing," *2011 International Conference on Body Sensor Networks*, Dallas, TX, 2011.

[14] Amato, F., C. Occhiuzzi, and G. Marrocco, "Performances of a 3.6 GHz Epidermal Loop for Future 5G-RFID Communications," *14th European Conference on Antennas and Propagation*, Copenhagen, Denmark, 2020.

[15] Shilkrot, R., *Digital Digits: Designing Assistive Finger Augmentation Devices*, PhD thesis, Massachusetts Institute of Technology, Cambridge, MA, 2015.

[16] Amendola, S., V. Greco, G. M. Bianco, E. Daprati, and G. Marrocco, "Application of Radio-Finger Augmented Devices to Cognitive Neural Remapping," *IEEE International Conference on RFID Technology and Applications*, Pisa, Italy, 2019.

[17] Amendola, S., V. Di Cecco, and G. Marrocco, "Numerical and Experimental Characterization of Wrist-Fingers Communication Link for RFID-Based Finger Augmented Devices," *IEEE Transactions on Antennas and Propagation*, Vol. 67, No. 1, 2019, pp. 531–540.

[18] Di Cecco, V., S. Amendola, P. P. Valentini, and G. Marrocco, "Finger-Augmented RFID System to Restore Peripheral Thermal Feeling," *IEEE International Conference on RFID*, Phoenix, AZ, 2017.

[19] Orfanidis, S., *Electromagnetic Waves and Antennas*, New Brunswick, NJ: Rutgers University Press, 2010.

[20] Bianco, G. M., S. Amendola, and G. Marrocco, "Near-Field Modeling of Self-Tuning Antennas for the Tactile Internet," *33rd General Assembly and Scientific Symposium of the International Union of Radio Science*, Rome, Italy, 2020.

[21] Amendola, S., and G. Marrocco, "Optimal Performance of Epidermal Antennas for UHF Radio Frequency Identification and Sensing," *IEEE Transactions on Antennas and Propagation*, Vol. 65, No. 2, 2017, pp. 473–481.

[22] Ferreira, D., P. Pires, R. Rodrigues, and R. F. S. Caldeirinha, "Wearable Textile Antennas: Examining the Effect of Bending on Their Performance," *IEEE Antennas and Propagation Magazine*, Vol. 59, No. 3, 2017 pp. 54–59.

[23] Caccami, M. C., and G. Marrocco, "Electromagnetic Modeling of Self-Tuning RFID Sensor Antennas in Linear and Nonlinear Regimes," *IEEE Transactions on Antennas and Propagation*, Vol. 66, No. 6, 2018, pp. 2779–2787.

[24] Bianco, G. M., S. Amendola, and G. Marrocco, "Near-Field Constrained Design for Self-Tuning UHF-RFID Antennas," *IEEE Transactions on Antennas and Propagation*, Vol. 68, No. 10, 2020, pp. 6906–6911.

[25] Marrocco, G., "The art of UHF RFID Antenna Design: Impedance-Matching and Size-Reduction Techniques," *IEEE Antennas and Propagation Magazine*, Vol. 50, No. 1, 2008, pp. 66–79.

[26] GS1, "Regulatory Status for Using RFID in the EPC Gen2 (860 to 960 MHz) Band of UHF spectrum," March 13, 2021, https://www.gs1.org/docs/epc/uhf_regulations.pdf.

[27] Amendola, S., S. Milici, G. Marrocco, and C. Occhiuzzi, "On-Skin Tunable RFID Loop Tag for Epidermal Applications," *IEEE International Symposium on Antennas and Propagation & USNC/URSI National Radio Science Meeting*, Vancouver, Canada, 2015.

[28] Bianco, G. M., and G. Marrocco, "Fingertip Self-Tuning RFID Antennas for the Discrimination of Dielectric Objects," *Proceedings of the 13th European Conference on Antennas and Propagation*, Krakow, Poland, 2019.

[29] Naccarata, F., G. M. Bianco, and G. Marrocco, "Multi-Channel Radiofrequency Finger Augmentation Device for Tactile Internet," *34th URSI General Assembly and Scientific Symposium*, Rome, Italy, 2021.

[30] Miozzi, C., G. Diotallevi, M. Cirelli, P. P. Ventini, and G. Marrocco, "Radio-mechanical Characterization of Epidermal Antennas during Human Gestures," *IEEE Sensors Journal*, Vol. 20, No. 14, 2020, pp. 7588–7594.

[31] Barrell, K., and A. G. Smith, "Peripheral Neuropathy," *Medical Clinics of North America*, Vol. 103, No. 2, 2019, pp. 383–397.

[32] Jacob, E., V. Wong Chan, C. Hodge, L. Zeltzer, D. Zurakowsy and N. F. Sethna, "Sensory and Thermal Quantitative Testing in Children with Sickle Cell Disease," *Journal of Pediatric Hematology Oncology*, Vol. 37, No. 3, 2015, pp. 185–189.

[33] Blankenburg, M., et al., "Childhood Diabetic Neuropathy: Functional Impairment and Non-Invasive Screening Assessment," *Diabetic Medicine*, Vol. 29, No. 11, 2012, pp. 1425–1432.

[34] Griffith, K. A., et al., "Evaluation of Chemotherapy-Induced Peripheral Neuropathy Using Current Perception Threshold and Clinical Evaluations," *Support Care Cancer*, Vol. 22, No. 5, 2014, pp. 1161–1169.

[35] Pascolini, D., and S. P. Mariotti, "Global Estimates of Visual Impairment: 2010," *British Journal of Ophthalmology*, Vol. 96, 2012, No. 5, pp. 614–618.

[36] Pervasive Electromagnetics Lab, *RADIOFingerTip: a Finger-Augmented-Device to Restore Peripheral Neuropathy*, Rome: Italy, University of Rome Tor Vergata, 2016, https://www.youtube.com/watch?v=DGUkYqmt-5Q&t=1s.

[37] Pervasive Electromagnetics Lab, *RadioFingerTip: Estimation of the Temperature of the Objects Touched by the Fingertip Sensor*, Rome, Italy: University of Rome Tor Vergata, 2017, https://www.youtube.com/watch?v=6quuuEiNEBs.

[38] Bianco, G. M., C. Vivarelli, S. Amendola, and G. Marrocco, "Experimentation and Calibration of Near-Field UHF Epidermal Communication for Emerging Tactile Internet," *5th International Conference on Smart and Sustainable Technologies*, Split, Croatia, 2020.

[39] Bianco, G. M., and G. Marrocco, "Sensorized Facemask with Moisture-Sensitive RFID Antenna," *IEEE Sensors Letters*, Vol. 5, 2021, No. 3, pp. 1–4.

[40] Pervasive Electromagnetics Lab, *Recognizing Materials by Touch*, Rome, Italy: University of Rome Tor Vergata, 2020, https://www.youtube.com/watch?v=bQZOG3eVgm0.

[41] Barsalou, L. W., "Grounded Cognition: Past, Present, and Future," *Topics in Cognitive Science*, Vol. 2, No. 4, 2010, pp. 716–724.

[42] Kang, Y., L. E. Williams, M. S. Clark, J. R. Gray, and J. A. Bargh, "Physical Temperature Effects on Trust Behavior: The Role of Insula," *Social Cognitive and Affective Neuroscience*, Vol. 6, No. 4, 2011, pp. 507–511.

[43] MacLeod, C. M., "Half a Century of Research on the Stroop Effect: An Integrative Review," *Psychological Bulletin*, Vol. 109, No. 2, 1991 pp. 163–203.

CHAPTER 8

Wearable Imaging Techniques

Satheesh Bojja Venkatakrishnan and John L. Volakis

Preventive medicine requires noninvasive and continual monitoring of a person's vital signs and internal organs. Many chronic noncommunicable diseases, such as breast cancer and thyroid cancer, benefit from early detection of signs such as hypoxia, angiogenesis, and melanoma formation [1]. Moreover, numerous medical applications necessitate periodic monitoring of blood arteries or organ activities [2]. As a result, medical sensors that allow continuous, unattended, and uninterrupted monitoring of a person's vital signs, including imaging of internal organs, are in high demand [3–5]. To this end, academics and healthcare pioneers are very interested in building a tailored medical imaging system that focuses on early detection and treatment [6]. The patient benefits from the sensor being worn because it allows for flexible multisite scanning and comfortable long-term monitoring for easy tailored operation. As a result, the wearable sensor enables for continued monitoring outside of the hospital without interfering with daily activities. These types of medical sensors should be tiny, portable to the backend hardware, and incorporate fast processing algorithms.

Many procedures and their associated hardware are often bulky, and diagnosis frequently necessitates hospitalization of patients. Furthermore, because they are not worn on the body, they do not allow for continuous monitoring. For example, X-rays are unable to determine whether tumors are benign or malignant without requiring surgery [7]. On the other hand, probe methods provide an accurate approximation of the dielectric constant [8, 9], but they require direct access to the specimen and are hence limited to in vitro applications. Another significant disadvantage is that typical RF sensors can only penetrate 3 cm into the body, and therefore, deep tissue monitoring is not possible [10].

Using ultrasound transducers instead of RF sensors allows for deeper penetration into the body. Due to its low cost and lack of ionizing radiation [2], ultrasound imaging has become one of the most used forms of imaging in the medical industry. However, modern ultrasound probes are bulky and impractical for continuous and periodic monitoring.

In contrast, photoacoustic imaging is an emerging modality employing a combination of optical excitation and acoustic detection that has been used to image vascular, functional, and molecular changes within living tissue [11]. Unlike optical imaging technologies such as optical coherence tomography, which uses ballistic photons, this technique uses diffused photons, which allows for far deeper penetration [12]. When it comes to functional photoacoustic imaging, however, the optical parametric oscillator (OPO) system is essential. The interconnected cooling system and stability control system (e.g., vibration-isolated optical table) result in a bulky system configuration, not to mention the extra size expansion produced by the interconnected cooling system and stability control system (e.g., vibration-isolated optical table) [13]. To acoustically pursue high-resolution and real-time imaging capabilities in 3-D space, a large-aperture matrix array is preferred.

As a result, miniaturizing photoacoustic devices is particularly difficult with such a large-scale acoustic implementation. Furthermore, acoustic coupling between the photoacoustic probe and the imaging target remains a difficult job. An intermediary coupling medium is required to fill the spherical cavity [14], which impedes the development of photoacoustic techniques for wearable implementation and tailored healthcare applications.

As a result, there are a variety of medical imaging approaches available, and the underlying question is how to select the best imaging methodology using wearable electronics. Nonetheless, each method has its own set of benefits and drawbacks. A suitable imaging modality can be chosen based on the medical application.

In this chapter, we discuss four different medical imaging modalities: (1) microwave tomography, (2) ultrasound imaging techniques, (3) optical topography, and (4) photoacoustic hybrid imaging including photoacoustic microscopy, as well as their importance in medical imaging. It should be emphasized that this chapter simply serves as a review of the most commonly used wearable medical imaging techniques.

1. Radar-based microwave imaging (MWI) technology is a low-cost, safe, and relatively simple system that can be used in a variety of portable and wearable applications, according to [15]. This method can also be used to assess tissue properties by looking at the difference between dielectric permittivity and conductivity across a broad frequency range [16]. One of the widely explored applications over microwave imaging is the early detection of a breast tumor since early tumor diagnosis and treatment is considered the best chance for surviving breast cancer [17].
2. As an emerging alternative, capacitive micromachined ultrasound transducers (CMUTs) offer wider bandwidth, better integration with electronics, and ease of fabricating large arrays [18]. The CMUTs outperformed piezoelectric probes in terms of signal-to-noise ratio when phased-array steering was used; this was attributed to the larger acoustic radiation field experimentally measured from CMUTs.
3. On the other hand, optical sensing exploits the wavelength selectivity to image a specific location with specific properties in the tissue. For example, functional near-infrared spectroscopy (fNIRS) is an effective optical topography and noninvasive neuroimaging tool for monitoring oxygenation and cerebral hemodynamics [19].
4. Photoacoustic imaging, a multiwave imaging modality, has been demonstrated to be a powerful tool in early cancer detection [20]. This method exploits the advantages of both optics and ultrasound technologies: wavelength selectivity and adequate penetration depth. By illuminating the

region of interest (ROI) with a high-energy laser pulse and then detecting the generated ultrasound wave, the human vasculature structure and other photoabsorbers (e.g., melanin) can be photoacoustically visualized in a noninvasive manner. A miniaturized multispectral photoacoustic imager can perform wearable 3-D imaging of human arm vasculature and work as a handheld scanner for flexible multisite diagnosis. On the other hand, photoacoustic microscopy (PAM) imaging has undergone tremendous advancements and has been used for several biomedical applications since its first demonstration for in vivo tissue imaging in 2005 [21–23]. Photoacoustic spectroscopy allows mapping of a wide variety of endogenous molecules (e.g., hemoglobin, melanin, DNA, RNA, lipids, and water) and exogenous contrast agents (e.g., small molecules and nanoparticles) with scalable spatial and temporal resolutions and penetration depths [24].

8.1 Wearable Imaging Algorithms

Wearable imaging approaches can be categorized into radar imaging, sonar-based (ultrasound), and optical imaging (photoacoustics). Approaches pertaining to each category will be discussed in detail below.

8.1.1 Radar-Based RF and THz Imaging

8.1.1.1 Employing Backscattering Imaging

Numerous techniques employing RF backscattering have been used to monitor tissue and organ properties by exploiting the variation in the dielectric permittivity and conductivity recorded using backscattered fields. For example, the authors in [7] showed that the dielectric constant of the underlying tissues can be accurately estimated within predicted limits. The following approach was used to accomplish this.

The sensor used in [7] is composed of a set of N electrodes (antennas) fed by $N - 1$ ports as shown in Figure 8.1. A monotonic RF signal excites the first port, which is situated between the first two electrodes. The others are only used to receive the radiated fringing field after it has propagated and dissipated through the tissues. The scattering parameters measured at these ports are denoted as S_{i1}, where $i = 1, 2, 3, ..., N - 1$, with port #1 being active. As implied from Figure 8.2(a), the signal at the farthest (leftmost) electrodes is more dependent on the deep layer properties. This observation is critical to the premise that a multiprobe sensor can extract the dielectric constant (ε_r) of deeper layers/tissues. With the proper calibration, the sensor can potentially provide an accurate determination of ε_r as a function of depth.

As indicated in Figure 8.2(b), by plotting the port scattering parameters versus port number, the curves corresponding to the dielectric constants of deeper tissue layers can be distinguished. Two properties in this graph are of interest: (1) the overall signal levels (port readouts or S_{i1} values) are sufficiently large and varying from probe to probe, and (2) the dynamic range of the S_{i1} values is wide enough for

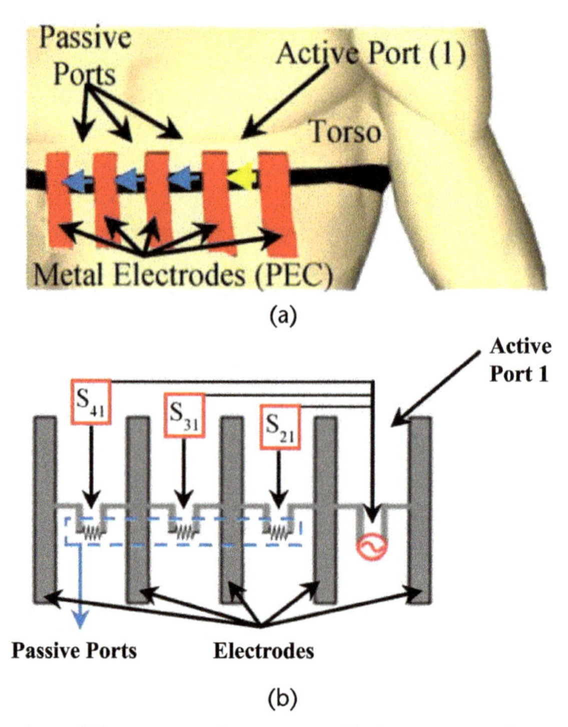

Figure 8.1 (a) Overview of the proposed organ monitoring sensor setup, and (b) port readout setup. (©IEEE [7].)

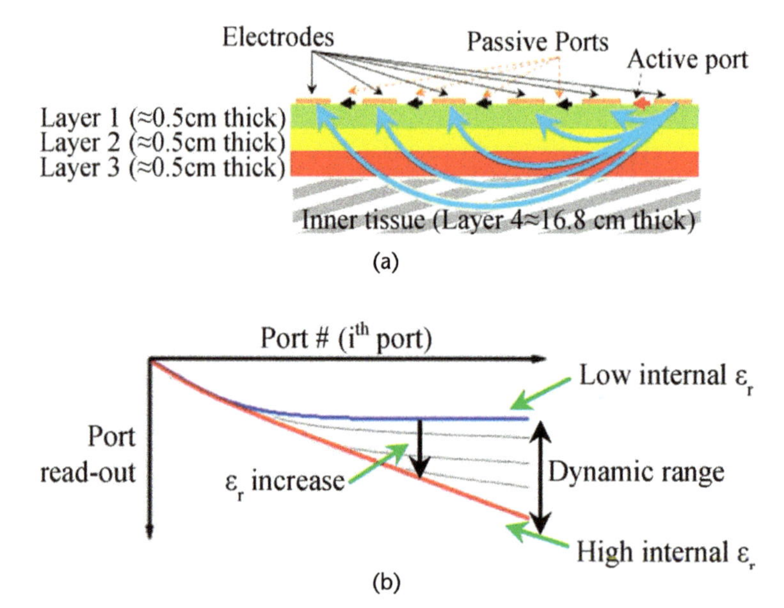

Figure 8.2 (a) Illustrations of the electrodes on the layered human torso with corresponding ports for measuring the voltages between the electrodes, and (b) port readout versus port number for different inner tissue dielectric constants. (©IEEE [7].)

accurate mapping of the tissues' dielectric properties. The dynamic range is associated with the farthest port from the active one as the S_{i1} value of this port is most affected by the inner layers. The larger the dynamic range, the more precisely one can detect finer changes in ε_r; an essential aspect of monitoring recovery/illness. The operating frequency was selected to provide better differentiation in the electrical properties of the biological tissues. As can be surmised from [25], tissues show significant variation at frequencies below 100 MHz, and hence the sensors were designed to operate at 40 MHz.

The goal is to extract the dielectric constant of deep biological tissues (see Figure 8.2) using the S_{i1} values. To do so, first it is necessary to remove the impact of the outer layers (skin, fat, and muscle) and only focus on the inner tissue layer(s). Toward this, ε_r is represented as a weighted sum of the scattering parameters measured from the multiprobe sensor at ports $i = 1, 2, 3, ..., N$, where N refers to the total number of measurement ports. That is:

$$\varepsilon_r = \sum_{i=1}^{N} w_{i-1} S_{i1} \tag{8.1}$$

where i is the port number, w_{i-1} refers to the weights (to be calculated) at each port, and S_{i1} is the scattering parameters read across the ith port. In effect, the weights in (8.1) need to be trained to minimize the error in extracting ε_r, basically calibrating the system.

Validation of the sensor's functionality was done in two steps. First, measurements of a small electrode set were carried out to validate the analysis. In all these measurements, a mannequin filled with a dielectric to emulate the human body was employed. Specifically, data was collected using an empty mannequin and one filled with a liquid of $\varepsilon_r = 56$ to emulate the human body. The first prototype consisted of nine electrodes and four ports. As noted, the rightmost port was active, and the three other ports were used to collect the penetrating field through the torso. The fabricated nine-electrode sensor is depicted in Figure 8.3(a). The mannequin's skin was 3 mm thick and was associated with $\varepsilon_r \approx 3$. Of importance is that good agreement was achieved between measurements and simulations, validating our representation of the torso as an elliptic cylinder having minor and major radius of 10.17 and 16.425 cm. Figure 8.3(b) shows a maximum error of 6.8% due to using a simplified torso for simulations.

Table 8.1 shows a comparison between the exact values (attained using an Agilent 85070D dielectric probe kit) and those attained using the S_{i1} values from the probe sensor. From Table 8.1, it was observed that the computed ε_r values are fairly accurate. Specifically, when $\varepsilon_r = 111.52$, the error is only 2.98%, and for $\varepsilon_r = 60.48$, the error is 6.69%. These errors are rather small and acceptable for practical use.

To better demonstrate the capability of the sensor, a second set of measurements was involved using a block of minced pork hollowed out to encase a single foam block. This experiment was repeated with foam blocks of different sizes. These foam blocks were intended to represent air volumes within the lung. For each foam block experiment, the average of the inner layers' ε_r was calculated. The results in Table 8.2 show that the computed ε_r values are always less than 8% in error.

The authors extended the same principle to a different medical imaging application. The authors in [26] developed a method that images the differential permittivity of the human torso's cross section by modeling each pixel as a tiny cylinder whose scattering is known. Moreover, to overcome the signal attenuation due to

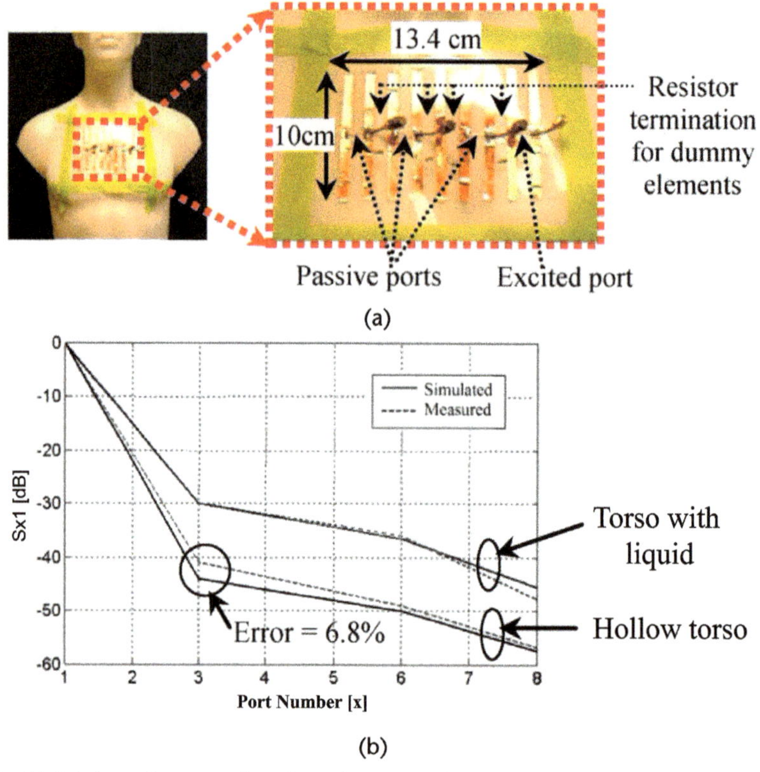

Figure 8.3 (a) Fabricated sensor placed on the torso of a mannequin filled with liquid to emulate the human tissue, and (b) simulated versus measured data. (©IEEE [7].)

Table 8.1 Measured ε_r Versus Its Exact Value

Case #	*Inner Layer*	*Outer Layer*	*Exact* ε_r	*Calculated* ε_r	*% Error*
1	Distilled water	Steak	80.40	83.56	3.93
2	Distilled water + salt	Steak	111.52	108.20	2.98
3	Ground beef	Ground beef	60.84	64.91	6.69
4	Ground beef + air balloon	Ground beef	56	58.85	5.09

From: [7].

Table 8.2 Measured ε_r Versus Its Exact Value

Foam Block Size	*Volume of Block*	*Exact* ε_r	*Calculated* ε_r	*% Error*
Small	$9 \times 7 \times 7$ cm^3	75.60	81.33	7.58
Medium	$8 \times 11.5 \times 10$ cm^3	58.51	60.21	2.90
Large	$12.5 \times 12.5 \times 12.5$ cm^3	21.65	20.00	7.62

highly lossy nature of the biological medium, a novel loss-compensation technique was introduced. Notably, the employed imaging method is different from typical backpropagation methods [27, 28], where plane wave approximation was adopted and the inverse Fourier transform of the collected data was performed to reconstruct the image.

The proposed method in [26] presents an ingenious derivation of Green's function of the medium with a new loss-compensation methodology by using electromagnetic reciprocity of forward and backward waves (not necessarily plane waves) to recover the image. Hence, the drawbacks of iterative microwave imaging methods that rely on ill-conditioned matrix inversions are completely avoided [29, 30]. Also, the proposed method renders almost real-time image recovery as opposed to those in [29, 30].

The proposed imaging method is referred to as the loss-compensated back-propagation (LC-BP) method. This proposed wearable sensor overcomes a number of challenges with body-worn imaging. These include (1) uncertainty in the torso's shape, (2) reduced SNR of the measured data, and (3) issues of mismatch between the sensor and torso/body [31–33]. The above issues have often led conventional microwave imaging methods to poor image recovery for body-worn sensors.

As shown in Figure 8.4(a), the proposed sensor consists of several probes placed conformally around the body. Its operation is based on exciting one of the dipole-like probes by a 0.8-GHz signal (source), while the rest remain in the receiving mode (measurement points) as shown in Figure 8.4(b). Similarly, every dipole is excited sequentially and the rest will act as receivers. It should be noted that although a higher frequency may yield a better image resolution, a low enough frequency was chosen to ensure sufficient penetration (using dipoles) into this layered lossy torso model. Notably, exact knowledge of the torso's shape is challenging to extract and any uncertainty in this regard makes the image reconstruction mathematically challenging [34]. The mismatch between the actual and estimated positions of the probes (x_{ac}, y_{ac}) and (x_{est}, y_{est}), respectively) placed around the torso cross section is expressed by the boundary mismatch parameter (BMP) defined by

$$BMP = \frac{\sum_{i=1}^{Q}\left(\left(x_{ac,i} - x_{est,i}\right)^2 + \left(y_{ac,i} - y_{est,i}\right)^2\right)}{L_{maj}^2} \tag{8.2}$$

where L_{maj} is the major axis length of the closest ellipse that can be drawn to minimize BMP (see Figure 8.4(c)) and subscript i refers to the ith probe with $L_{maj} = 35$ cm in [34].

The human torso will be modeled to have four outer layers corresponding to skin, fat, muscle, and bone. Table 8.2 gives the nominal values of permittivity and loss tangent both at 0.8 GHz and thicknesses of the outer layers [35]. We note that the input power at the source was set to 1 mW. As a result, the corresponding maximum SAR averaged over 1g of tissue was $SAR_{1g,\max} \approx 0.2$ W/kg. This value is far below the limit 1.6 W/kg based on the IEEE C95.1-1999 and FCC safety exposure guidelines.

To obtain the imaging algorithm, the formulation is derived exploiting the solution of forward scattering from a PEC cylinder due to line source radiator.

Let a circular PEC cylinder of radius a and infinite length be located near a line source. The origin of the coordinate system was placed at the center of the cylinder. The line source (along the z-axis) is located at (ρ', ϕ') and any measurement point is located at (ρ, φ) where $\rho > a$. The electric line source carries a constant current A_μ

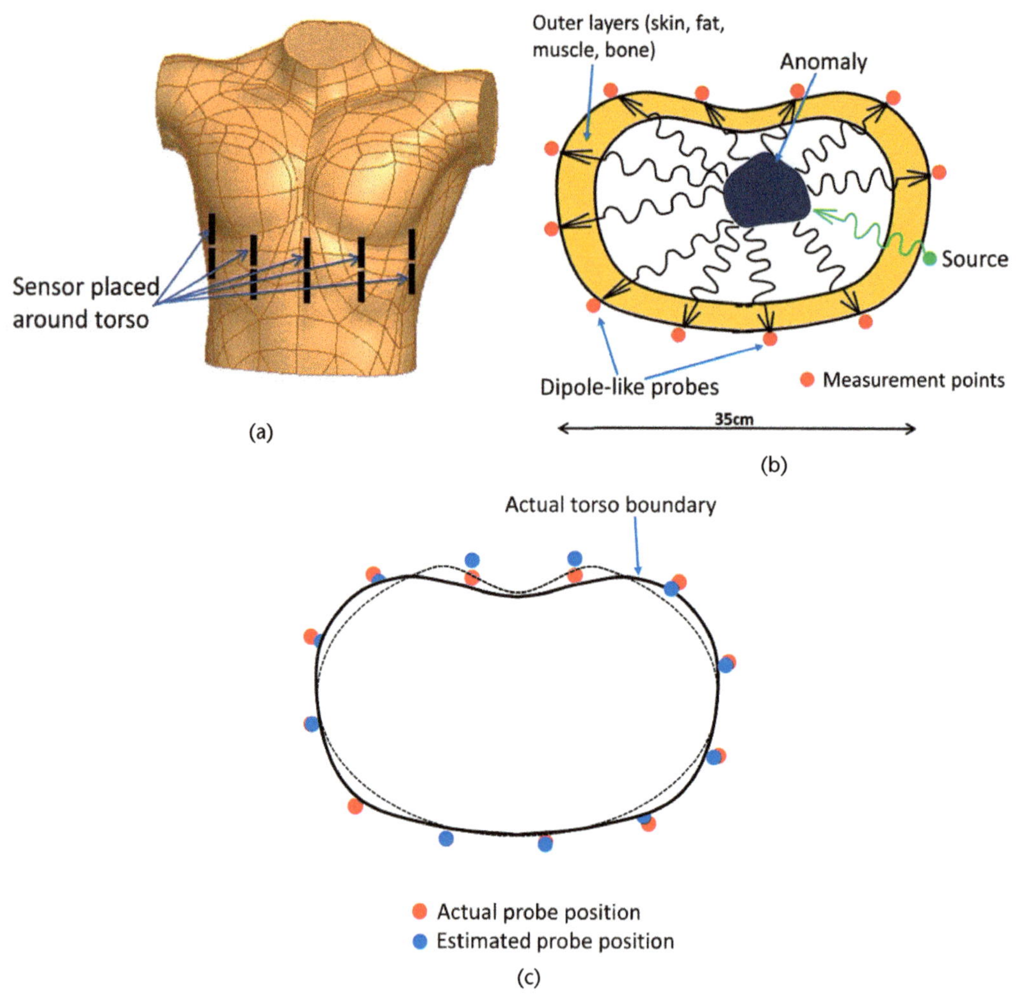

Figure 8.4 (a) Proposed body-worn imaging sensor placed around a human torso model, (b) scattered EM field measured at the receiver probes, and (c) actual (red) and estimated (blue) probe positions around the torso boundary [26].

and radiates electromagnetic waves. For this excitation, the scattered field in the presence of the PEC cylinder is [36],

$$E_z^s = \frac{\beta^2 A_m}{4\omega\varepsilon} \sum_{n=-\infty}^{+\infty} \left[H_n^{(2)}(\beta\rho') \frac{J_n(\beta a)}{H_n^{(2)}(\beta a)} H_n^{(2)}(\beta\rho) e^{jn(\phi-\phi')} \right] \tag{8.3}$$

where $H_n^{(2)}(.)$ refers to nth order Hankel function of the second kind, $J_\nu(\cdot)$ refers to nth order Bessel function, ω is the frequency in rad/s, $\varepsilon = \varepsilon_0\, \varepsilon_r$ is the permittivity of the medium, and $\beta = \dfrac{2\pi}{\lambda}$ is the propagation constant of the medium. ε represents any homogeneous medium including a typical lossy biological tissue. Looking at (8.3), the 2-D Green's function of the medium can be written as

Table 8.3 Nominal Values of Tissue Electrical Properties at 0.8 Ghz and Thicknesses

Tissue	*Permittivity* (ε_r)	*Loss Tangent (tanδ)*	*Thickness (h) [cm]*
Skin	46.52	0.39	0.3
Fat	5.48	0.20	1.5
Muscle	55.29	0.37	1.5
Bone	12.55	0.24	2.0
Body fluid	68.93	0.52	-
Blood	61.7	0.54	-
Heart	60.61	0.44	-
Lung (moderately inflated)	37.0	0.40	-

From: [26].

$$G_{2D}(\rho,\rho',\phi,\phi') = \frac{\beta^2 A_m}{4\omega\varepsilon}\sum_{n=-\infty}^{+\infty}\left[H_n^{(2)}(\beta\rho')\frac{J_n(\beta a)}{H_n^{(2)}(\beta a)}H_n^{(2)}(\beta\rho)e^{jn(\phi-\phi')}\right] \tag{8.4}$$

Hence, (8.3) can be written as

$$E_z^s(\rho,\rho',\phi,\phi') = A_m G_{2D}(\rho,\rho',\phi,\phi') \tag{8.5}$$

Notably, G_{2D} in (8.4) refers to the 2-D Green's function of the forward-scattering problem.

Here, a conventional multistatic measurement scheme of microwave tomography is employed where one antenna is excited at a time while the others are used as receivers. This process is repeated until all antennas are excited one by one. Thus, a measured scattered field vector, $E_{z,meas}^s = (E_{z,12}^s, E_{z,13}^s, \ldots, E_{z(Q-1)Q}^s)^T$, is obtained where q refers to the qth line source, m refers to the mth measurement point, Q is the total number of line sources, and the superscript T refers to the transpose of the vector. Each element of the measurement vector, $E_{z,meas}^s$, is the difference between two measurements ((total field – incident field) as discussed in [26]). As the measurement is typically carried out using a vector network analyzer (VNA), the electric field, $E_{z,qm}^s$ can be replaced with the scattering parameter, $S_{z,qm}^s$ (the transmission coefficient between probes q and m), measured using VNA. Equation (8.5) now becomes

$$S_z^s(\rho,\rho',\phi,\phi') = A_m G_{2D}(\rho,\rho',\phi,\phi') \tag{8.6}$$

Equation (8.6) gives the forward solution of the scattered field from a PEC cylinder buried in a homogeneous medium. Now, the domain of interest (DOI) is subdivided into N_E small pixels (as shown in Figure 8.4(b)) and apply (8.6) to calculate the scattered fields around the DOI, (8.6) will give nonzero values for the pixels where the PEC target resides and zero values for the pixels where the PEC target is not present. Hence, for a pixelated DOI, summing the contributions of all the pixels, the scattered field at the mth measurement point due to the unity excitation ($A_m = 1$) of the qth line source, can be obtained as

$$S_z^s(\rho,\rho',\phi,\phi')=\sum_{n=1}^{N_E} G_{2D}^n(\rho,\rho',\phi,\phi')I_{2D,n} \tag{8.7}$$

A factor $I_{2D,n}$ is introduced in (8.7), which takes a value of either 1 (when the PEC target is located at the nth pixel) or 0 (when the same is not located at that pixel).

If $[G_{2D}]$ is considered to be a linear mapping from the $[I_{2D}]$ vector space to the $[S]$ vector space, $[G_{2D}]^*$ (conjugate response) can be considered as the inverse of G, resulting in an approximated solution for pixel located at (x_{pix}, y_{pix}):

$$I_{2D}\left(x_{pix},y_{pix}\right)=\left|\sum_{q=1}^{Q}\sum_{\substack{m=1\\m\neq q}}^{Q} G_{2D}^*(\rho,\rho',\phi,\phi')S_{z,qm}^s(\rho',\rho,\phi,\phi')\right| \tag{8.8}$$

It can be deduced from Figure 8.5 (rightmost column) that the proposed method can reconstruct the approximate shape of the anomalies. For comparison, actual shapes are overlaid using dotted lines on the reconstructed shapes, in Figure 8.5 (rightmost column).

8.1.1.2 Employing Fractional Parameters

The approach in [37] uses fraction parameters instead of the complex permittivity directly. These fraction parameters represent the weights of the preselected permittivities to reconstruct the imaging domain. The goal of fraction imaging is to mitigate the imbalance between the real and imaginary portion of the complex permittivity. Imbalance refers to the difference between the actual versus estimated permittivities. To do so, an iterative algorithm was developed based on the Gauss-Newton (GN) method to calculate the fraction parameters. Both the relative permittivity and conductivity images are then obtained from the fraction parameters once they are available.

Based on Figure 8.6, from the set of dipole transceivers placed around the domain, one antenna is excited at a time while the others are used as receivers. This process is repeated until all antennas are excited one by one. Scattering parameters are collected from all receiver antennas and stacked in one vector, $\{S\}_{meas}$, to form the M X 1 measurement vector, M being the number of measurements. The complex relative permittivity of the DOI shown in Figure 8.6 is defined as:

$$\{\varepsilon\}=\{\varepsilon_n : n=1,2,3,\ldots,N_E\} \tag{8.9}$$

In this, N_E refers to the total number of pixels subdividing the DOI. A vital aspect of the fraction method is to express the permittivity of the nth pixel as:

$$\varepsilon_n=\sum_{t=1}^{N_{dielectric}} r_{t,n}e_t \tag{8.10}$$

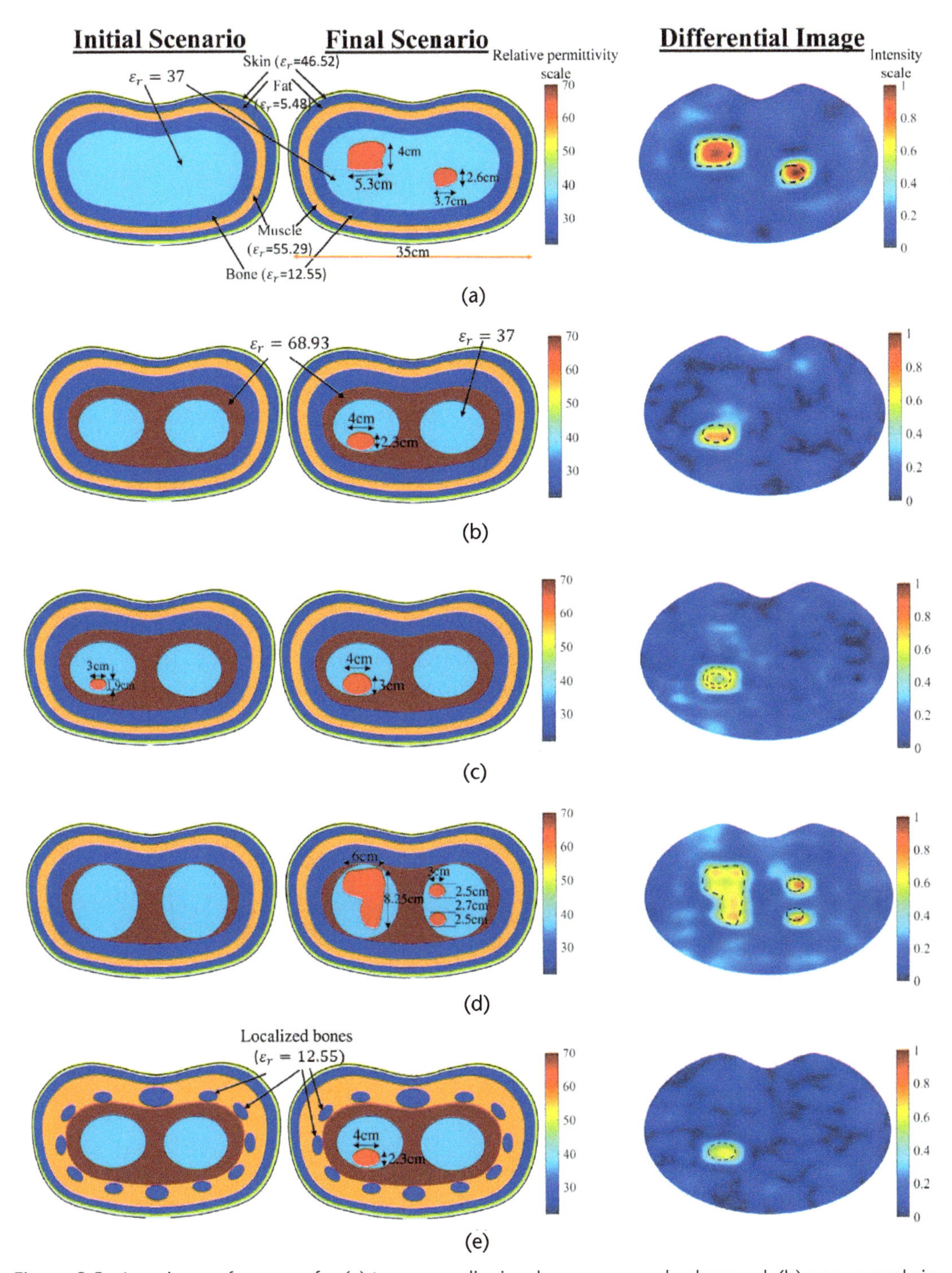

Figure 8.5 Imaging performance for (a) two anomalies in a homogeneous background, (b) one anomaly in an inhomogeneous background, (c) one annular-shaped anomaly in a homogeneous background, (d) three anomalies in an inhomogeneous background, and (e) one anomaly in an inhomogeneous background (bones modeled as localized elements). (©IEEE [26].)

In this, $r_{1,n}, r_{2,n}, \ldots, r_{N_{dielectric},n}$ represent the permittivity fractions selected a priori to synthesize the pixel's permittivity, ε_n. That is, ε_n is a weighted linear sum of the

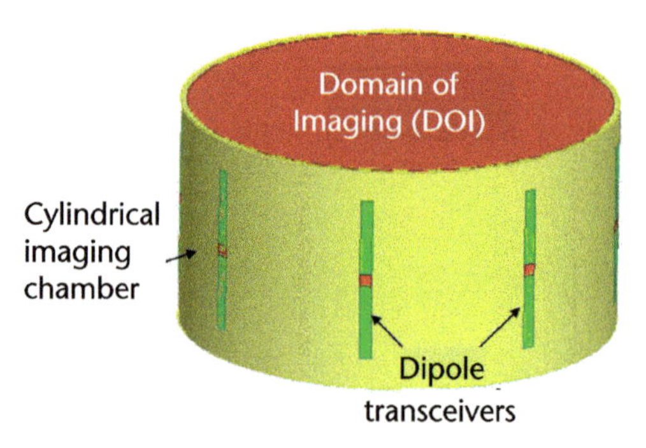

Figure 8.6 Geometrical configuration of the imaging domain. (©IEEE [37].)

preselected permittivities $e_1, e_2, ..., e_{N_{dielectric}}$ where $N_{dielectric}$ = the number of preselected permittivities. Obviously, as r's are fractions:

$$\sum_{t=1}^{N_{dielectric}} r_{t,n} = 1, \quad (0 \leq r_{t,n} \leq 1) \tag{8.11}$$

Notably, the permittivity values $e_1, e_2, ..., e_{N_{dielectric}}$ are selected a priori and the inverse problem involves finding the unknown fractions $r_{1,n}, r_{2,n}, ..., r_{N_{dielectric},n}$ for each pixel. Notably, this kind of a priori information is typically available (or at least can be estimated) for practical biomedical domains.

In conventional microwave imaging methods, the goal is to find $\{\varepsilon\}$ using the measured scattering data, $\{S\}_{meas}$. To do so, one proceeds to solve the following optimization problem [38]:

$$C(\{\varepsilon\}) = argmin_{\{\varepsilon\}} \frac{1}{2}(\| \{S\}_{meas} - F(\{\varepsilon\}) \|^2 + \mu \| [R](\{\varepsilon\} - \{\varepsilon^0\} \|^2) \tag{8.12}$$

where $\{S\}_{meas}$ is the measurement vector and $F(\{\varepsilon\})$ is the forward problem solution for a set of given $\{\varepsilon\}$. That is, $F(\{\varepsilon\}) = \{S\}_{calc}$ for the given $\{\varepsilon\}$ and $\|\cdot\|$ is the L_2 norm of a vector. The second term in (8.12) is referred to as the regularization functional with $[R]$ denoting an appropriate regularization matrix, μ is a chosen scalar controlling the level of regularization (user specified), and $\{\varepsilon^0\}$ is an assumed initial permittivity distribution.

Instead of solving for the desired complex permittivity $\{\varepsilon\}$, only the fraction parameters are solved. To do so, the optimization process is revised as follows:

$$C(\{\tilde{r}\}) = argmin_{\{r\}} \frac{1}{2}(\| \{S\}_{meas} - F(\{\tilde{r}\}) \|^2 + \mu \| [R](\{\tilde{r}\} - \{\widetilde{r^0}\} \|^2) \tag{8.13}$$

with

$$\{\tilde{r}\} = \{\tilde{r}\}_t \tag{8.14}$$

From (8.11), $\{\tilde{r}\}_t$ represents an $N_E \times 1$ vector of the fraction parameters for all the pixel elements in the DOI. For the sake of avoiding repetition, the subscript "t" will be omitted hereafter. As was the case of $\{\varepsilon^0\}$, $\{\widetilde{r^0}\}$ is any assumed initial value (typically set to the same real number for all the pixels) of the fraction parameters.

The optimization problem in (8.11) can be solved via the conventional Gauss-Newton method. As reported in [38], a fast implementation of the Gauss-Newton method can be carried out, avoiding the forward solution in each iteration, by introducing the Taylor series expansion:

$$F\left(\left\{\widetilde{r^k}\right\}\right) \approx F\left(\left\{\widetilde{r^0}\right\}\right) + [J_{\tilde{r}}]\left(\left\{\widetilde{r^k}\right\} - \left\{\widetilde{r^0}\right\}\right) \tag{8.15}$$

Here, $[J_{\tilde{r}}]$ is the $M \times N_E$ Jacobian matrix with the superscript "k" denoting the kth iteration. It is noted that unlike typical microwave imaging where the Jacobian matrix is calculated with respect to the permittivities [38], the Jacobian matrix is calculated for the fraction parameters.

By using the Gauss-Newton iterative method to solve (8.13),

$$\begin{aligned} \left\{r_t^{k+1}\right\} &= \left\{r_t^k\right\} - \alpha\left(B_t^k\right)^{-1} G_t^k, t = 1,2,\ldots,N_{dielectric} - 1, \\ \left\{r_{N_{dielectric}}^{\;\;k+1}\right\} &= 1 - \sum_{t=1}^{N_{dielectric}-1} \left\{r_t^k\right\} \end{aligned} \tag{8.16}$$

In the above, $\left[G_t^{\,k}\right]$ and $\left[B_t^{\,k}\right]$ are the gradient and Hessian matrices calculated from (8.13) with respect to r_t at the kth iteration and α is the step size. Notably, the last equation of (8.16) is due to the condition (8.11). Another critical condition enforced at each iteration is (8.11). This condition ensures the stability of the iterative algorithm. Notably, conventional algorithms used to reconstruct the permittivity do not typically employ such boundary checks.

The process of obtaining relative permittivity and conductivity from the reconstructed fraction parameters is explained in Figure 8.7. For this example, a set of three ($N_{dielectric}$ =3) known permittivities e_1, e_2, and e_3 (i.e., t = 1, 2, 3) to construct pixel permittivities in the DOI were used. Initially, $e_1 = 60 - j8.7$ (conductivity = 0.6 S/m), $e_2 = 20 - j2.88$ (conductivity = 0.2 S/m), and $e_3 = 40 - j5.75$ (conductivity = 0.4 S/m) was set. Hence, (8.13) needs to be solved for the unknown fraction parameters $\{\tilde{r}\}_1$, $\{\tilde{r}\}_2$ and $\{\tilde{r}\}_3$. Once $\{\tilde{r}\}_1$, $\{\tilde{r}\}_2$ and $\{\tilde{r}\}_3$ are reconstructed, as shown in Figure 8.7, they can be used to obtain the complex permittivity across the DOI from

$$e_1\{\tilde{r}\}_1 + e_2\{\tilde{r}\}_2 + e_3\{\tilde{r}\}_3 = \{\tilde{\varepsilon}\}_{re} - j\{\tilde{\varepsilon}\}_{im} \tag{8.17}$$

In the above, $\{\tilde{\varepsilon}\}_{re}$, $\{\tilde{\varepsilon}\}_{im}$ are $N_E \times 1$ vectors of the real and imaginary parts of the complex permittivity, respectively, for each pixel. In Figure 8.7, all the $N_E \times 1$ vectors $\{\tilde{r}\}_1$, $\{\tilde{r}\}_2$, $\{\tilde{r}\}_3$ and $\{\tilde{\varepsilon}\}_{re}$, $\{\tilde{\varepsilon}\}_{im}$ are plotted as 2-D images to better explain the concept.

It should be noted that the accuracy of reconstruction for both the relative permittivity and conductivity parts of the complex permittivity is excellent. This is due

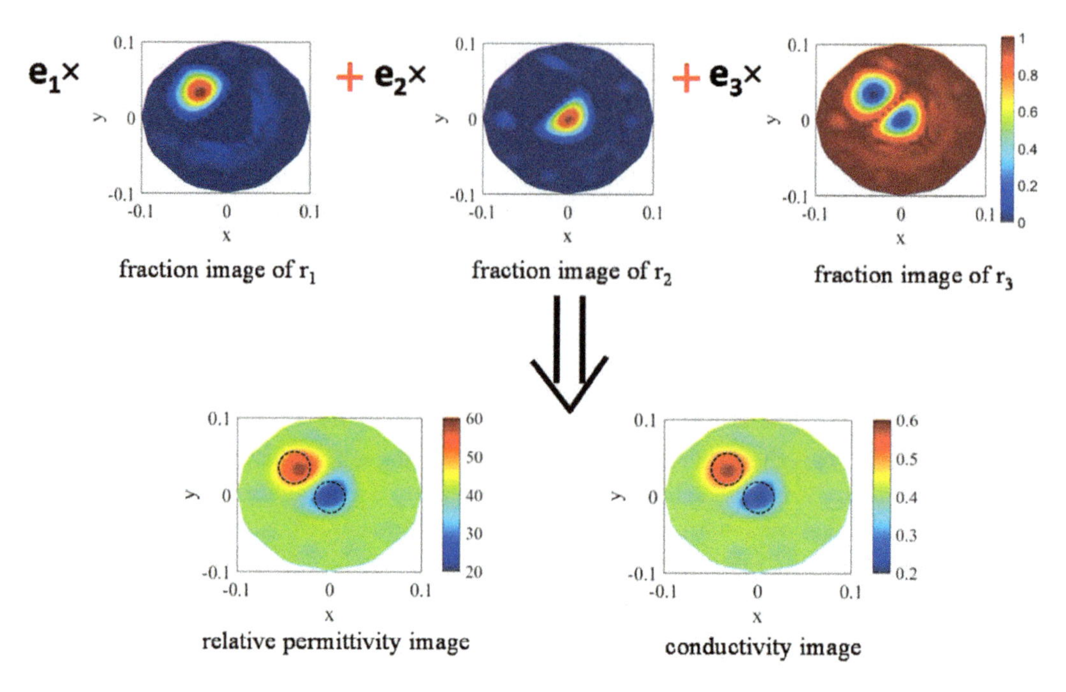

Figure 8.7 Image reconstruction process when three fraction parameters ($N_{\delta\iota\varepsilon\lambda\varepsilon\chi\tau\rho\iota\chi}$ =3) are used. The *x*- and *y*-axes are in meters [37].

to two key reasons: (1) the use of the preselected permittivities to better predict the domain's permittivity (notably, if any one of the preselected permittivity elements is nonexisting in the actual imaging domain, the fraction parameter of that element will be zero), and (2) the significant reduction of the search space for finding the fraction parameters, $\{\tilde{r}\}$, implying a more robust solution.

8.1.1.3 Employing Radar-Based Time Domain Approach

While the previous sections focused on frequency domain techniques, the time domain approaches based on radar are summarized here. Because this is a broadband technique, the tissue properties must be characterized over a wide range. According to the Cole-Cole dispersion model, the electrical characteristics of human tissue alter with frequency. The authors of [39] demonstrate the efficiency of this strategy at 12 GHz using a breast model. The important parameters for tissues described with the "Biological Property Definition" macro are shown in Table 8.4. The following Cole-Cole expressions are used to define the basic visual macro [39]:

$$\varepsilon(\omega) = \varepsilon_0 + \sum_{m=1}^{M} \frac{\Delta\varepsilon_m}{1 + (j\omega\tau_m)^{(1-\alpha_m)}} + \frac{\sigma_j}{j\omega\varepsilon_0} \tag{8.18}$$

where m is the order of the Cole-Cole dispersion model, ε_0 is the free-space permittivity (8.854 pF/m), and ω is the angular frequency. Table 8.4 lists the parameters for different types of tissues.

Table 8.4 Tissue Parameters for Single-Pole Cole-Cole Model

Tissue	(ε_0)	$(\Delta\varepsilon_m)$	τ_m *(ps)*	σ_i *(S/m)*
Fat, high	3.9870	7.5318	13.0000	0.0803
Fat, medium	3.1161	4.7077	13.0000	0.0496
Fat, low	2.8480	3.9521	13.0000	0.2514
Fibroglandular, high	14.2770	54.7922	13.0000	0.6381
Fibroglandular, medium	13.8053	49.3510	13.0000	0.7384
Fibroglandular, low	12.8485	37.4915	13.0000	0.2514
Skin	15.9300	39.7600	13.0000	0.8310
Malignant tumor	20.2800	45.5000	13.0000	1.3000

From: [39].

For evaluating the proposed methodology, the authors in [39] utilized a comparable measurement approach. More specifically, the antenna array was linked to a switch RF matrix controller, which was then linked to a VNA. The human tissue being investigated, such as the breast, receives a short pulse from a transmitting antenna. The other receiving antennas receive backscattering signals from human tissues. This process is repeated with the other antennas, and the data is then sent to a computational device, such as a personal computer (PC), to determine whether a tumor is present. For radar-based microwave imaging applications, computation often entails using simple and reliable methods such as the delay and sum algorithm. Furthermore, using multiantenna array units can improve the accuracy.

The proposed conformal antenna arrays for microwave breast imaging applications were validated in two configurations by the authors in [39]. The gaps between the antenna array and the breast phantom are filled with a dispersion-free coupling medium that has the same dielectric constant as the mixture mold. The proposed single-level antenna array covers the imported voxel-based breast model in Figure 8.8(a and c), while the other is a four-level antenna array with the dielectric spacer cover and coupling medium hidden. Using the CST Design Studio, the data collecting process was carried out in a simulated environment. For the two antenna array configurations, the total 36 and 576 signals in the time domain were subsequently processed using the delay and sum algorithm to produce pictures.

8.1.1.4 Monostatic Radar-Based Fidelity Factor

An on-body wearable antenna's EM radiation and performance analysis differs dramatically from that of other commercial free-space antennas. Furthermore, wearable on-body antennas, which are usually encased in lossy biological tissues, are often designed for near-field applications [40]. Before using wearable on-body antennas for biomedical imaging applications, it is critical to examine and analyze their strength and quality [40, 41].

Estimating the fidelity factor (FF), which determines the normalized cross-correlation coefficient between the antenna's input signal $T_s(t)$ and the received signal $R_s(t)$ [42], is one such method.

To do so, the authors in [40] used a Gaussian pulse signal $T_s(t)$ to excite the antenna, and E-field probes to detect the received E-field signal $R_s(t)$. The virtual

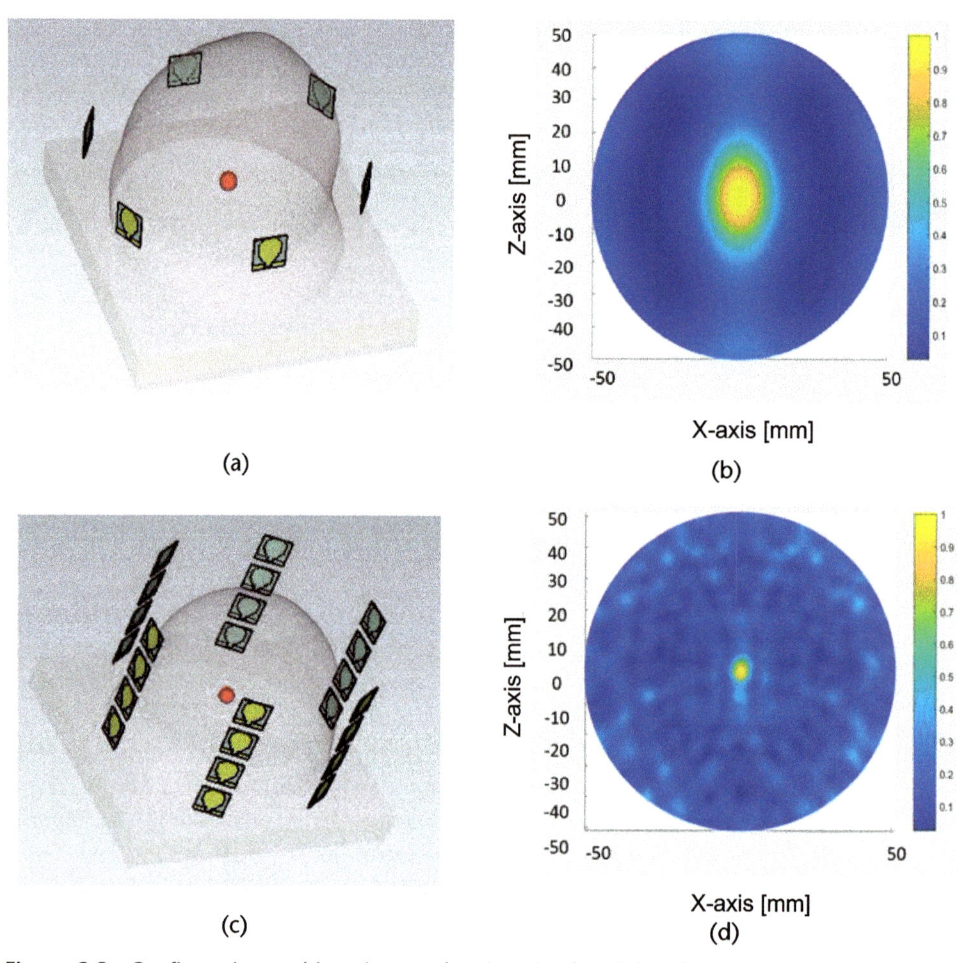

Figure 8.8 Configuration and imaging result using simulated data from CST (a) single-level antenna array mounting on breast model, (b) reconstructed image with single-level antenna array, (c) four-level antenna array mounted on breast model, and (d) reconstructed image with four-level antenna array. (©IEEE [39].)

probes' input Gaussian pulse signal $T_s(t)$ and receiving E-field signal $R_s(t)$ are then normalized as [42]:

$$T_s(t) = \frac{T_s(t)}{\sqrt{\int_{-\infty}^{\infty} \left|T_s(t)\right|^2 dt}} \tag{8.19}$$

and

$$R_s(t) = \frac{R_s(t)}{\sqrt{\int_{-\infty}^{\infty} \left|R_s(t)\right|^2 dt}} \tag{8.20}$$

The similarity between the two signals is estimated using the FF defined as [42]

$$FF = max\left|\int_{-\infty}^{\infty} T_s(t) R_s(t+\tau)dt\right| \tag{8.21}$$

The authors of [40] fabricated and tested an array of 13 antennas on a homogenous phantom to experimentally evaluate the suggested antenna's imaging capacity. An elliptical head phantom with a circumference of 52 cm, 13 antenna elements, and a high-performance VNA make up the system. The antenna array is placed directly on the head phantom, as shown in Figure 8.9.

After that, two measurements were taken: (1) with an empty imaging domain (healthy), and (2) with a target (unhealthy) that represented the abnormality within the imaging domain. As a result, the measured data with a target is calibrated by subtracting the data from the healthy domain, which eliminates all changes save those caused by the anomaly inside the imaging domain. The confocal multistatic delay-multiply-and-sum beamforming technique described in [43] is then utilized to reassemble the image. The time-domain delays from the dispersed signals are used in this approach to focus on places in the imaging domain where there is significant contrast in the dielectric characteristics. The contrast is caused by changes in the dielectric characteristics of normal and pathological tissues within the human cranium. The exact placements of the antenna elements surrounding the head phantom are used to calculate the picture domain's outer boundary. Figure 8.10 shows the results of the reconstructed images at various positions inside the head phantom.

Notably, the aforementioned five approaches use the same measurement setup, in which many antennas are connected to the VNA and one antenna transmits while the others receive. The approaches in Sections 8.1.1.1 and 8.1.1.2 are based on frequency domain response, while the radar-based approaches in Sections 8.1.1.3 and 8.1.1.4 are based on time-domain response.

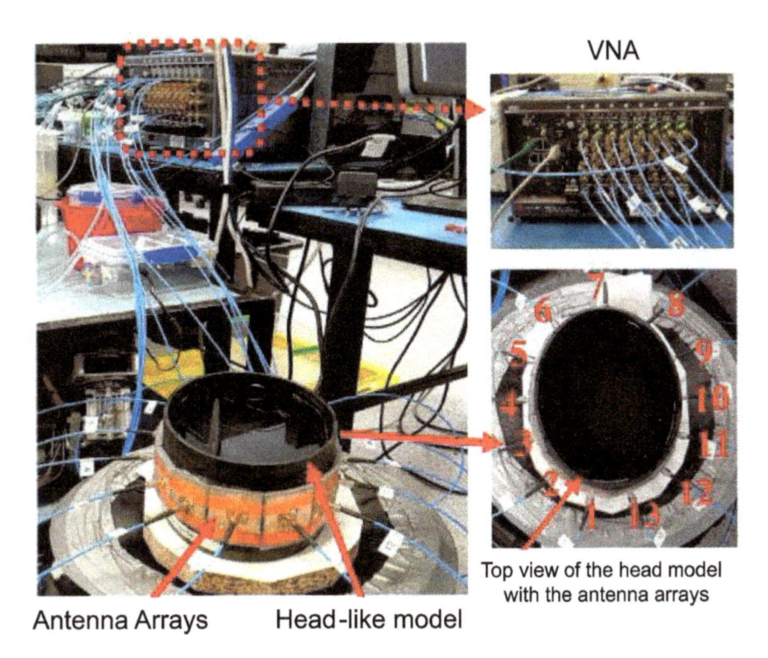

Figure 8.9 Experimental setup of the head imaging system. (©IEEE [40].)

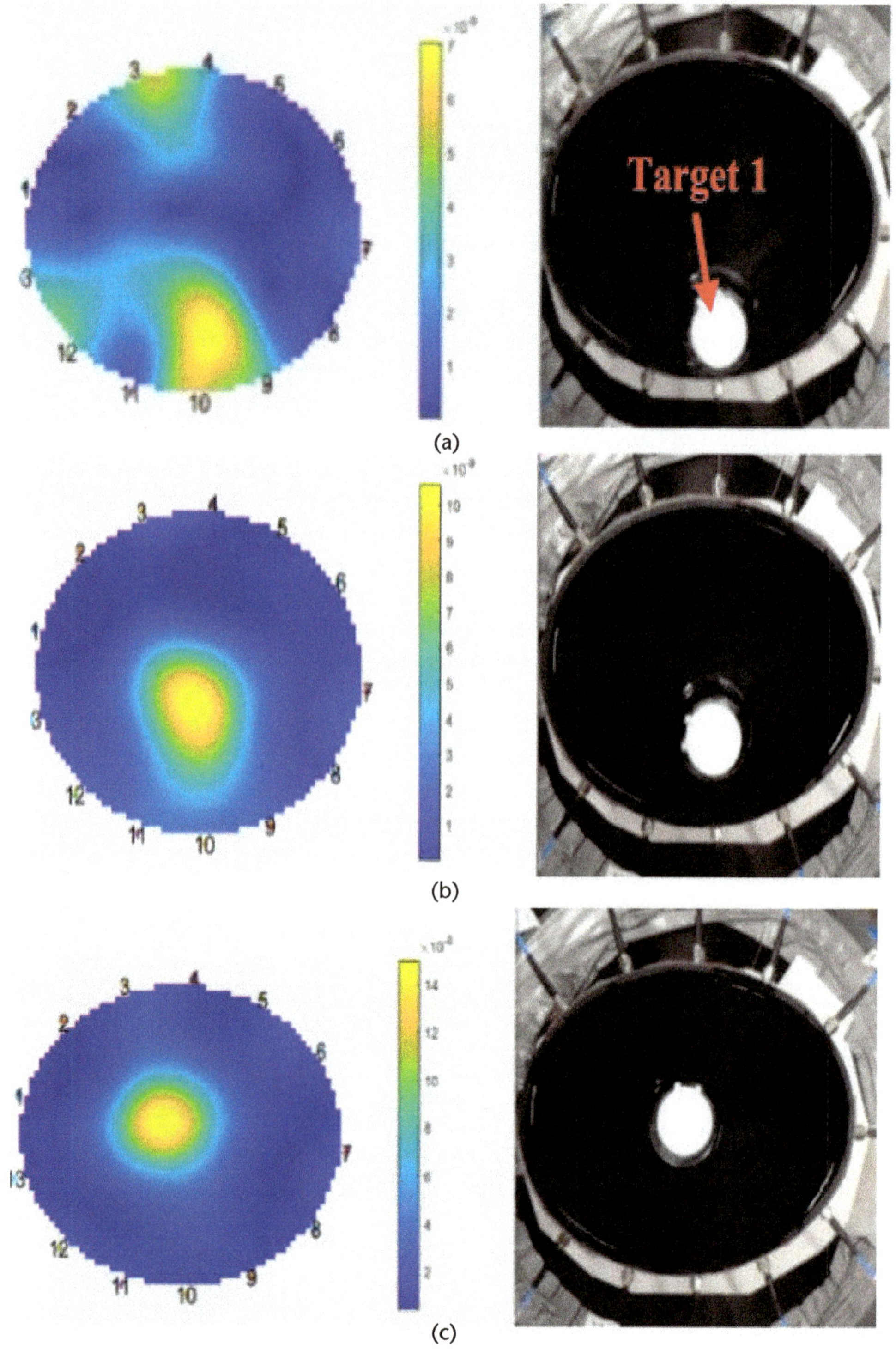

Figure 8.10 Reconstructed images for Target 1 at different locations inside the head phantom. (©IEEE [40].)

8.1.1.5 Electrical Impedance Tomography

Alternately, electrical impedance tomography has recently opened up numerous applications in health care monitoring. As mentioned by the authors in [44], one of the typical approaches used to image the conductivity distribution in electrical impedance tomography (EIT) is the Jacobian matrix, which discretizes the sensing domain inside the sensor into a mesh using the infinite element method (FEM) as a linearization of distinct conductivities. In practice, the Jacobian matrix is computed beforehand under the assumption of a uniform conductivity distribution inside the fixed boundary shape sensor [44].

The wearable EIT sensors used by the authors in [44] are shown in Figure 8.11. The true boundary shape Ω_{true} is represented by the black dotted line, the estimated standard boundary shape Ω_{std} is represented by the solid blue line, and the estimated flexible boundary shape $\Omega(t)$ is represented by the solid black line in Figure 8.12. The results show that the boundary shape estimated by $\Omega(t)$ is qualitatively more accurate than the boundary shape estimated by Ω_{std}.

The authors in [45] hypothesized that changes in the medium's impedance may be used to detect biological irregularities. When a cell is damaged or dies, the membrane structure and integrity are lost, enabling ions and current to pass through. As a result, damaged cells will have a higher electrical conductance through the membrane and a lower charge storage capacity [46]. In other words, the cell behaves more like a resistor than a capacitor. This impedance measurement shows up as a phase angle θ that is closer to zero (or equivalently, a smaller reactance, X), in vivo impedance measurements of tissues have also been linked to better tissue health: bioelectrical impedance analysis of patients at high risk of developing pressure sores revealed lower reactance and phase angle than the control group [47].

The authors of [45] conceived and built an electronic sensing device that uses a multiplexed electrode array to monitor spatially correlated complex impedance

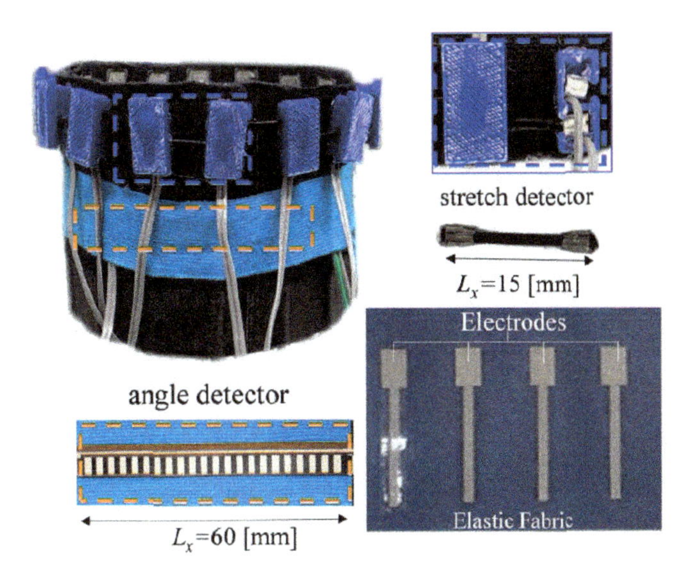

Figure 8.11 Experimental setup of the Wearable EIT. (©IEEE [44].)

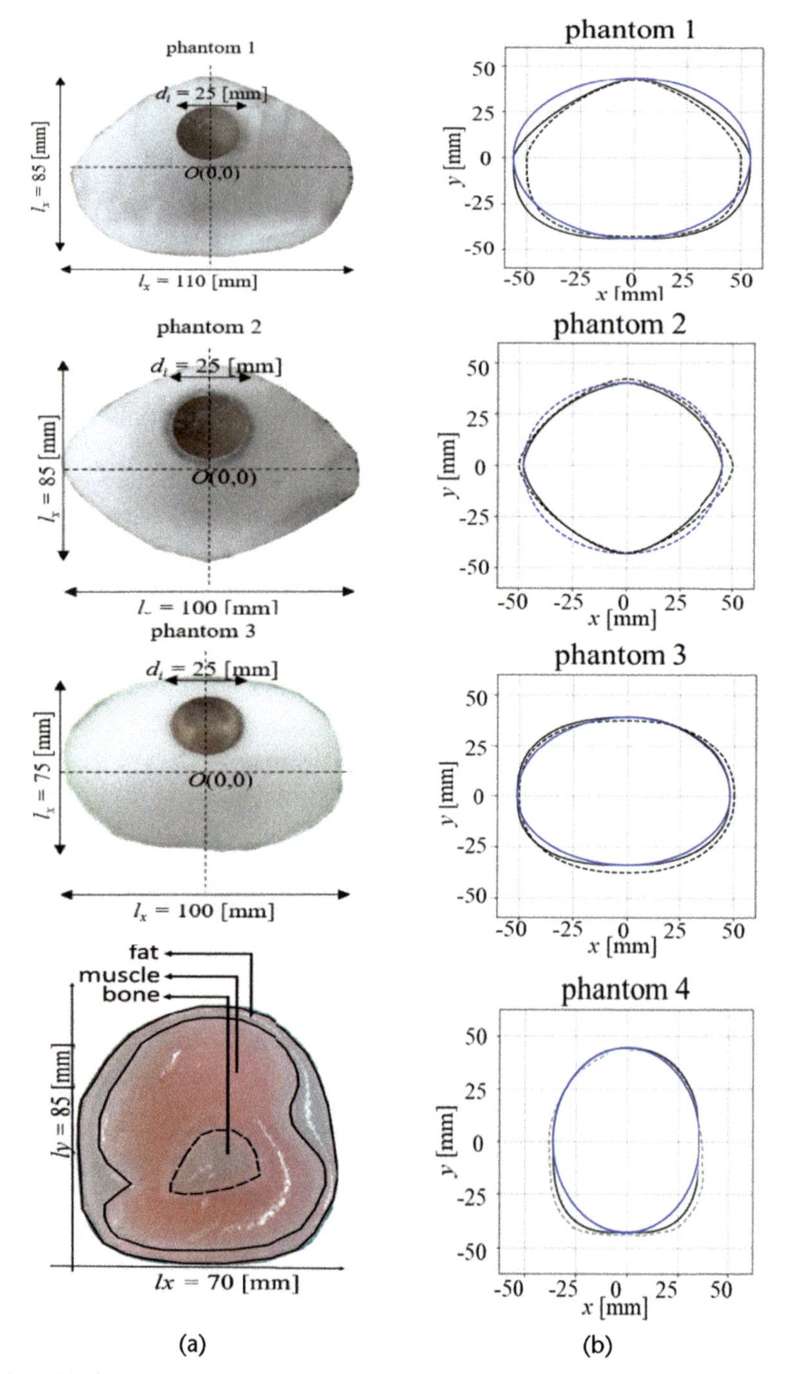

Figure 8.12 (a) Phantoms with various boundary shapes, and (b) comparison of the standard $\Omega_{\sigma\tau\alpha}$ (blue solid) and the novel boundary shape $\Omega(t)$ (black solid) estimation of true boundary shape $\Omega_{\tau\rho\upsilon\epsilon}$ (black dotted). (©IEEE [44].)

in vivo. An electrode array that contacts the skin and control gear that performs impedance spectroscopy across the array make up the device. The electrode array was constructed in two versions: (1) a commercial rigid printed circuit board with gold-plated electrodes, and (2) a flexible bandage-like array printed on a plastic substrate utilizing inkjet printing. The rigid gadget served as a stable calibration platform for detecting pressure ulcers with impedance spectroscopy while also developing the flexible device to use that method in vivo.

8.2 Ultrasound Imaging

Ultrasound imaging, unlike RF imaging, is based on the transducer's acoustic performance. Based on the ultrasonic detection technique, transducers can be classified as physical ultrasound transducers or optical ultrasound detectors. There are two types of physical transducers: (1) traditional piezoelectric, and (2) micromachined (capacitive or piezoelectric).

CMUTs have a better sensitivity, wider bandwidth, easier integration with electronics, and the flexibility to construct bigger arrays than piezoelectric transducers. The wideband nature of CMUT aids in picture resolution, while seamless interaction with electronics aids in overall image SNR preservation. Furthermore, employing CMUTs, fabrication of tiny transparent transducer arrays with desired shapes that are also portable and low cost is possible [2, 11, 18]. As a result, this section concentrates on CMUT-based ultrasonic imaging.

The efficiency of an ultrasound transducer is measured by the electromechanical coupling coefficient k. A large k in CMUTs results in increased transducer sensitivity, increased bandwidth, and hence greater image resolution. The efficiency of CMUTs has been reported to be as high as 82% [48].

The electromechanical coupling coefficient k characterizes the efficiency of an ultrasound transducer. In CMUTs, a high k leads to higher transducer sensitivity, improved bandwidth, and therefore improved image resolution. Efficiency levels reported for CMUTs can be as high as 82% [48]. The CMUT cell structure also contributes to the wide bandwidth; the thin membrane constricted at the rim has a low mechanical impedance, allowing for improved acoustic matching to the medium [49].

Small, compact, and portable CMUTS have been realized in a variety of ways [2, 11, 18, 50, 51]. In this section, we will look at two of these realizations. The authors in [2] created a probe that included frontend ICs and the CMUT, all of which were combined on a flexible PCB. The placement and assembly of these several components are critical to achieving the small factor size. Figure 8.13 depicts the probe and its components. The entire structure (Figure 8.13(b)) was then enclosed in a biocompatible lens made of a PDMS)to provide insulation and focusing in the elevation direction. Finally, to protect the traces from external interference, the flexible wire was electrically insulated with copper tape.

For picture reconstruction, the authors in [2] employed the traditional phased array technique [52]. The imaging code was created to capture, reconstruct, and display real-time images of a commercially available phantom (Gammex 406 LE, Gammex, Inc., Middleton, WI) up to a depth of 5 cm [2]. The phantom is made out of 0.1-mm nylon wires that are encased in tissue-mimicking gel. Finally, photos of the human neck, notably the jugular vein, were presented. The real-time neck pictures collected with the CMUT probe are shown in Figure 8.14. The pulsing of the vein was plainly visible at 30 frames per second.

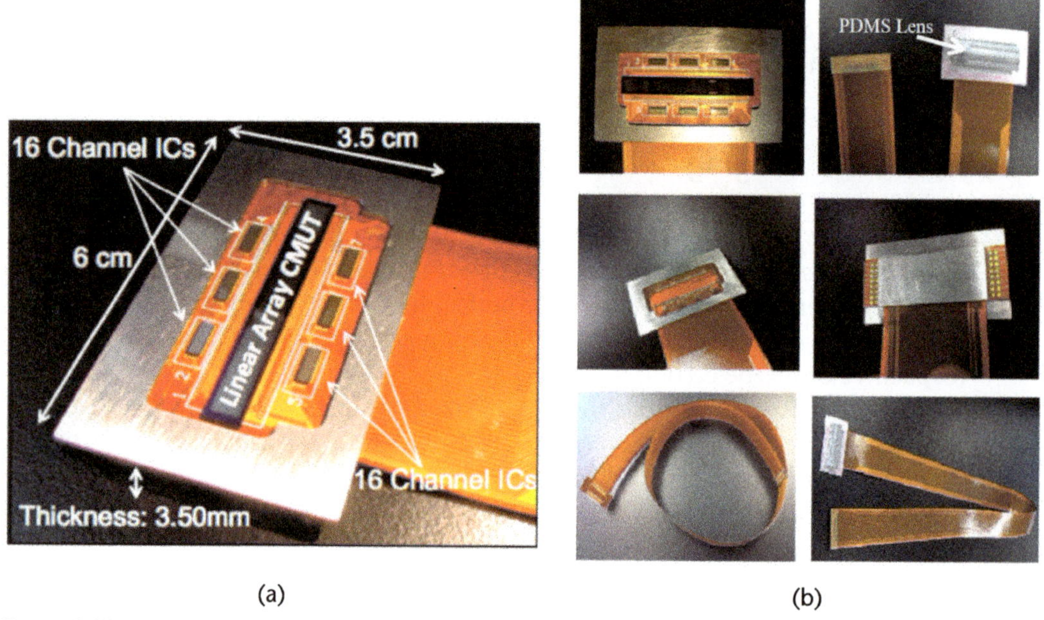

(a) (b)

Figure 8.13 (a) Integration of CMUT and ICs on flexible PCB, and (b) assembly photos of CMUT probe. (©IEEE [2].)

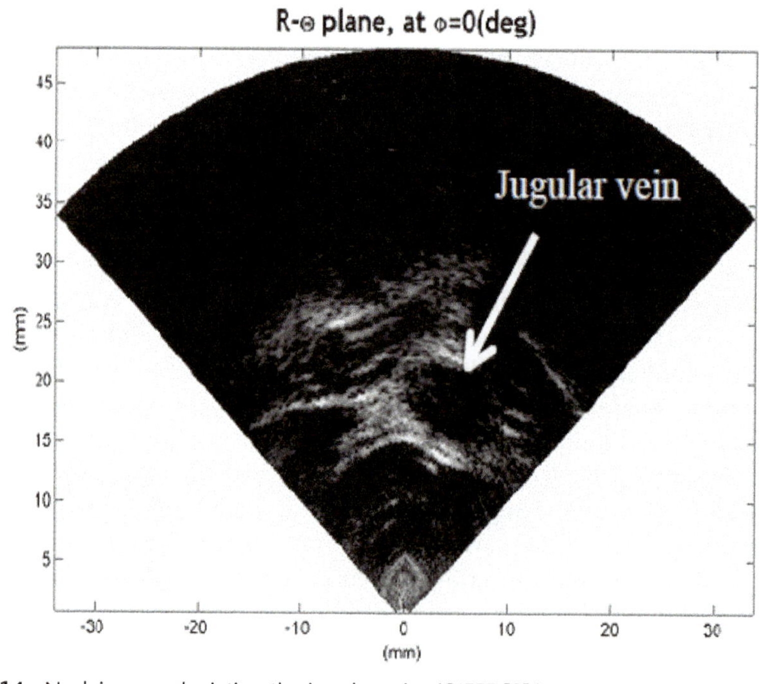

Figure 8.14 Neck images depicting the jugular vein. (©IEEE [2].)

Along these lines, investigators in [50] used two 128-element linear CMUT arrays for a carotid ultrasound, which detects plaque accumulation in these arteries. As seen in Figure 8.15, the device is designed like a neck brace. To monitor both

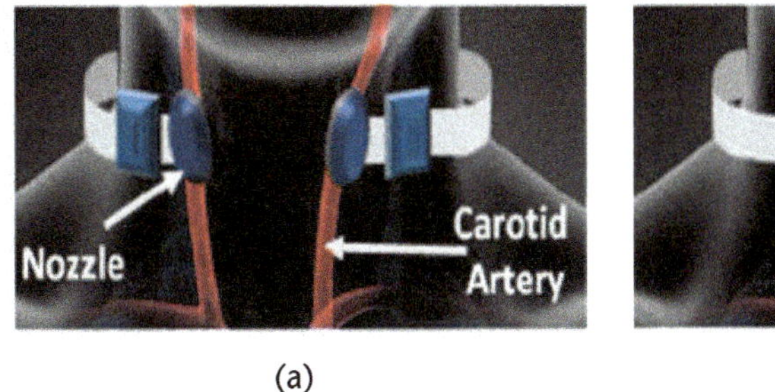

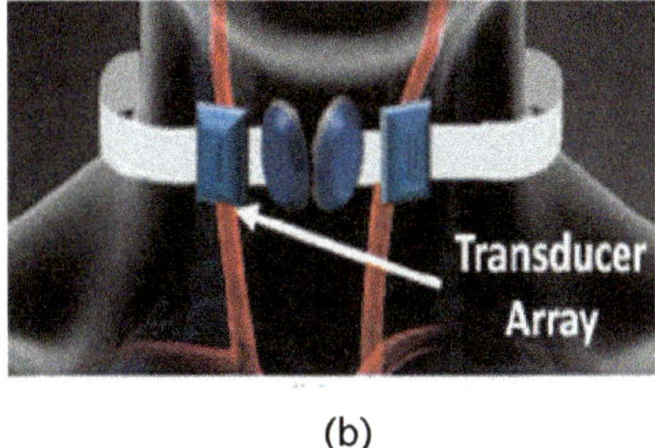

(a) (b)

Figure 8.15 (a) Nozzle placement while releasing ultrasound gel, and (b) transducer array placement on strategic locations during diagnosis. (©IEEE [50, 51].)

carotid arteries, two ultrasonic transducer array probes are strategically inserted on the brace, by the authors in [50]. The images are then evaluated to see if there are any abnormalities in the thickness of the carotid intima-media (CIMT). The intima-lumen and media-adventitia interfaces are where CIMT is assessed. The growth of plaque within the inner artery lining causes CIMT to rise over the typical range seen in asymptomatic people. When the gadget detects abnormally high CIMT levels, it gives a warning and advises the user to see a clinic or hospital for a comprehensive cardiovascular examination. As a result, it allows for early detection of CVDs.

Instead of blood flow measurements, the device suggested in [50, 51] uses CIMT and lumen diameter measurements as surrogate markers to predict cardiovascular disease. It employs B-mode ultrasonography, a widely accepted approach for CIMT measurements that can be carried out using small and low-cost equipment (see Figure 8.16).

8.3 Optical Tomography

Optical tomography (OT) is a form of computed tomography that creates a digital volumetric model of an object by constructing images made from light transmitted and scattered through an object. Optical tomography relies on the object under study being partially light-transmitting or translucent, so it works best on soft tissue, such as breast and brain tissue.

The high scatter-based attenuation is typically addressed by employing powerful, sometimes pulsed or intensity-modulated, light sources and high sensitivity light sensors, as well as the use of infrared light at frequencies where biological tissues are most transmissive. Because soft tissues scatter strongly but absorb weakly in the near-infrared and red sections of the spectrum, this is the wavelength range most commonly used.

OT can be classified into four types based on their principle of operation: (1) diffuse optical tomography, (2) time-of-flight diffuse optical tomography, (3) fluorescence molecular tomography, and (4) confocal diffuse tomography. In near-infrared (NIR) diffuse optical tomography (DOT), transmitted diffuse photons are collected and a diffusion equation is used to reconstruct an image from them.

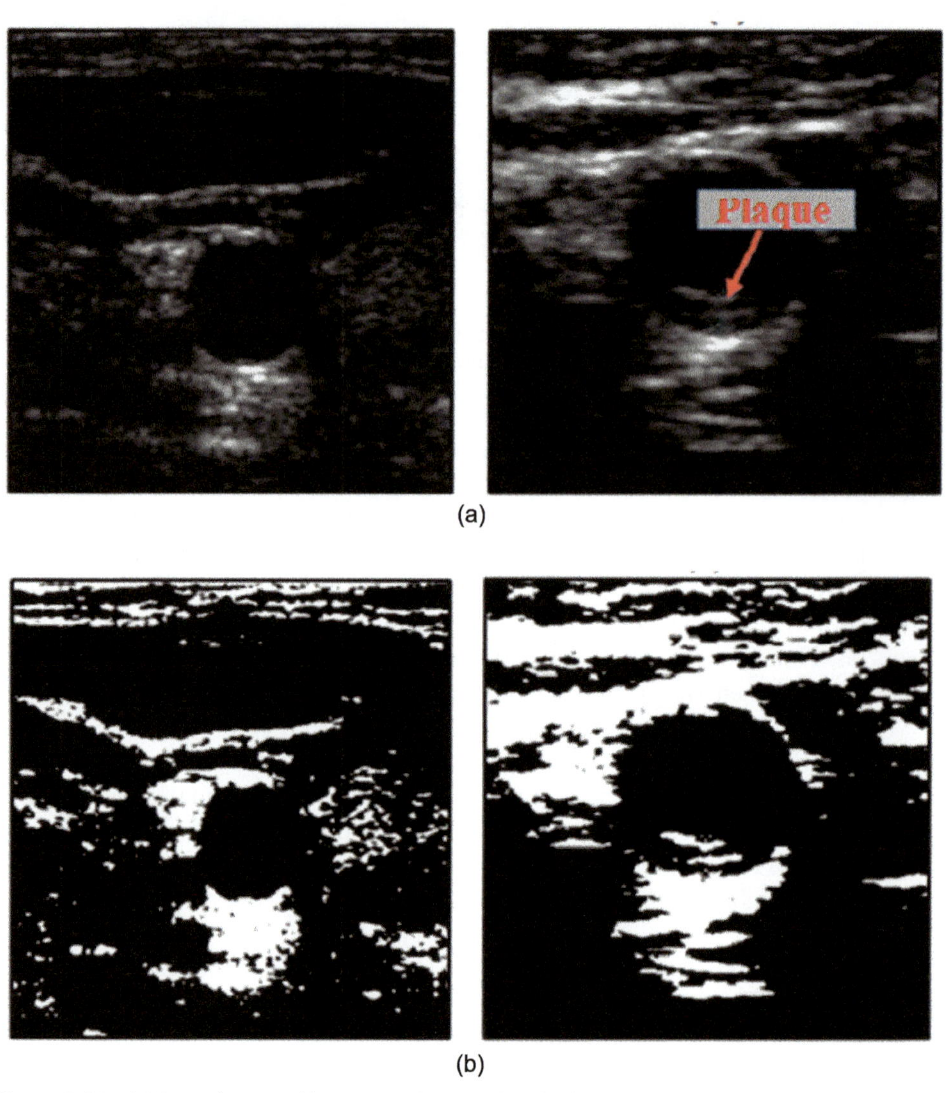

Figure 8.16 (a) Raw ultrasound images without and with plaque, (b) processed versions of raw images, (c) locations of carotid arteries, and (d) surface plots for detecting location of carotid arteries. (©IEEE [50, 51].)

A variant of optical tomography uses optical time-of-flight sampling to distinguish transmitted light from scattered light [53] for breast cancer imaging and cerebral measurement. The utilization of either time-resolved or frequency domain data, which is then paired with a diffusion theory-based estimate of how light traveled through the tissue, is the key to distinguishing absorption from scatter. The measurement of time-of-flight or frequency domain phase shift is essential to allow separation of absorption from scatter with reasonable accuracy. In fluorescence molecular tomography, the fluorescence signal transmitted through the tissue is normalized by the excitation signal transmitted through the tissue, thereby not requiring the use of time-resolved or frequency domain data. Since the applications of fluorescent molecules in humans are fairly limited, most of the work in fluorescence tomography has been in the realm of preclinical cancer research and tracking tumor proteins. Confocal diffuse tomography uses a powerful laser to illuminate a

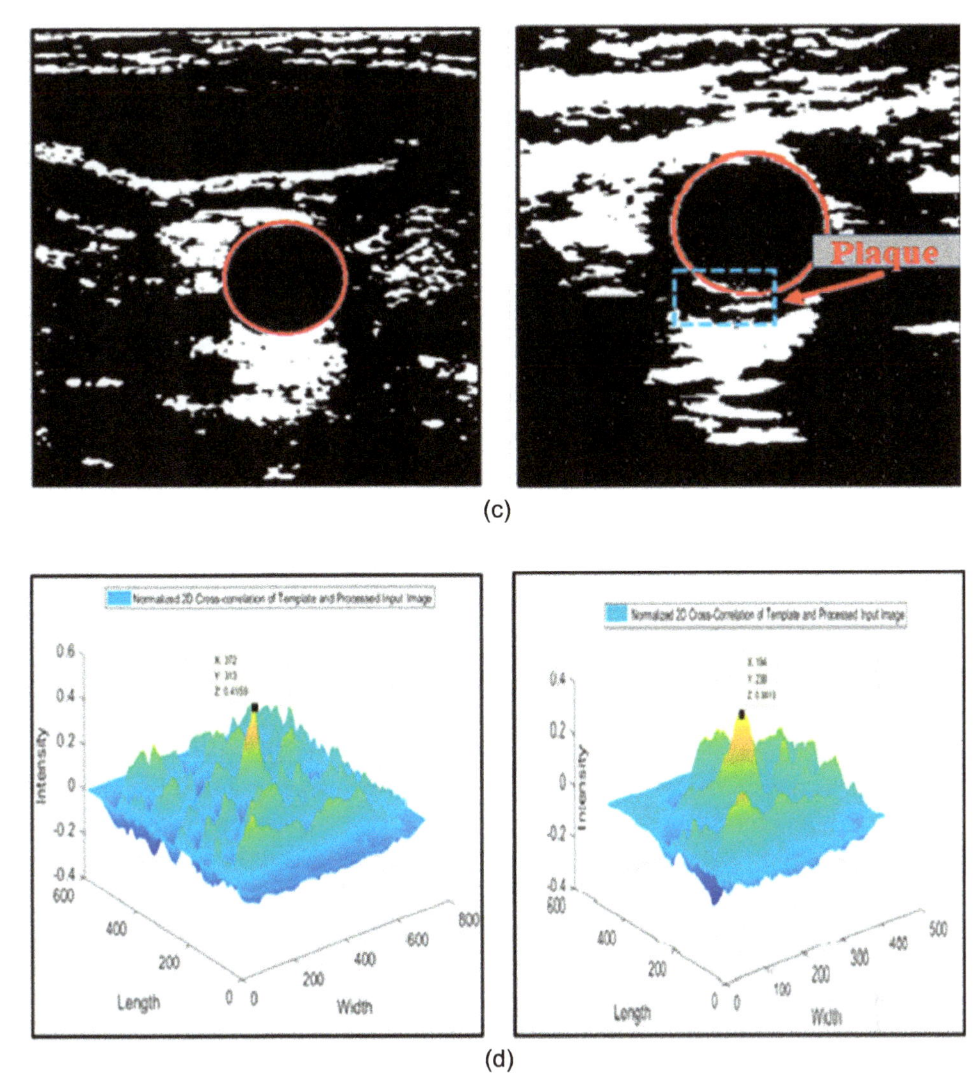

Figure 8.16 (continued)

sample through a scattering medium, followed by deconvolution with a calibrated diffusion operator to estimate a volume without the effects of diffusive scattering and subsequent application of a confocal inverse filter to recover the sample image [54].

In this section, we present examples of only DOT. For instance, researchers in [19] presented a portable diffuse optical NIR imaging system that allows multichannel brain imaging in freely moving subjects and lends itself readily to general-purpose large-area imaging of brain activity. Notably, fNIRS is an effective and noninvasive neuroimaging tool for monitoring oxygenation and cerebral hemodynamics.

A typical wearable fNIRS instrument utilizes dual-wavelength LEDs for direct skin illumination [19]. Time-multiplexing of the source positions with simultaneous frequency-encoded dual-wavelength illumination is employed. Each LED has two emitters with spectral half-widths of 25 and 30 nm, respectively, at 760 and

850 nm. During the "on" state, the emitters are intensity-modulated at 1.0 and 1.1 kHz and emit around 10 mW of average optical power. Optical detection is achieved using photoelectrical receivers, containing a silicon photodiode ((SiPD), BPW34, Siemens, Germany) followed by a transimpedance amplifier.

In [19], optodes were set in two groups (four sources and four detectors each) centered at different positions to cover the principal motor regions of both hemispheres, resulting in 20 NIRS measurement channels for each of the two wavelengths. Notably, a bandage and a dark woolen cap worn over the optical probes served to improve probe-tissue-contact, shield ambient light, and provide stability against motion artifacts. To estimate the signal-to-noise performance of a data channel, the relative coefficient of variation (CV, in %) was calculated for the unfiltered raw data (intensity changes), which is a standard procedure for multichannel NIRS measurements [19, 55, 56]:

$$CV = \frac{\sigma}{\mu}.100\% \tag{8.22}$$

Here, σ is the temporal standard deviation for a data channel, and μ is the corresponding mean value. Possible sources of reduced signal-to-noise ratio (= increased CV) due to physical exercise include physical artifacts such as motion-induced instabilities of the coupling efficiency at the tissue-optical interfaces as well as physiological artifacts such as blood-pressure–induced hemodynamics. The hemodynamic signals of the significant channel show the prototypical increase in ΔhbO_2 and decrease in ΔHbR in response to neuronal activation throughout all conditions.

In similar work, the authors in [57, 58] developed a wearable optical topography (WOT) to investigate cognitive functions when a subject performs a cognitive task while moving. It is essential to evaluate task-related changes in measurement signals without artifacts related to the subject's physical movement. For this examination, the subjects in the study performed an attention-demanding (AD) task that was designed to simplify human behavior in daily life, such as fully carrying something while walking [57, 58].

The WOT system was designed to be tiny, light, and comfortable to wear. A probe unit, a processing unit, and the WOT's computer make up the system. The probe unit has a 2 × 8 alternating arrangement of irradiation and detection positions covering the entire forehead, with 22 measurement points. This arrangement enables monitoring of the cortical activations mainly in the dorsolateral prefrontal cortex (DLPFC) and the rostral prefrontal area. The probe unit has a flexible pad with irradiation and detection positions to fit the subject's head. Vertical-cavity surface-emitting laser diodes with two wavelengths—790 and 850 nm—were used as light sources to emit light onto the scalp at each irradiation position, and silicon photodiodes were located at each detection position to detect multiply scattered light in the biological tissue. In [57, 58], the authors presented the mean estimate for the location of measurement points for 10 volunteers, and its associated variation of oxy-Hb and deoxy-Hb signals ($\Delta C'_{oxy}$ and $\Delta C'_{deoxy}$, respectively) obtained from channel 8 over a course of time as a moving-averaged data for 5s. During the task condition periods, $\Delta C'_{oxy}$ increased, and $\Delta C'_{deoxy}$ decreased slightly, which shows task-related and reproducible changes of $\Delta C'_{oxy}$.

8.4 Photoacoustics Imaging

Photoacoustic imaging is another technique that directly benefits from ultrasound transducer technology (PAI). PAI is a new imaging technique that combines optical excitation and acoustic detection to visualize vascular, functional, and molecular changes in living tissue [11]. As opposed to the optical imaging modalities such as optical coherence tomography [12] that employs ballistic photons, PAI uses diffused photons providing significantly deeper penetration. When tissue chromophores are treated with a nanosecond-pulsed laser, thermoelastic expansion occurs, resulting in the production of acoustic waves, which are subsequently detected by ultrasonic transducers for image generation. The need for an efficient and effective ultrasonic detection method is emphasized by the smaller generated pressure range of the waves in PAI compared to ultrasound imaging [59]. The detection unit in PAI must have a high sensitivity and a large acceptance angle over a wide range of spectral bandwidth since the optically induced ultrasonic pressure is inherently weak.

The commonly utilized piezoelectric ultrasonic transducers' large and optically opaque construction causes technical issues in several biological applications. Optical ultrasound detection methods offer a technique that could be a potential solution. To detect incident elastic waves, this approach uses high-finesse optical resonators. Optical ultrasonic detection techniques can be divided into two categories based on their configurations and detecting parameters: (1) interferometric method, and (2) refractometric approach.

Interferometric detection can be realized using Michelson interferometry (MI) [61–62], Mach-Zehnder interferometry (MZI) [63, 64], Doppler [65, 66], or resonator [67–69]. A two-beam approach is used in MI or MZI, in which a laser beam is split into two optical channels, one of which is disrupted by the ultrasonic wave and the other acts as a reference (see Figure 8.17(a)). The received pressure waves affect the optical path, causing proportionate variations in the intensity of the beam at the interferometer output. In contrast to two-beam interferometers, the Doppler method captures ultrasound waves by detecting Doppler shift (see Figure 8.17(b)). In resonator-based techniques, a micron-scale optical resonator detects ultrasound waves (see Figure 8.17(d)). Fabry–Pérot (FP) interferometers, microring resonators (MRRs), and π-phase-shifted fiber Bragg gratings (π-FBGs) are the most common optical resonator geometries utilized in photoacoustic imaging.

There are three types of refractometric methods: intensity sensitive, beam deflectometry, and phase sensitive [60]. When ultrasonic waves flow through the interface of two mediums with differing refractive indices, the intensity of the beam incident on that contact varies in the intensity sensitive method (see Figure 8.18(a)). In the beam deflectometry method (see Figure 8.18(b)), the interaction of the received acoustic waves with the medium modifies the refractive index of the medium, which in turn deflects the probe beam that is eventually detected using a position-sensitive detector such as a quadrant photodiode [60]. In phase-sensitive method, a collimated light beam passes through an acoustic field; the beam is deflected from

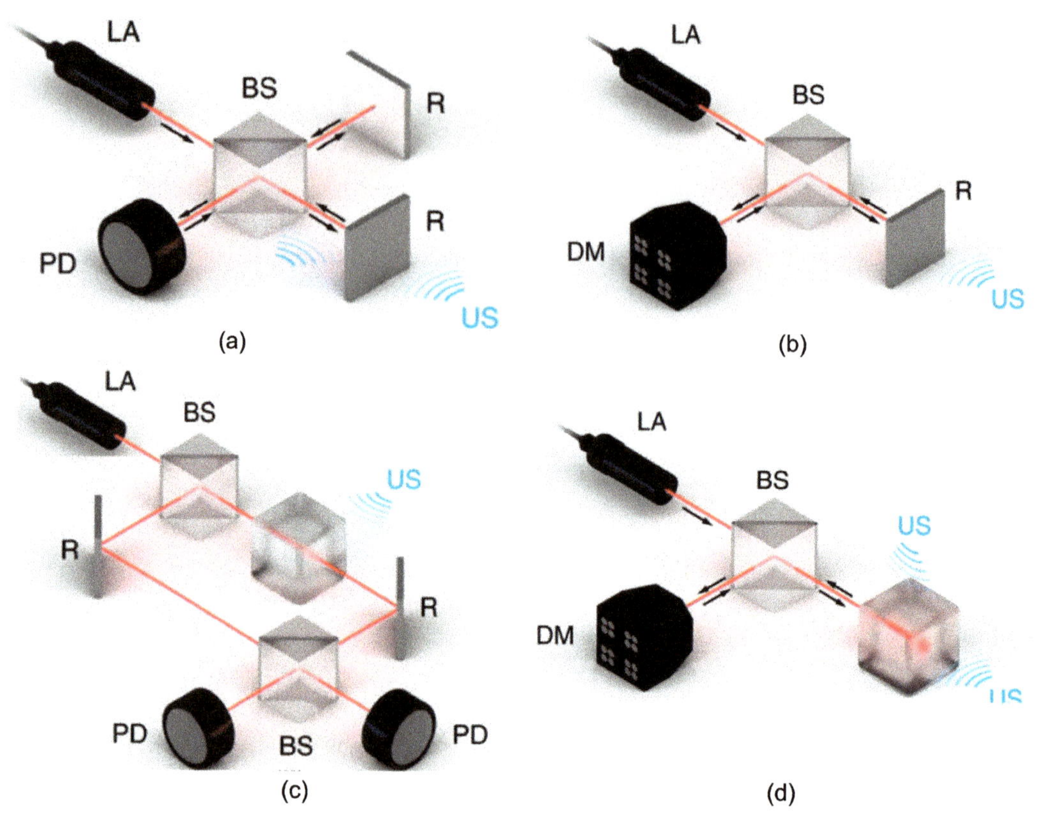

Figure 8.17 Interferometric-based optical ultrasound detection techniques: (a) Michelson, (b) Mach-Zehnder, (c) Doppler-based sensing, and (d) resonator-based sensing. AL: acoustic lens, BS: beam splitter, D: detector, DM: demodulator, US: ultrasound, LA: laser, R: reflector. ([11], reused under creative commons CC BY license, 60.)

the original path and perturbed, and this beam is then focused through a spatial filter (see Figure 8.18(c)). The resultant beam is collimated and detected by a charge-coupled device (CCD) or complementary metal-oxide-semiconductor (CMOS) camera; the image produced by the camera is the intensity map of the acoustic field.

Photoacoustic imaging, a multiwave imaging technique, has been shown to be an effective technique in the early diagnosis of cancer [20]. The human vascular structure and other photoabsorbers (e.g., melanin) can be photoacoustically seen in a noninvasive manner by illuminating the ROI with a high-energy laser pulse and then measuring the produced ultrasonic wave.

The majority of photoacoustic systems are implemented in this regime due to the commonly used intermediate penetration depth achieved by IR light, increased sensitivity of IR devices, and the rapid improvement of IR optical components. Both preclinical and clinical applications of infrared photoacoustic technology have proved effective.

A miniaturized photoacoustic 3-D imager for superficial medical imaging is presented by [1]. By employing the compact continuous-wave laser diode based optical irradiation and an ultrathin 2-D matrix array based photoacoustic detection in the coherent frequency domain, a wearable imaging probe with a size of about $80 \times 25 \times 24$ mm^3 and a weight of 21 grams was developed by the authors

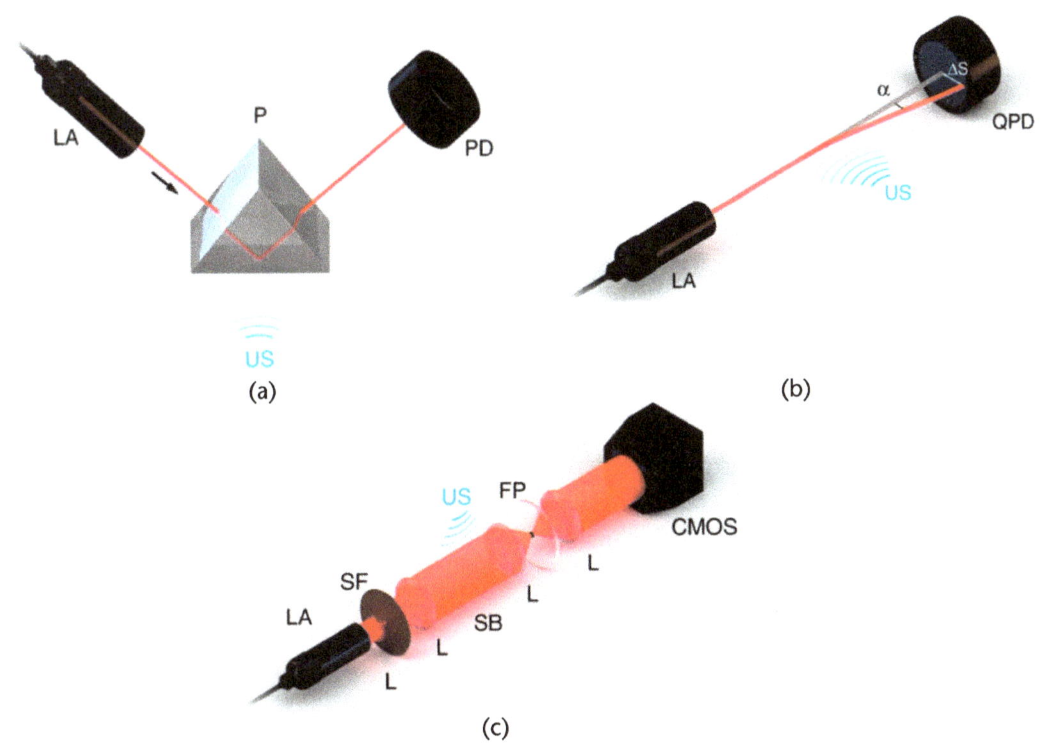

Figure 8.18 Refractometric-based optical ultrasound detection techniques: (a) intensity-sensitive detection of refractive index, (b) single-beam reflectometry, and (c) phase-sensitive ultrasound detection. CMOS: complementary metal-oxide-semiconductor, FP: Fourier plane, L: lense, LA: laser, P: prism, PD: photo diode, QPD: quadrant photodiode, SB: Schlieren beam, SF: spatial filter, US: ultrasound. ([11], reused under creative commons CC BY licenese, 60.)

in [1], as shown in Figure 8.19. A current source controlled by an FPGA drives the laser diodes in the backend to excite linear frequency modulated optical irradiation. The produced photoacoustic responses are first compressed with the coherent frequency domain photoacoustic approach and then extrapolated in the wavenumber-frequency domain for quick image reconstruction using data from a portable multichannel data-gathering device. Multispectral photoacoustic imaging can be achieved using three-wavelength (450, 638, and 808 nm) laser irradiation, giving the developed imager functional imaging capability in 3-D space.

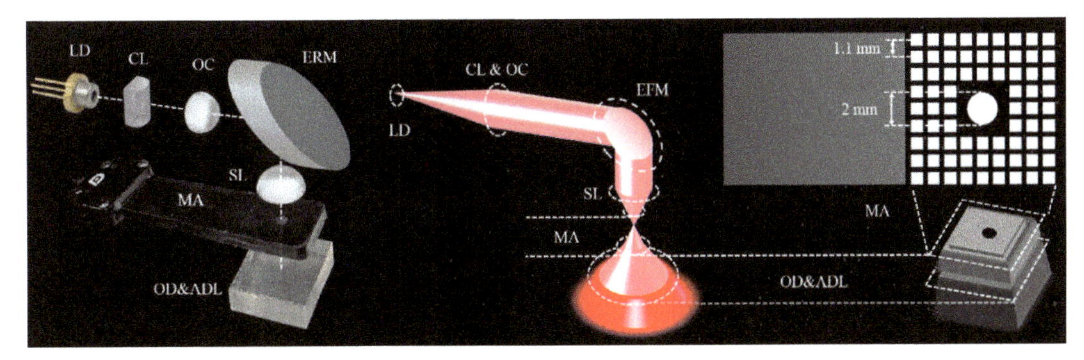

Figure 8.19 Photoacoustic imager with its schematics. (©IEEE [1].)

Complete data processing and image reconstruction employed by the authors in [1] are accomplished in three steps. First, cross-correlation (CCR) based pulse compression is performed using the Fourier transform of the $(m, n)^{th}$ acoustic signal $S_{m,n}(\omega)$ and the Fourier transform of the linear frequency modulated (LFM) waveform $R(\omega)$, as:

$$\begin{aligned} ccr(x,y,t-\tau) &= \frac{1}{2\pi}\int EB(\omega)R(\omega)S^{*}_{m,n}(\omega)e^{i\omega(t-\tau)}d\omega \\ m &= 1,2,\ldots,9 \\ n &= 1,2,\ldots,9 \end{aligned} \tag{8.23}$$

where $ccr(x,y,t-\tau)$ is the CCR function of the raw acoustic signal $S_{m,n}(t)$ and $EB(\omega)$ is the exact-Blackman window for noise suppression. Second, after proper data truncation and volumetric interpolation, field extrapolation is performed in the wavenumber-frequency domain at different imaging depths:

$$CCR(k_x,k_y,z,\omega) = e^{ik_z z}\iiint ccr(x,y,t)e^{-ik_x x}e^{-ik_y y}e^{i\omega t}dxdydt \tag{8.24}$$

where k_x and k_y are the wavenumber in the lateral directions, and k_ς is the wavenumber in depth direction and can be calculated as $k_z = ((\omega / c)^2 - k_x^2 - k_y^2)^{-1/2}$. Finally, extrapolated wave field is transformed back into the space domain by first integration over ω, followed by inverse Fourier transform over k_x and k_y, as:

$$p(x,y,z) = \iiint CCR(k_x,k_y,z,\omega)e^{ik_x x}e^{ik_y y}dk_x dk_y d\omega \tag{8.25}$$

The second and third steps sequentially compose the fast phase shift migration (PSM) method for image reconstruction over the imaging ROI. Overall reconstruction time for a $10 \times 10 \times 10$ mm^3 data structure is less than 0.05 seconds.

For testing the proposed wearable imaging probes, the authors of [1] gently strapped the probes on the forearm with two soft ligatures (Figure 8.20(a)). By simply substituting the diode tubes, photoacoustic images at different wavelengths can be obtained, as shown in Figure 8.20(b, c, and d). Oxyhemoglobin and deoxyhemoglobin have nearly equal molar extinction coefficients at 450 and 808 nm (around the isosbestic point). As a result, both superficial venous and arterial blood vessels emit visible photoacoustic emission, as seen in Figures 8.20(b, d). These photoacoustically reconstructed images match the anatomic structure of the arm vasculature quite well (Figure 8.20e). Figure 8.20(f, g) illustrate the contour images of 35% and 72% SO_2 levels, which correspond to the outline of the cephalic vein and the radial artery, respectively. Because the wearable imager is securely fastened to the forearm, long-term monitoring of SO_2 is possible. Figure 8.20(h) depicts the photoacoustically measured SO_2 levels at two specified points in the superficial cephalic vein and radial artery, respectively.

Alternatively, matched filtering approach can be employed in place of CCR to recover the image. In the presence of strong directly coupled optoacoustic signal (scattered light-induced simultaneously optoacoustic generation at the transducer aperture surface [1, 71, 72]) and the background noise (including electronic noise and acoustic noise), the effective optoacoustic wave is submerged and cannot be

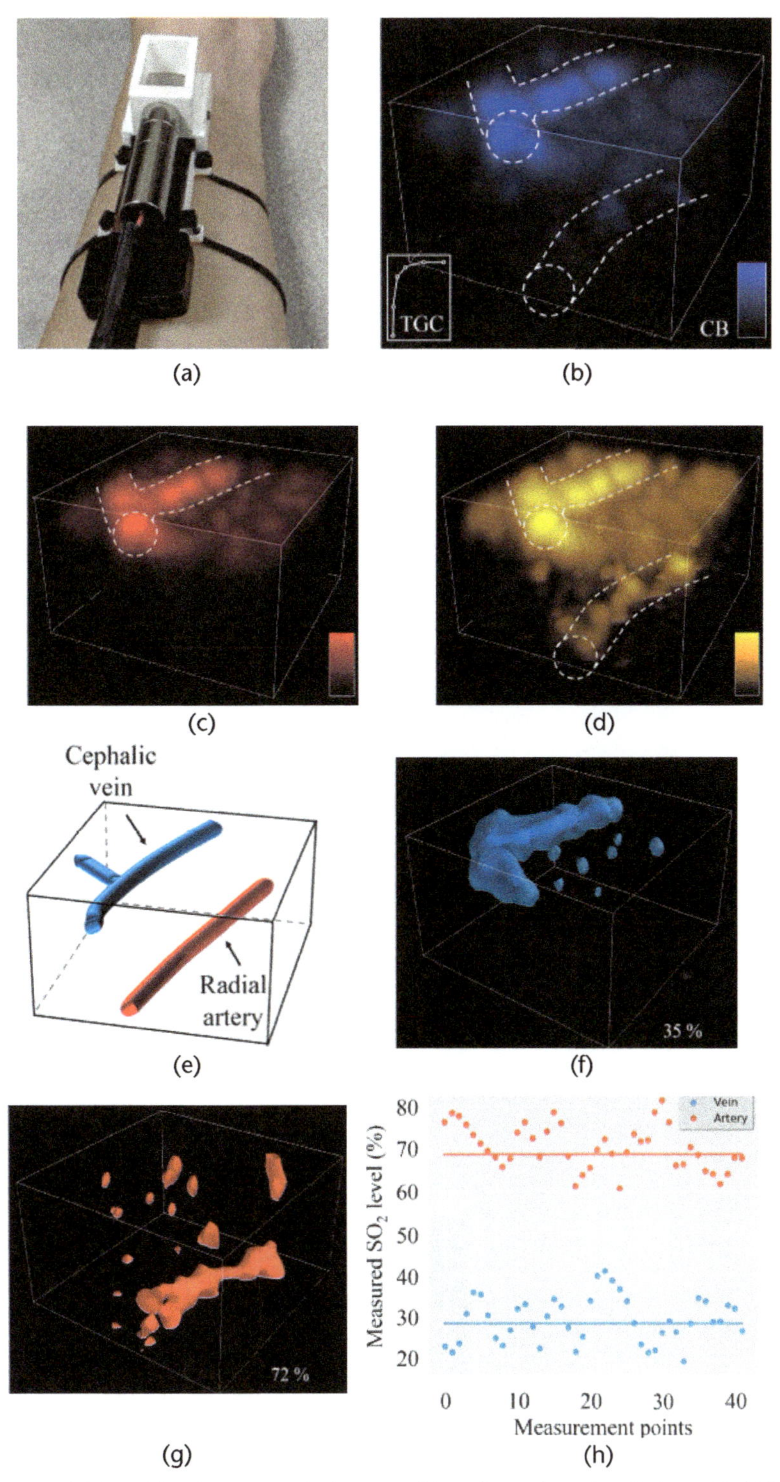

Figure 8.20 (a) Wearable photoacoustic imager, (b) 450-nm laser irradiation, (c) 638-nm laser irradiation, (d) 808-nm laser irradiation, (e) arm vasculature distribution, (f) contour 3-D image of 35% SO_2 level, (g) contour 3-D image of 72% SO_2 level, and (h) measured average SO_2 level in artery and vein. (©IEEE [1].)

directly observed, even after signal average. Therefore, matched filtering based pulse suppression is performed on the collected acoustic signal from the $(i, j)^{th}$ piezoelectric element with the reference chirp waveform $R(t)$:

$$pi,j(t-\tau)=\frac{1}{2\pi}\int R(\omega)S_{i,j}^{*}(\omega)e^{i\omega(t-\tau)}d\omega \tag{8.26}$$

With the uniform backprojection algorithm in a planar plane [71, 73], the source distribution $P(x, y, z)$ can be retrieved as:

$$p(x,y,z)=\sum_{i=1}^{n}\sum_{j=1}^{n}\omega(i,j)P_{i,j}\left(\frac{\sqrt{Z^2+(x-x_i)^2+(y-y_i)^2}}{c}\right) \tag{8.27}$$

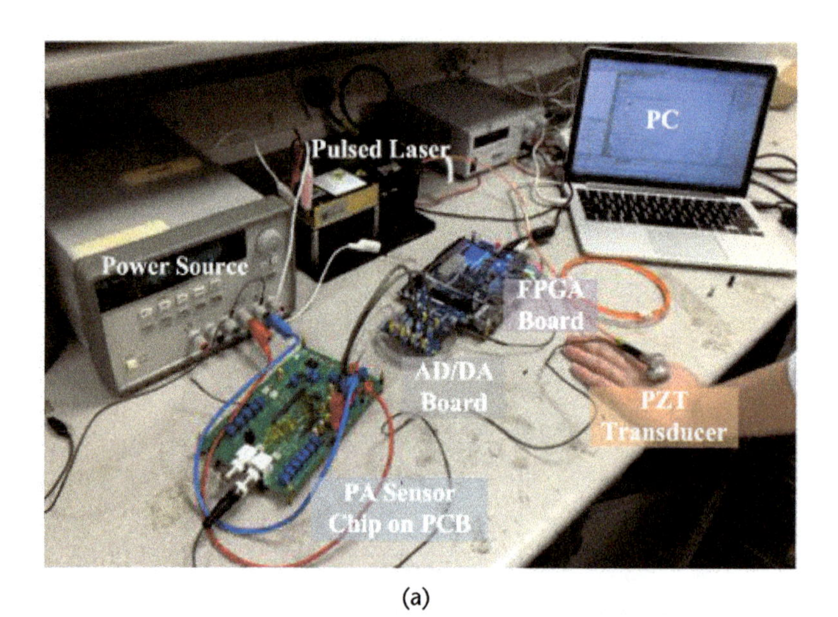

(a)

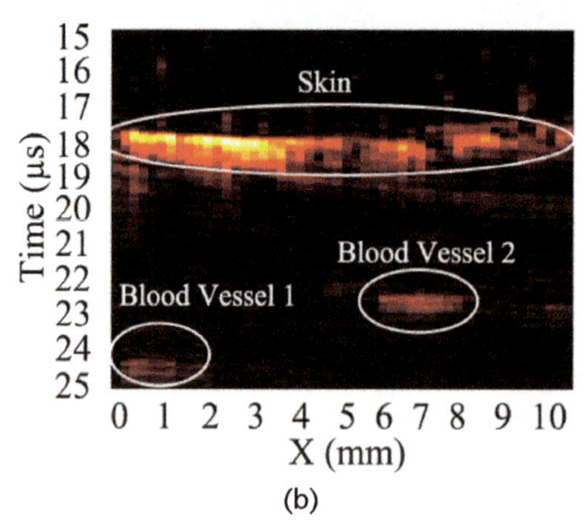

(b)

Figure 8.21 (a) Measurement setup, and (b) in vivo vessel imaging. (©IEEE [70].)

where $\omega(i,j)$ is the weighting factor for optimized image reconstruction, $(x, y, 0)$ is the position of the $(i,j)^{th}$ piezoelectric element and (x, y, z) is the position of performing the backprojection reconstruction.

In [70], the authors developed a photoacoustic (PA) sensor that is compact in size, lightweight, and has low power consumption, which is ideal to implement on wearable medical devices for in vivo blood metabolic sensing and imaging. A reconstructed image based on the developed prototype is shown in Figure 8.21.

Another aspect of photoacoustic imaging involves photoacoustic microscopy. Optical resolution photoacoustic microscopy (ORPAM) is a new imaging technology that has been used to investigate various brain activity and pathologies in anesthetized/restricted mice, with a focus on morphological and functional visualization of cerebral cortex. ORPAM is a noninvasive and label-free imaging method capable of monitoring brain hemodynamics at the microscale level [74]. It has excellent spatial resolution, deep penetration depth, and a broad field of view.

The authors in [74] enhanced the contrast of blood vessels by using an adaptive histogram equalization (AHE) algorithm, sharpened the edge of blood vessels with a high-pass filter, and improved the SNR by employing a Gabor wavelet transformation, top-hat operators, and a Hessian matrix. The total hemoglobin and total vascular numbers were calculated and normalized.

Due to their distinct characteristics, we provided four different medical imaging modalities: (1) microwave tomography, (2) ultrasonic imaging techniques, (3) optical topography, and (4) photoacoustic imaging. It should be noted that this chapter has merely offered a brief overview of the most commonly used wearable medical imaging methods. There are alternative approaches that could be employed for different biomedical applications.

References

[1] Liu, S., K. Tang, X. Feng, H. Jin, F. Gao, and Y. Zheng, "Toward Wearable Healthcare: A Miniaturized 3D Imager with Coherent Frequency-Domain Photoacoustics," *IEEE Transactions on Biomedical Circuits and Systems,* Vol. 13, No. 6, 2019, pp. 1417–1424.

[2] Bhuyan, A., J. W. Choe, B. C. Lee, P. Cristman, O. Oralkan, and B. T. Khuri-Yakub, "Miniaturized, Wearable, Ultrasound Probe for On-Demand Ultrasound Screening," *2011 IEEE International Ultrasonics Symposium,* 2011, pp. 1060–1063.

[3] Salman, S., A. Kiourti, and J. L. Volakis, "Rudimentary Deep Tissue Imaging through a Wearable Realtime Monitoring System," *2015 IEEE International Symposium on Antennas and Propagation USNC/URSI National Radio Science Meeting,* 2015, pp. 729–729.

[4] Bonfiglio, A., and D. De Rossi (eds.), *Wearable Monitoring Systems,* New York: Springer Science & Business Media, 2010.

[5] Paradiso, R., G. Loriga, and N. Taccini, "A Wearable Health Care System Based on Knitted Integrated Sensors," *IEEE Transactions on Information Technology in Biomedicine,* Vol. 9, No. 3, 2005, pp. 337–344.

[6] Teng, X., Y. Zhang, C. C. Y. Poon, and P. Bonato, "Wearable Medical Systems for P-Health," *IEEE Reviews in Biomedical Engineering,* Vol. 1, 2008, pp. 62–74.

[7] Salman, S., D. Psychoudakis, and J. L. Volakis, "Determining the Relative Permittivity of Deep Embedded Biological Tissues," *IEEE Antennas and Wireless Propagation Letters,* Vol. 11, 2012, pp.1694–1697.

[8] Kuhn, M. J., M. Awida, M. R. Mahfouz, and A. E. Fathy "Open-Ended Coaxial Probe Measurements for Breast Cancer Detection," *2010 IEEE Radio and Wireless Symposium (RWS)*, 2010, pp. 512–515.

[9] Raabe, M. E., and C. Davis, "Measuring Dielectric Properties of Simulants for Biological Tissue," *Merit Fair*, 2011, pp. 1–8.

[10] Yilmaz ,T. and Y. Hao, "Compact Resonators for Permittivity Reconstruction of Biological Tissues," in *2011 XXXth URSI General Assembly and Scientific Symposium*, 2011, pp. 1–4.

[11] Manwar, R., K. Kratkiewicz, and K. Avanaki, "Overview of ultrasound detection technologies for photoacoustic imaging," Micromachines, Vol. 11, No. 7, 2020, p. 692.

[12] Turani, Z., E. Fatemizadeh, T. Blumetti, et al., "Optical Radiomic Signatures Derived from Optical Coherence Tomography Images Improve Identification of Melanoma," *Cancer Research*, Vol. 79, No. 8, 2019, pp. 2021–2030.

[13] Maslov, K., H. F. Zhang, S. Hu, and L. V. Wang, "Optical-Resolution Photoacoustic Microscopy for In Vivo Imaging of Single Capillaries," *Optics Letters*, Vol. 33, No. 9, 2008, pp. 929–931.

[14] Deán-Ben, X. L., and D. Razansky, "Portable Spherical Array Probe for Volumetric Real-Time Optoacoustic Imaging at Centimeter-Scale Depths," *Optics Express*, Vol. 21, No. 23, 2013, pp. 28062–28071.

[15] Bahramiabarghouei, H., E. Porter, A. Santorelli, B. Gosselin, M. Popovic, and L. A. Rusch, "Flexible 16 Antenna Array for Microwave Breast Cancer Detection," *IEEE Transactions on Biomedical Engineering*, Vol. 62, No. 10, 2015, pp. 2516–2525.

[16] Martellosio, A., M. Pasian, M. Bozzi, et al., "Dielectric Properties Characterization from 0.5 to 50 GHz of Breast Cancer Tissues," *IEEE Transactions on Microwave Theory and Techniques*, Vol. 65, No. 3, 2017, pp. 998–1011.

[17] Jotwani, A. C., and J. R. Gralow, "Early Detection of Breast Cancer," *Molecular Diagnosis & Therapy*, Vol. 13, No. 6, 2009, pp. 349–357.

[18] Gerardo, C. D., E. Cretu, and R. Rohling, "Fabrication and Testing of Polymer-Based Capacitive Micromachined Ultrasound Transducers for Medical Imaging," *Microsystems & Nanoengineering*, Vol. 4, No. 1, 2018, pp. 1–12.

[19] Piper, S. K., A. Krueger, S. P. Koch, , et al., "A Wearable Multi-Channel fNIRS System for Brain Imaging in Freely Moving Subjects," *Neuroimage*, Vol. 85, 2014, pp. 64–71.

[20] Mehrmohammadi, M., S. J. Yoon, D. Yeager, and S. Y. Emelianov, "Photoacoustic imaging for cancer detection and staging," *Current Molecular Imaging* (discontinued), Vol. 2, No. 1, 2013, pp. 89–105.

[21] Dangi, A., S. Agrawal, G. R. Datta, V. Srinivasan, and S. R. Kothapalli, "Towards a Low-Cost and Portable Photoacoustic Microscope for Point-of-Care and Wearable Applications," *IEEE Sensors Journal*, Vol. 20, No. 13, 2020, pp. 6881–6888.

[22] Maslov, K., G. Stoica, and L. V. Wang, "In Vivo Dark-Field Reflection-Mode Photoacoustic Microscopy," *Optics Letters*, Vol. 30, No. 6, 2005, pp. 625–627.

[23] Zhang, H. F., K. Maslov, G. Stoica, and L. V. Wang, "Functional Photoacoustic Microscopy for High-Resolution and Noninvasive In Vivo Imaging," *Nature Biotechnology*, Vol. 24, No. 7, 2006, pp. 848– 851.

[24] Cox, B. T., J. G. Laufer, P. C. Beard, and S. R. Arridge, "Quantitative Spectroscopic Photoacoustic Imaging: A Review," *Journal of Biomedical Optics*, Vol. 17, No. 6, 2012, 061202.

[25] Gabriel, S., R. W. Lau, and C. Gabriel, "The Dielectric Properties of Biological Tissues: II. Measurements in the Frequency Range 10 Hz to 20 GHz," *Physics in Medicine & Biology*, Vol. 41, No. 11, 1996, pp. 2251.

[26] Islam, M. A., and J. L. Volakis, "Wearable Microwave Imaging Sensor for Deep Tissue Real-Time Monitoring Using a New Loss-Compensated Backpropagation Technique," *IEEE Sensors Journal*, Vol. 21, No. 3, 2021, pp. 3324–3334.

[27] Devaney, A. J., "A Filtered Backpropagation Algorithm for Diffraction Tomography," *Ultrasonic Imaging*, Vol. 4, No. 4, 1982, pp. 336–350.

[28] Tabbara, W., B. Duchêne, C. Pichot, D. Lesselier, L. Chommeloux, and N. Joachimowicz, "Diffraction Tomography: Contribution to the Analysis of Some Applications in Microwaves and Ultrasonics," *Inverse Problems,* Vol. 4, No. 2, 1988, p. 305.

[29] Meaney, P. M., M. W. Fanning, D. Li, S. P. Poplack, and K. D. Paulsen, "A Clinical Prototype for Active Microwave Imaging of the Breast," *IEEE Transactions on Microwave Theory and Techniques,* Vol. 48, No. 11, 2000, pp. 1841–1853.

[30] Rubaek, T., O. S. Kim, and P. Meincke, "Computational Validation of a 3-D Microwave Imaging System for Breast-Cancer Screening," *IEEE Transactions on Antennas and Propagation,* Vol. 57, No. 7, 2009, pp. 2105–2115.

[31] Ostadrahimi, M., A. Zakaria, J. LoVetri, and L. Shafai, "A Near-Field Dual Polarized (TE–TM) Microwave Imaging System," *IEEE Transactions on Microwave Theory and Techniques,* Vol. 61, No. 3, 2013, pp. 1376–1384.

[32] Haynes, M., J. Stang, and M. Moghaddam, "Real-Time Microwave Imaging Of Differential Temperature for Thermal Therapy Monitoring," *IEEE Transactions on Biomedical Engineering,* Vol. 61, No. 6, 2014, pp. 1787–1797.

[33] Chen, G., J. Stang, M. Haynes, E. Leuthardt, and M. Moghaddam, "Real-Time Three-Dimensional Microwave Monitoring of Interstitial Thermal Therapy," *IEEE Transactions on Biomedical Engineering,* Vol. 65, No. 3, 2018, pp. 528–538.

[34] Kolehmainen, V., M. Lassas, P. Ola, and S. Siltanen, "Recovering Boundary Shape and Conductivity in Electrical Impedance Tomography," *Inverse Problems & Imaging,* Vol. 7, No. 1, 2013, p. 217.

[35] Islam, M. A., A. Kiourti, and J. L. Volakis, "A Novel Method of Deep Tissue Biomedical Imaging Using a Wearable Sensor," *IEEE Sensors Journal,* Vol. 16, No. 1, 2016, pp. 265–270.

[36] Balanis, C. A., *Advanced Electromagnetic Engineering,* New York: John Wiley & Sons, 1989.

[37] Islam, M. A., A. Kiourti, and J. L. Volakis, "A Novel Method to Mitigate Real–Imaginary Image Imbalance in Microwave Tomography," *IEEE Transactions on Biomedical Engineering,* Vol. 67, No. 5, 2020, pp. 1328–1337.

[38] Islam, M. A., A. Kiourti, and J. L. Volakis, "A Modified Gauss-Newton Algorithm for Fast Microwave Imaging Using Near-Field Probes," *Microwave and Optical Technology Letters,* Vol. 59, No. 6, 2017, pp. 1394–1400.

[39] Wang, F., T. Arslan, and G. Wang, "Breast Cancer Detection with Microwave Imaging System Using Wearable Conformal Antenna Arrays," *2017 IEEE International Conference on Imaging Systems and Techniques (IST),* 2017, pp. 1–6.

[40] Alqadami, A. S. M., N. Nguyen-Trong, B. Mohammed, A. E. Stancombe, M. T. Heitzmann, and A. Abbosh, "Compact Unidirectional Conformal Antenna Based on Flexible High-Permittivity Custom-Made Substrate for Wearable Wideband Electromagnetic Head Imaging System," *IEEE Transactions on Antennas and Propagation,* Vol. 68, No. 1, 2020, pp. 183–194.

[41] Mobashsher A. T., and A. M. Abbosh, "Compact 3-D Slot-Loaded Folded Dipole Antenna with Unidirectional Radiation and Low Impulse Distortion for Head Imaging Applications," *IEEE Transactions on Antennas and Propagation,* Vol. 64, No. 7, 2016, pp. 3245–3250.

[42] Lamensdorf, D., and L. Susman, "Baseband-Pulse-Antenna Techniques," *IEEE Antennas and Propagation Magazine,* Vol. 36, No. 1, 1994, pp. 20–30.

[43] Stancombe, A. E., K. S. Bialkowski, and A. M. Abbosh, "Portable Microwave Head Imaging System Using Software-Defined Radio and Switching Network," *IEEE Journal of Electromagnetics, RF and Microwaves in Medicine and Biology,* Vol. 3, No. 4, 2019, pp. 284–291.

[44] Darma, P. N., M. R. Baidillah, M. W. Sifuna, and M. Takei, "Real-Time Dynamic Imaging Method for Flexible Boundary Sensor in Wearable Electrical Impedance Tomography," *IEEE Sensors Journal,* Vol. 20, No. 16, 2020, pp. 9469–9479.

[45] Swisher, S. L., M. C. Lin, A. Liao, et al., "Impedance Sensing Device Enables Early Detection of Pressure Ulcers In Vivo," *Nature Communications,* Vol. 6, No. 1, 2015, pp. 1–10.

[46] Patel, P., and G. H. Markx, "Dielectric Measurement of Cell Death," *Enzyme and Microbial Technology,* Vol. 43, No. 7, 2008, pp. 463–470.

[47] Wagner, D. R., K. F Jeter, T. Tintle, M. S. Martin, and J. M. Long 3rd, "Bioelectrical Impedance as a Discriminator of Pressure Ulcer Risk," *Advances in Wound Care: The Journal for Prevention and Healing,* Vol. 9, No. 2, 1996, pp. 30–37.

[48] Guldiken, R. O., J. Zahorian, F. Y. Yamaner, and F. L. Degertekin, "Dual-Electrode CMUT with Non-Uniform Membranes for High Electromechanical Coupling Coefficient and High Bandwidth Operation," *IEEE Transactions on Ultrasonics, Ferroelectrics, and Frequency Control,* Vol. 56, No. 6, 2009, pp. 1270–1276.

[49] Park, K. K., M. Kupnik, H. J.. Lee, B. T. Khuri-Yakub, and I. O. Wygant, "Modeling and Measuring the Effects of Mutual Impedance on Multi-Cell CMUT Configurations," *2010 IEEE International Ultrasonics Symposium,* IEEE, 2010, pp. 431–434.

[50] Shomaji, S., A. Basak, S. Mandai, R. Karam, and S. Bhunia, "A Wearable Carotid Ultrasound Assembly for Early Detection of Cardiovascular Diseases," *2016 IEEE Healthcare Innovation Point-Of-Care Technologies Conference (HI-POCT),* 2016, pp. 17–20.

[51] Shomaji, S., P. Dehghanzadeh, A. Roman, D. Forte, S. Bhunia, and S. Mandal, "Early Detection of Cardiovascular Diseases Using Wearable Ultrasound Device," *IEEE Consumer Electronics Magazine,* Vol. 8, No. 6, 2019, pp. 12–21.

[52] Johnson, J. A., M. Karaman, and B. T. Khuri-Yakub, "Coherent-Array Imaging Using Phased Subarrays. Part I: Basic Principles," *IEEE Transactions on Ultrasonics, Ferroelectrics, and Frequency Control,* Vol. 52, No. 1, 2005, pp. 37–50.

[53] Lyons, A., F. Tonolini, A. Boccolini, et al., "Computational Time-of-Flight Diffuse Optical Tomography," *Nature Photonics,* Vol. 13, No. 8, 2019, pp. 575–579.

[54] Lindell, D. B., and G. Wetzstein, "Three-Dimensional Imaging Through Scattering Media Based on Confocal Diffuse Tomography," *Nature Communications,* Vol. 11, No. 1, 2020, pp. 1–8.

[55] Schmitz, C. H., D. P. Klemer, R. Hardin, et al., "Design and Implementation of Dynamic Near-Infrared Optical Tomographic Imaging Instrumentation for Simultaneous Dual-Breast Measurements," *Applied Optics,* Vol. 44, No. 11, 2005, pp. 2140–2153.

[56] Schneider, S., P. Piper, C. H. Schmitz, et al., "Fast 3D Near-Infrared Breast Imaging Using Indocyanine Green for Detection and Characterization of Breast Lesions," *RöFo-Fortschritte auf dem Gebiet der Röntgenstrahlen und der bildgebenden Verfahren,* Vol. 183, 2011, Stuttgart: Georg Thieme Verlag KG, pp. 956–963.

[57] Atsumori, H., M. Kiguchi, T. Katura, et al., "Noninvasive Imaging of Prefrontal Activation During Attention-Demanding Tasks Performed While Walking Using a Wearable Optical Topography System," *Journal of Biomedical Optics,* Vol. 15, No. 4, 2010, p. 046002.

[58] Atsumori, H., M. Kiguchi, A. Obata, et al., Development of Wearable Optical Topography System for Mapping the Prefrontal Cortex Activation," *Review of Scientific Instruments,* Vol. 80, No. 4, 2009, p. 043704.

[59] Nasiriavanaki, M., J. Xia, H. Wan, A. Q. Bauer, J. P. Culver, and L. V Wang, High-resolution photoacoustic tomography of resting-state functional connectivity in the mouse brain. *Proceedings of the National Academy of Sciences,* 111(1):21– 26, 2014.

[60] Wissmeyer, G., M. A. Pleitez, A. Rosenthal, and V. Ntziachristos, "Looking at Sound: Optoacoustics with All-Optical Ultrasound Detection" *Light: Science & Applications,* Vol. 7 No. 1, 2018, pp. 1–16.

[61] Deferrari, H. A., R. A. Darby, and F. A. Andrews, "Vibrational Displacement and Mode-Shape Measurement by a Laser Interferometer," *The Journal of the Acoustical Society of America,* Vol. 42, No. 5, 1967, pp. 982–990.

[62] Deferrari, H. A., and F. A. Andrews, "Laser Interferometric Technique for Measuring Small-Order Vibration Displacements," *The Journal of the Acoustical Society of America,* Vol. 39, No. 5A, 1966, pp. 979–980.

[63] Paltauf, G., R. Nuster, M. Haltmeier, and P. Burgholzer, "Photoacoustic Tomography Using a Mach-Zehnder Interferometer as an Acoustic Line Detector," *Applied Optics,* Vol. 46, No. 16, 2007, pp. 3352–3358.

[64] Bauer-Marschallinger, J., K. Felbermayer, A. Hochreiner, et al., "Low-Cost Parallelization of Optical Fiber Based Detectors for Photoacoustic Imaging," *Photons Plus Ultrasound: Imaging and Sensing 2013,* Vol. 8581, p. 85812M.

[65] Thomson, J. K., H. K. Wickramasinghe, and E. A. Ash, "A Fabry-Perot Acoustic Surface Vibration Detectorapplication to Acoustic Holography," *Journal of Physics D: Applied Physics,* Vol. 6, No. 6, 1973, p. 677.

[66] Wickramasinghe, H. K., "High Frequency Acoustic Holography in Solids," *Acoustical Holography,* P. Green (ed.), Springer, 1974, pp. 121–132.

[67] JA Bucaro, HD Dardy, and EF Carome. Fiber-optic hydrophone. *The Journal of the Acoustical Society of America,* 62(5):1302–1304, 1977.

[68] Shajenko, P., J. P. Flatley, and M. B. Moffett, "On Fiber-Optic Hydrophone Sensitivity," *The Journal of the Acoustical Society of America,* Vol. 64, No. 5, 1978, pp. 1286–1288.

[69] Layton, M. R., and J. A. Bucaro, "Optical Fiber Acoustic Sensor Utilizing Mode-Mode Interference," *Applied Optics,* Vol. 18, No. 5, 1979, pp. 666–670.

[70] Fang, Z., C. Yang, H. Jin, et al., A Digital Enhanced Chip-Scale Photoacoustic Sensor System for Blood Core Temperature Monitoring and In Vivo Imaging," *IEEE Transactions on Biomedical Circuits and Systems,* Vol. 13, No. 6, 2019, pp. 1405–1416.

[71] Liu, S., K. Tang, H. Jin, R. Zhang, T. T. Kim, and Y. Zheng, "Continuous Wave Laser Excitation Based Portable Optoacoustic Imaging System for Melanoma Detection," in *2019 IEEE Biomedical Circuits and Systems Conference (BioCAS),* 2019, pp. 1–4.

[72] Siyu Liu, Ruochong Zhang, Yunqi Luo, and Yuanjin Zheng. Magnetoacoustic microscopic imaging of conductive objects and nanoparticles distribution. *Journal of Applied Physics,* 122(12):124502, 2017.

[73] Liu, S., X. Feng, F. Gao, et al., "GPU-Accelerated Two Dimensional Synthetic Aperture Focusing for Photoacoustic Microscopy," *APL Photonics,* Vol. 3, No. 2, 2018, 026101.

[74] Chen, Q., H. Xie, and L. Xi, "Wearable Optical Resolution Photoacoustic Microscopy," *Journal of Biophotonics,* Vol. 12, No. 8, 2019, e201900066.

CHAPTER 9

Wearable Wireless Power Transfer Systems

Stavros Georgakopoulos, Juan Barreto, and Abdul-Sattar Kaddour

9.1 Introduction

The usage of wireless power transfer (WPT) systems expected to grow significantly, with forecasts predicting that the global wireless power transmission market will reach 12 billion USD by the end of 2022 and with most of the market share in North America [1]. WPT technologies eliminate wires, thereby enabling new applications where wires are either not feasible or potentially dangerous. Recent technological advancements in wireless communications and RF devices have paved the way for various wireless power applications, including but not limited to electric vehicle (EV) charging [2, 3], implantable medical devices (IMDs) [4, 5], wearable devices [6, 7], mobile devices [8], consumer electronics [9,10], unmanned aerial vehicles (UAVs) [11, 12], sensors [13, 14], and RFID [15, 16].

Another growing application of WPT systems is wearable technologies. Notably, forecasts project that by the end of 2020 the wearable healthcare device market will be worth over 18 billion USD in North America, Europe, Asia, Latin America, Middle East, and Africa, and will continue to grow beyond 45 billion USD by the end of 2025 [17]. This rapid growth can be attributed to applications that include diagnostic/monitoring devices (e.g., heart rate monitors, blood pressure monitors, and neuromonitors) and therapeutic devices (e.g., insulin pumps, pain management devices, and respiratory therapy devices).

For various biomedical applications, devices must be implantable or wearable, and they are potentially lifesaving to their users. For example, devices such as deep brain neurostimulators, cochlear implants, gastric stimulators, cardiac defibrillators, insulin pumps, foot drop implants, and pacemakers keep certain body parts from malfunctioning and/or inform users if any problems arise. However, most of these devices have a finite operational life because of their limited battery storage. It is also typically very difficult to replace such devices once their batteries are depleted and such replacements usually require risky and costly surgeries. Alternatively, WPT systems can be used to recharge the batteries of biomedical devices wirelessly and extend their operational life, thereby eliminating the risks of surgeries associated with replacements. In this chapter, we focus on WPT systems for wearable applications, as shown in Figure 9.1.

While wearable devices with WPT systems can be very beneficial, the electromagnetic field (EMF) absorption by human tissues can be detrimental to human health and affect WPT system performance. In fact, nonprofit organizations such as the International Committee of Non-Ionizing Radiation Protection (ICNIRP) have set standards to ensure RF devices are safe for their operation near the human body. Therefore, in this chapter, we also study the performance of WPT systems on the human body and examine the SAR at numerous body locations to ensure that high-efficiency power transfer can be safely achieved.

9.2 WPT Methods

Currently, many different WPT methods have been proposed for various applications. Here, we discuss the most commonly used WPT methods, including inductive power transfer (IPT), magnetic resonance coupling (MRC), and strongly coupled magnetic resonance (SCMR).

An ideal WPT system will wirelessly transfer power with no losses (i.e., 100% efficiency) at long ranges while maintaining physically small footprints for the transmitter (TX) and receiver (RX) elements. In addition, WPT systems should be insensitive to misalignment and operate safely and efficiently on or inside the human body. However, in practical implementations, WPT systems cannot offer ideal performance, and performance trade-offs must be considered. Figure 9.2 shows a typical WPT system. It consists of a DC voltage supply that feeds the transmitter with power. An oscillator and power amplifier are used to convert the DC signal

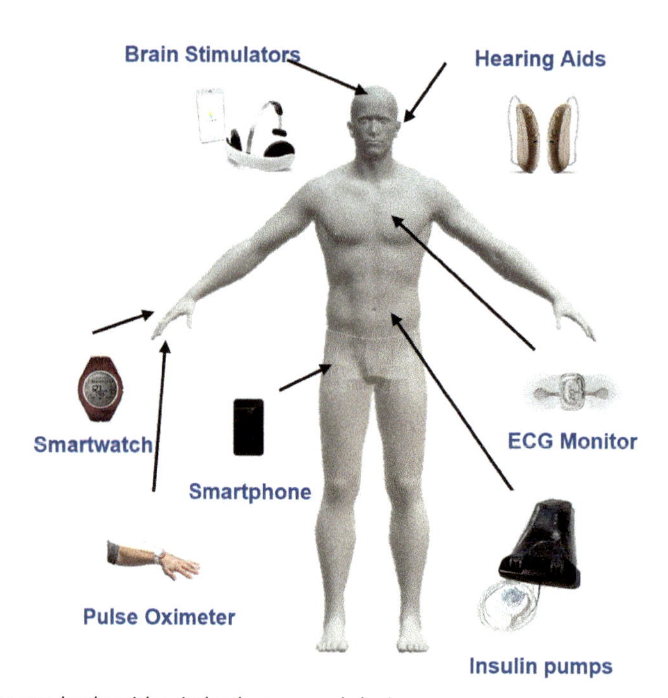

Figure 9.1 Human body with wirelessly powered devices.

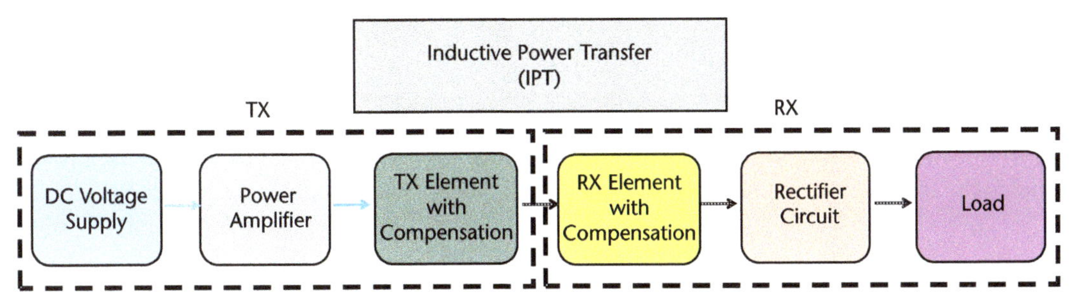

Figure 9.2 Wireless power transfer system.

to the frequency where the TX/RX coils resonate. Also, a capacitor in series or parallel is utilized to compensate for the reactive nature of the IPT system and provide optimal efficiency. The TX element wirelessly transmits the power to the RX element. Finally, a rectifying circuit is used to convert the RX's received high-frequency signal to a DC signal suitable for a desired load.

While IPT, MRC, and SCMR are different WPT methods, they all fundamentally obey electromagnetics laws. Therefore, these methods share similar theoretical foundations that will be discussed here. Ampere's law and Faraday's law of induction describe why power transfer is possible. Suppose a pair of not necessarily resonant coils are placed near each other (i.e., within the nonradiative near-field region) with an exciting current I_1 flowing through coil one with N_1 number of turns. If I_1 is a time-varying current, an induced EMF will be generated in the second coil due to the magnetic flux change. The induced EMF is given by (9.1) as:

$$\varepsilon_{21} = -N_1 \frac{d\Phi}{dt} = \frac{-d}{dt}_{coil\,1} \vec{B}_1 \bullet d\vec{A}_2 \tag{9.1}$$

This produces a current flowing through coil two. The rate of change of magnetic flux going through coil two is proportional to the rate of change of the current in coil one, as given by (9.2):

$$N_2 \frac{d\Phi_{21}}{dt} = M_{21} \frac{dI_1}{dt} \tag{9.2}$$

Therefore, the mutual inductance (M_{21}) is rewritten as:

$$M_{21} = \frac{N_2 \Phi_{21}}{I_1} \tag{9.3}$$

The mutual inductance due to an exciting current in coil one and corresponding magnetic flux going through coil two is a function of the number of turns, the magnetic flux and exciting current as seen in (9.3). A similar analysis may be done, where an exciting current in coil two generates a magnetic flux that goes through coil one to produce a mutually induced link M_{12} given by (9.4):

$$M_{12} = \frac{N_1 \Phi_{12}}{I_2} \tag{9.4}$$

Due to the reciprocity theorem that combines Ampere's and Biot-Savart's laws, the mutual inductance constants can be shown to be equal as seen in (9.5):

$$M_{12} = M_{21} = M \tag{9.5}$$

The analysis above shows that mutual inductance is simply the ratio of the magnetic flux going through the number of turns in one coil and the exciting current in the opposite coil that generates the magnetic field. Therefore, mutual inductance can be thought of as the effect that one coil has on the other based on the current changing on one coil. Also, the coupling coefficient can be written as:

$$\kappa_{12} = \frac{M_{12}}{\sqrt{L_1 L_2}} = \frac{N_1 \Phi_{12}}{I_2 \sqrt{L_1 L_2}} \tag{9.6}$$

The above equations and concepts apply to all near-field wireless power transfer (i.e., IPT, MRC, and SCMR methods) and they describe how electromagnetic coupling enables the wireless transfer of power.

9.2.1 Inductive Power Transfer

IPT or inductive coupling is the traditional method of delivering power wirelessly, which was first explored by Nikola Tesla [18], and it has been used in medical devices since the 1960s. The IPT method provides sufficient efficiencies at short range (i.e., distances between TX and RX resonators that are within 3 cm), which is beneficial for some mobile device applications and IMDs. However, IPT methods cannot provide high efficiencies at midrange distances (i.e., distances that are equal or greater than the maximum dimensions of TX/RX resonators [19]). In summary, IPT designs are simple, and they provide high efficiency at short distances and relatively low operating frequencies. Specifically, typical IPT transmitters and receivers are composed of a coil with parasitic inductance and capacitance that determines its operating resonant frequency. Also, high efficiency is desired at every stage of the WPT system to provide maximal efficiency (higher efficiency means less power is lost). Therefore, significant research has been done to optimize the different parts of WPT systems. For instance, high efficiency and high-power class E amplifiers are of great interest and have been subject of many studies. Furthermore, compensation topologies at the TX and RX circuit stages have been rigorously studied, and they are particularly important for IPT methods because coils are typically loosely coupled with coupling coefficients less than 1. When this is the case, the power transmitted to the load can be approximated as:

$$P_{L,k_{12}} = \frac{L_2}{L_1} |V_1|^2 \frac{R_L}{(\omega L_2 + X_L)^2 + R_L^2} \kappa_{12}^2 \tag{9.7}$$

The previous equation demonstrates the need for a compensation network for IPT systems. Both primary and secondary circuits require a capacitive element to eliminate the reactive part appearing in (9.7). We can see that the power delivered at the load can be significantly increased by reducing the $(\omega L_2 + X_L)^2$ term, thereby increasing the WPT system's efficiency.

9.2.2 Resonant Inductive Coupling

The resonant inductive coupling (RIC) method is similar to the IPT method. However, they are distinguished by one significant difference: the RIC method utilizes a pair of coupled resonant LC coils at TX and RX elements. This resonant condition allows RIC systems to increase the magnetic flux between TX and RX elements, thereby increasing their efficiency at slightly longer distances than the ones achieved by IPT. A popular wireless charging standard based on the resonant inductive coupling is the Qi standard [20]. The Qi standard also specifies three different power ranges for wireless chargers.

9.2.3 Strongly Coupled Magnetic Resonance

In 2006, researchers at the Massachusetts Institute of Technology showed that wireless power transfer systems with high efficiencies at midrange can be achieved using strongly coupled resonators [21]. These researchers introduced the SCMR method that shows great promise for next-generation WPT systems. Similar to RIC, the SCMR utilizes resonant conditions to improve efficiency. However, SCMR uses an additional source loop at the TX element and a load loop at the RX element to maximize efficiency and range. To understand why SCMR systems can provide high efficiency at midrange, circuit theory analysis is used here. In Figure 9.3, a SCMR system is shown. It consists of the four following components: (1) a source loop where an alternating current is fed into, (2) a TX resonant loop or coil to transmit power wirelessly, (3) an RX resonant loop or coil to receive the transmitted power wirelessly, and (4) a load loop where the power is delivered to. As previously mentioned, this transfer's premise begins with Faraday's law of induction. This occurs when the source loop is excited by an AC current that produces a magnetic field

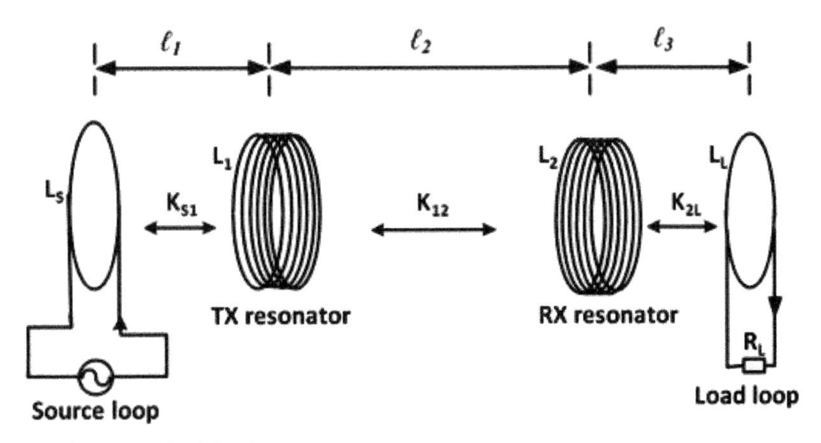

Figure 9.3 Schematic of a SCMR system.

that induces a voltage on the RX coil or loop due to the change in magnetic flux. The same process occurs throughout each loop or coil until the load loop absorbs the transfer of power.

Figure 9.4 shows the equivalent circuit of an SCMR system, where R_s and V_s are the internal source resistance and voltage, respectively. The source and load loop resistances and inductances are R_S' and L_s, and R_L', L_L respectively. RX and TX resonators have a capacitance, inductance, and resistance of R_1, L_1, C_1, and R_2, L_2, C_2 respectively. The coupling coefficient between source and TX resonator is denoted as K_{S1}, between TX and RX resonators as K_{12}, and between RX resonator and load loop as K_{2L}. Coupling terms between loops or coils not directly facing each other can be ignored as their values are negligible. The coupling between the resonators increases the overall mutual inductance between the TX and RX elements of the WPT system, allowing SCMR systems to work at extended ranges compared to traditional IPT methods. The transmission distance can be further extended by adding repeater coils [22].

Taking a closer look at each of the TX and RX resonators' equivalent circuits shows that they are composed of a resistance, inductance, and capacitance, which forms a resonant condition. For SCMR to achieve high efficiency, both RX and TX resonators must be operating at the same resonant frequency where they simultaneously exhibit their maximum Q-factor. The Q-factor can be defined as how well the resonators can store energy. In fact, it is the measurement of the peak energy stored (reactance) to the energy dissipated (resistance) during each cycle of oscillation. Therefore, high Q-factors mean that high amounts of energy can be stored within reactive components while maintaining minimal losses per cycle. This is shown clearly in (9.9) where it is seen that decreasing losses due to resistance in the system will allow for high Q-factors to be achieved. The energy in the resonant circuit oscillates between the inductor and the capacitor at the resonant frequency, f_r. The energy is dissipated in resistance R, and magnitudes of inductive reactance and capacitive reactance should be equal under resonant conditions, therefore $\omega_L = \frac{1}{\omega C}$ and the resonant frequency can be calculated as follows:

$$\omega_0 = \frac{1}{\sqrt{LC}} \tag{9.8}$$

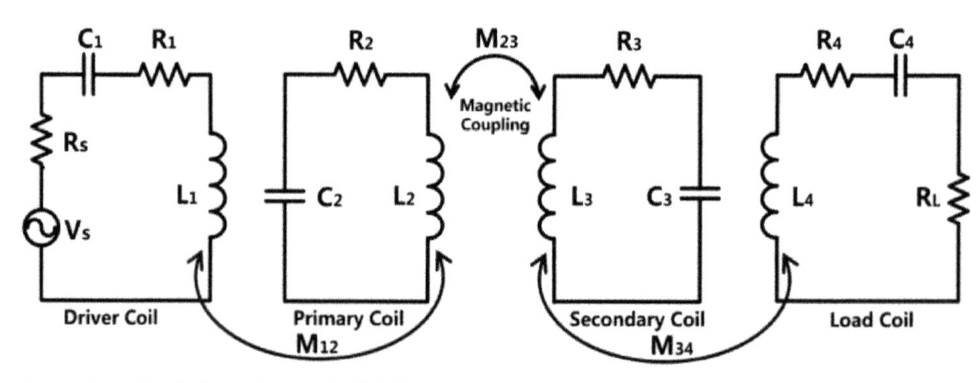

Figure 9.4 Equivalent circuit of a SCMR system.

where ω_0 is equal to $2\pi f_r$. Therefore, the Q-factor as described in [23] can be expressed as follows:

$$Q=\frac{\omega_r L}{R}=\frac{2\pi f_r L}{R} \tag{9.9}$$

If the resonator is a resonant helix, then the Q-factor may be written as [24]:

$$Q=\frac{2\pi f_r L_{helix}}{R_{ohm}+R_{rad}} \tag{9.10}$$

where L_{helix}, R_{rad}, and R_{ohm} are the self-inductance, radiation resistance, and ohmic resistance of the short helix or solenoid ($2r > h$) and are determined mathematically by (9.4), (9.12), and (9.13) [25–28]:

$$L_{helix}=\mu_0 r N^2\left[\ln\left(\frac{8r}{r_c}\right)-2\right] \tag{9.11}$$

$$R_{rad}=\left(\frac{\pi}{6}\right)\eta_0 N^2\left(\frac{2\pi f_r}{c}\right)^4 \tag{9.12}$$

$$R_{ohm(helix)}=\left(\sqrt{\mu_0 \rho \pi f_r}\right)\frac{Nr}{r_c} \tag{9.13}$$

The efficiency of an SCMR system with both resonators operating at the same resonant frequency f_r is written as follows [29]:

$$\eta(f_r)=\frac{\kappa^2_{(TX_{RX})}(f_r)Q_{TX}(f_r)Q_{RX(f_r)}}{1+\kappa^2_{(TX_{RX})}(f_r)Q_{TX}(f_r)Q_{RX(f_r)}} \tag{9.14}$$

Equation (9.14) shows that large Q-factors result in high efficiencies even when the mutual coupling $\kappa^2_{(TX_{RX})}$ between the resonators is not as large. In fact, this is the reason why SCMR provides higher efficiencies at greater distances. Under resonant conditions, $Q_{TX} = Q_{RX}$ and (9.14) may be rewritten as

$$\eta(f_r)=\frac{\kappa^2_{(TX_{RX})}(f_r)Q^2_{TX}(f_r)}{1+\kappa^2_{(TX_{RX})}(f_r)Q^2_{TX}(f_r)} \tag{9.15}$$

The limiting factor of the coupling coefficient is the magnetic flux. When the two coils are physically placed near each other, the coupling coefficient will always be highest since this is when the most flux generated by the transmitting coil is coupled to the receiving coil. The coupling coefficient ranges from values of zero to one, one meaning that all of the magnetic flux generated by the transmitting coil is going through the receiving coil, and zero meaning that none of the magnetic flux generated by the transmitting coil is going through the receiving coil. When the coils become farther apart, there is less flux going through the receiving coil, and the coupling coefficient lowers in value. This is fundamentally why WPT systems have a limited range.

Notably, the coupling coefficient of SCMR systems operating in the midrange is small, due to the distance between the coils, as explained above. In most cases, the value of the coupling coefficient is less than 0.1, thereby the WPT system operates in the loosely coupled regime. Therefore, for loosely coupled WPT systems to provide high WPT efficiency, they must exhibit high quality factors to compensate for their low coupling coefficients. Figure 9.5 shows the efficiency of an SCMR system for different coupling coefficients and quality factors to study their influence on the performance of SCMR systems.

In Figure 9.5, three different values of coupling coefficients of 0.1, 0.3, and 0.5 are considered along with quality factors of increasing values to obtain the maximum PTE. Two conclusions can be drawn from Figure 9.5. First, transmission efficiency increases as the coupling coefficient increases (this can be achieved by decreasing the transmission distance between transmitting and receiving coils). Another effect of high coupling coefficient or overcoupling values is the frequency splitting phenomena that the interested reader can further study [30]. Second, when the quality factor increases, that in turn increases the WPT efficiency. Therefore, our conclusions confirm the fundamental concept of SCMR systems by demonstrating that even though low coupling coefficients are expected in such systems, the high quality factors of their coil resonators can compensate for their low

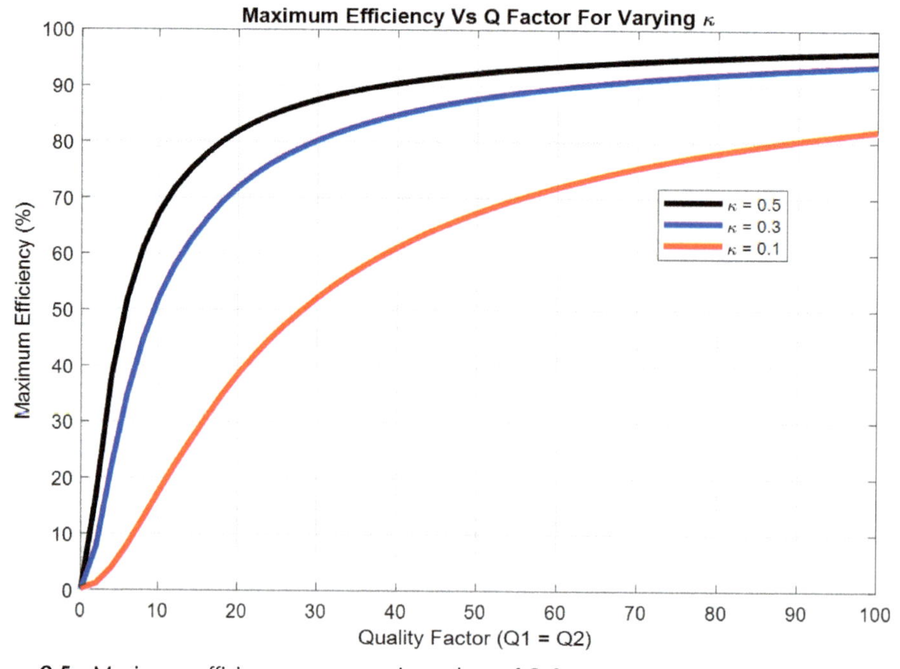

Figure 9.5 Maximum efficiency versus varying values of Q & κ.

coupling, which leads to high WPT efficiencies. For example, from the results in Figure 9.5, it is seen that if the coupling coefficient is less than 0.1 (which occurs in typical SCMR systems), the quality factor of the resonators must be larger than 100 in order to obtain high efficiencies.

Recently, the conformal strongly coupled magnetic resonance (CSCMR) method has been proposed [31]. CSCMR is similar to the SCMR method. However, CSCMR embeds the source and load loops coplanar to the TX and RX resonators, respectively, thereby providing WPT systems that are significantly more compact than SCMR ones [32, 33]. Therefore, CSCMR systems are better suited for numerous applications for which the size of wireless devices must be small, such as wearable devices [34, 35]. CSCMR systems are studied for wearable applications in the following section.

9.3 CSCMR Systems for Wearable Applications

9.3.1 CSCMR System Design

In this section, the performance of a CSCMR system is examined. A typical CSCMR system is shown in Figure 9.6, and it is formed from a traditional SCMR system by placing the source and load loops coplanar with the TX and RX resonator loops, respectively. The main advantages of CSCMR systems are their low profile and ease of integration in printed circuit boards. Therefore, CSCMR systems are very well suited for wearable applications as they can be conformal to the human body. This is the reason we propose and study their performance here and propose their use for wearable applications.

In our study, two CSCMR systems (see Figure 9.6) are designed to operate at the industrial, scientific, and medical (ISM) band of 27.12 MHz. In the first system, the TX and RX elements use an FR-4 substrate and in the second system they use a ferrite substrate. Ferrite substrates are considered due to their ability to isolate the effects of materials placed under the TX and RX elements of CSCMR systems

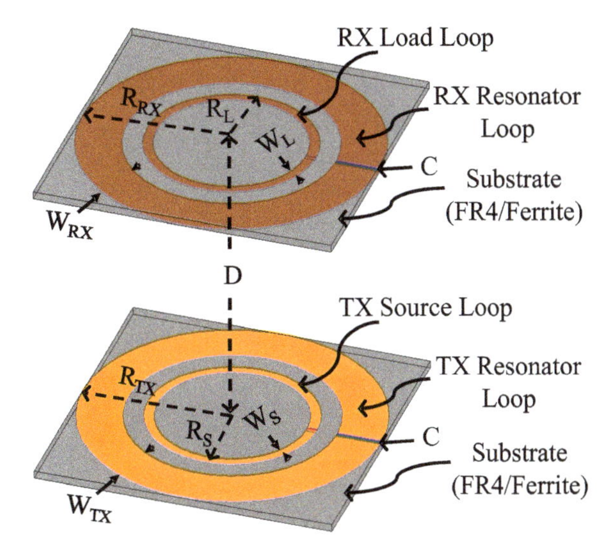

Figure 9.6 Geometry of the CSCMR WPT system [6].

[36]. These materials depend on the application and can often detrimentally affect the efficiency of CSCMR systems (e.g., in the case of wearable applications, RX elements are placed on the human body). Specifically, our system uses a capacitor, C, of 470 pF and 280 pF for FR-4 and ferrite substrates, respectively. Specifically, in this design, the width of the source/load loops is 2 mm. This allowed us to use resonant loops with larger widths that provide significantly higher PTE than previous CSCMR systems. Our CSCMR designs were simulated and analyzed using ANSYS HFSS. All the geometrical parameters of our optimized CSCMR system per Figure 9.6 are as follows: $R_S = R_L = 17$ mm, $W_S = W_L = 2$ mm, $R_{TX} = R_{RX} = 30$ mm, $W_{TX} = W_{RX} = 9$ mm, and $D = 60$ mm. Two sets of TX and RX CSCMR units were fabricated and measured in free space. The first pair was printed on a 1.5 mm thick FR-4 substrate and its performance is used as a benchmark. The second pair was printed on a flexible ferrite sheet [37] with $\mu_r = 45$, $\varepsilon_r = 12$, and tan $\delta_\mu = 0.$ 0.007 at 27.12 MHz. For a fair comparison, three stacked layers of 0.5 mm thick ferrite sheets were used to form a 1.5 mm thick ferrite substrate. The measurements and simulations of the CSCMR system on the FR-4 are compared in Figure 9.7. The CSCMR system on FR-4 exhibits a simulated and measured efficiency of 73.43% and 70.0%, respectively. This difference in PTE between measurements and simulations is attributed to the capacitor losses that are not modeled in ANSYS HFSS. It should be noted that the material properties of the ferrite substrate were not fully and accurately defined at 27.12 MHz in the specifications of the material. We chose to adjust our ferrite's material properties μ_r, ε_r, and tan δ_μ to the values defined above so that our simulation agreed well with the measurements. This is acceptable because the goal is to evaluate the relative effects that different parts of the human body have in the performance of CSCMR systems. Therefore, the simulation model of the CSCMR system with the ferrite substrate in free space (which we created and validated using these material properties) will serve as the basis for simulating the performance of this system on various parts of the human body. The simulated and measured results for the CSCMR system with the ferrite substrate are 75.27% and 71.0% PTE, respectively (see Figure 9.7).

The magnetic field distributions (H-field) of the CSCMR system on FR-4 and ferrite substrates are shown in Figure 9.8(a) and (b), respectively. It is clearly seen

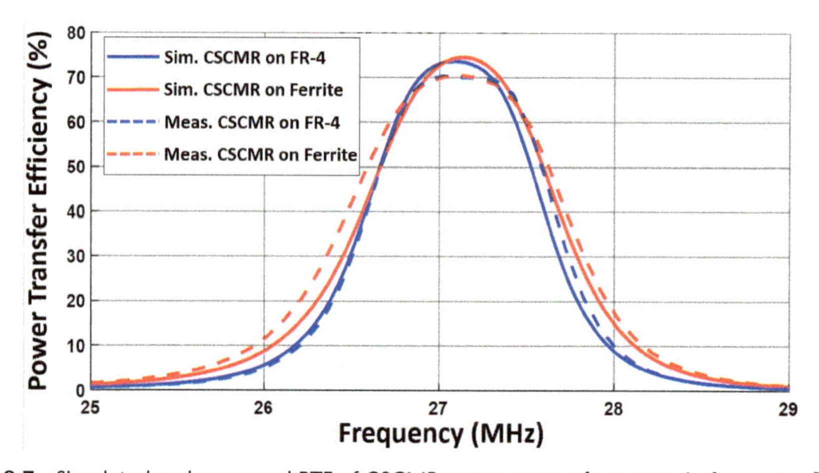

Figure 9.7 Simulated and measured PTE of CSCMR system versus frequency in free space [6].

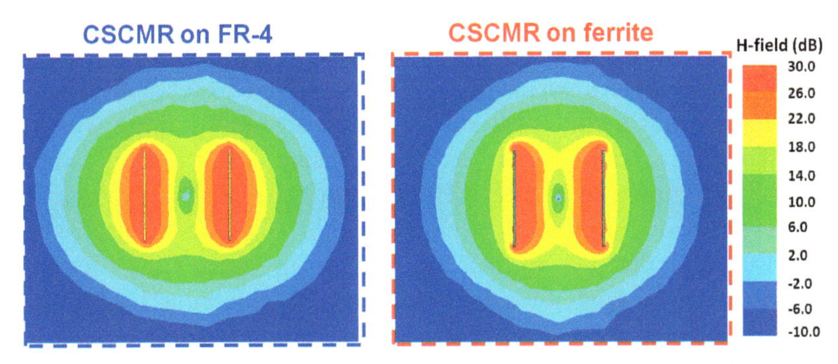

Figure 9.8 Magnetic field intensity of the CSCMR system in free space: (a) with an FR-4 substrate, and (b) with a ferrite substrate [6].

that the magnetic field distributions of these two CSCMR systems are different. The ferrite substrate confines the magnetic fields and reduces their intensity behind them. This behavior explains why ferromagnetic materials are well suited for isolating WPT systems from materials that are placed under their TX and RX elements.

9.3.2 Performance of CSCMR System on the Human Body

In this section, the performance of our CSCMR system was studied on 26 unique locations of the human body, as shown in Figure 9.9, [6]. The 26 locations were split into groups and categorized as either head (1–5), neck/bicep (6–10), arm (11–15), torso (16–21), and leg (22–26). The torso category is the only group with six different locations due to its wide coverage area compared to the other groups, which had only five locations each. Our simulation setups were simplified to save computational resources by including for each case only the relevant part of the

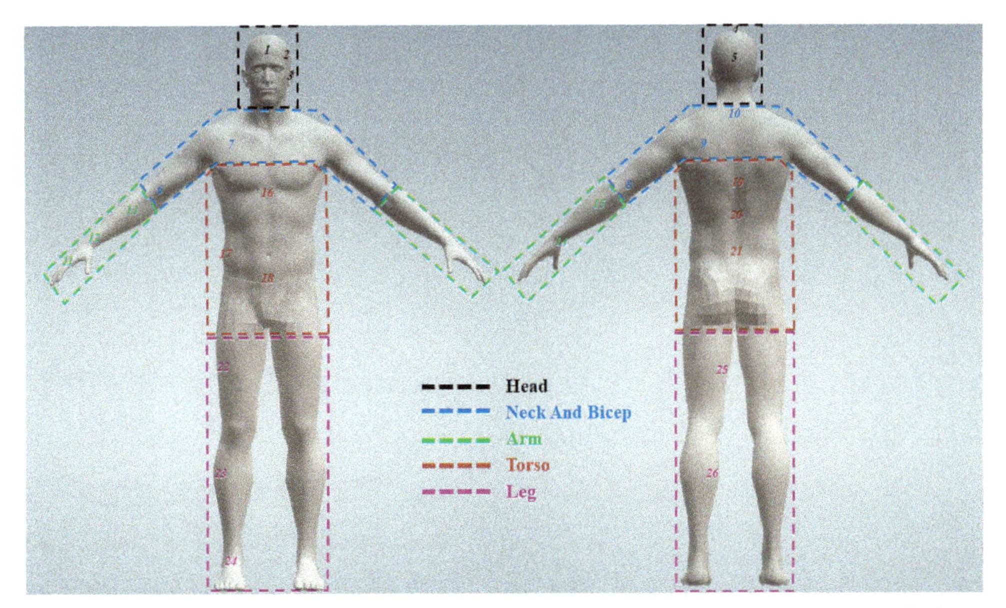

Figure 9.9 Human body model that outlines the five different groups of the CSCMR system's placement: (a) front view, and (b) back view [6].

human body (instead of having the entire body model), thereby ignoring the other parts that have negligible effects on the performance of our WPT system.

Simulations were performed using the ANSYS HFSS and the ANSYS human body model that includes the different human tissues along with their properties [38]. The measurements were conducted using a 3-D printed support for the TX and RX, as shown in Figure 9.10, and for the following two spacings: (a) flush against the body (i.e., 0 mm spacing), and (b) 10 mm away from the body. Specifically, Figure 9.10 shows our measurement setup for the CSCMR system with ferrite substrates placed flush on top of the hand (i.e., location 11). Also, when our 3-D

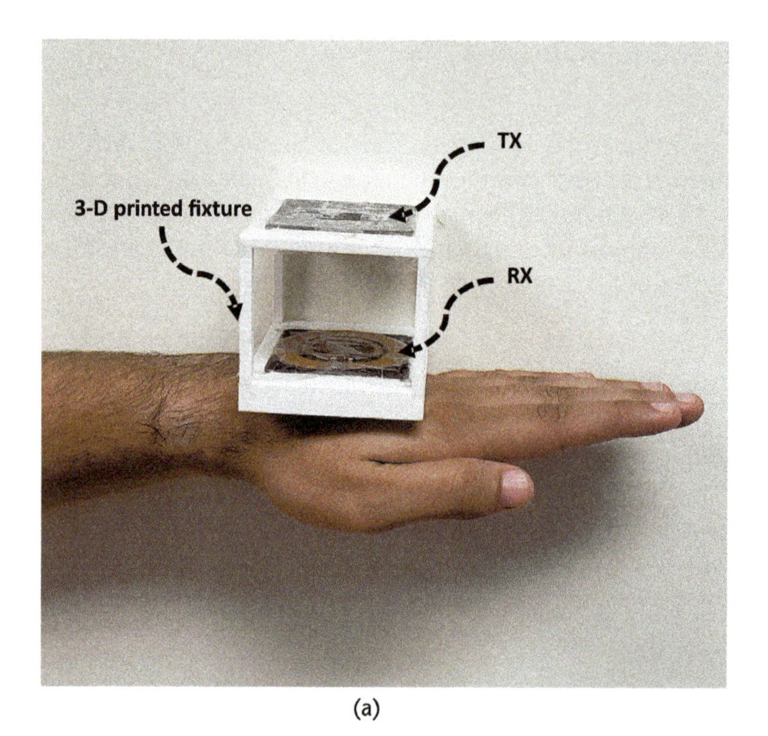

(a)

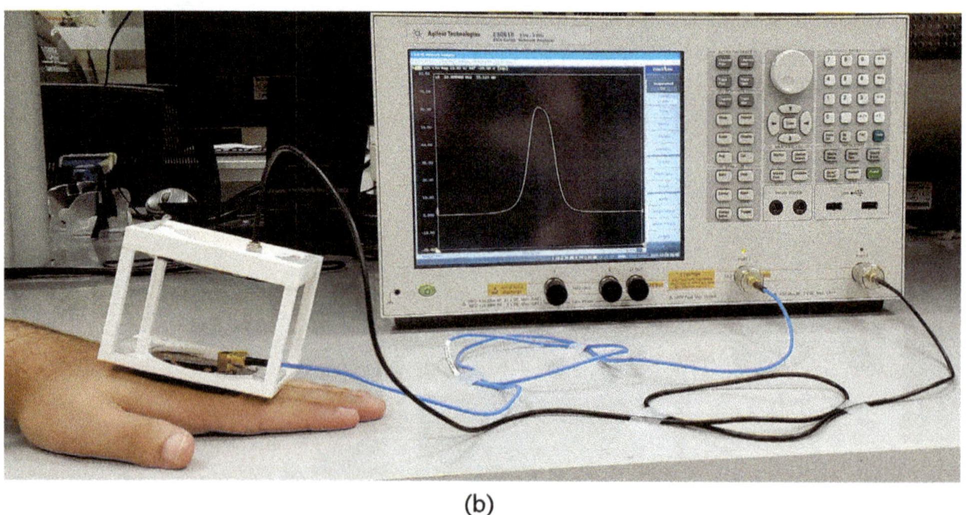

(b)

Figure.9.10 Measurement setup for the CSCMR system with ferrite substrates on top of the hand: (a) close-up of the setup, and (b) complete system connected to the VNA.

printed fixture is flipped, it supports a gap of 10 mm between the RX and the human body. An elbow connector was used to suitably obtain measurements on the body without the cable or connector obstructing the human body. Even though it seems that this connector setup might affect the results (as the connector and cable are directly between the transmission path), on the contrary, our tests indicate that using the elbow connector has negligible effects as compared to using two standard straight connectors. The PTE, η, for each scenario, was measured using a Keysight Vector Network Analyzer (VNA). The PTE was defined as $\eta = |S_{12}|^2$.

The simulated and measured results for all the placement locations on the human body (see Figure 9.9) are shown in Table 9.1. The efficiencies of the CSCMR systems for all five placement categories (i.e., head, neck/bicep, torso, arm, and leg) were calculated by averaging the efficiency values of all locations under each category. The free-space case was included as a benchmark case to quantify the losses caused by the human body. Also, Table 9.1 shows the efficiency of the CSCMR systems for the locations that experienced the highest (upper back torso 19) and lowest (top of the wrist 12) amount of losses (i.e., efficiency drop). Our measured results in Table 9.1 show that the CSCMR system with the ferromagnetic substrate exhibits higher efficiency for all the different placements on the body than the system with the FR-4 substrates. This can also be illustrated by finding the average losses (i.e., each loss is calculated based on the difference between the reference case in free space and the case that is on the human body) using the 26 measurements at all placement locations on the human body. Specifically, the WPT system with the ferromagnetic substrate exhibited average losses of 1.6% and 0.6%, at spacings of 0 and 10 mm, respectively. In contrast, the CSCMR with the FR-4 substrate exhibited average losses of 7.2% and 3.0% at spacings of 0 and 10 mm, respectively. Notably, while a 1.5 mm thick ferrite sheet was used here, it was observed that by using thicker ferrite substrates we can further increase the efficiency

Table 9.1 PTE of the CSCMR System on the Human Body

		Power Transfer Efficiency (%)											
		CSCMR on FR-4				*CSCMR on Ferrite*				*Increase from FR-4 to Ferrite*			
		Simulation		*Measurements*		*Simulation*		*Measurements*		*Simulation*		*Measurements*	
	Free Space	73.43		70.0		75.27		71.0		NA		NA	
	Distance from Human Body (mm)	0	10	0	10	0	10	0	10	0	10	0	10
Body Region	*Upper Back (19)*	65.6	69.7	60.1	66.2	71.9	73.5	69.0	70.4	6.3	3.8	8.9	4.2
	Top of Wrist (12)	71.8	73.2	64.6	68.0	75.0	75.3	69.0	69.6	3.2	1.8	4.4	1.6
	Head	68.8	71.6	64.3	66.6	73.5	74.3	69.0	69.7	4.7	2.7	4.7	3.1
	Neck/Bicep	66.9	70.0	62.4	67.2	72.1	73.0	69.4	69.7	5.2	3.0	7.0	2.5
	Torso	66.8	70.4	61.6	66.4	72.5	73.7	69.4	69.6	5.7	3.3	7.8	3.2
	Arms	70.8	72.8	63.7	67.8	74.7	75.2	69.3	69.7	3.9	2.4	5.6	1.9
	Legs	68.4	71.3	61.8	66.8	72.9	73.8	69.6	69.7	4.5	2.5	7.8	2.9

of our WPT systems. Finally, Table 9.1 also shows for each location on the human body the increase of the efficiency when a ferrite substrate is used instead of FR-4.

9.3.3 Magnetic Field Distributions

Our WPT system's magnetic field intensities on a FR-4 substrate are compared here at the locations where they exhibited the highest and lowest simulated PTE in Figure 9.11. Specifically, the highest and lowest simulated PTEs occurred at the top of the wrist (see location 12 in Figure 9.9) and the upper back of the torso (see location 19 in Figure 9.9). The field plots of Figure 9.11 justify why the simulated PTE achieves its highest and lowest values at these locations. This happens due to two reasons: (a) the upper torso is thicker and is a wider area of the human body compared to the top of the wrist, which is also narrower than the size of our WPT TX and RX; therefore, higher WPT losses due to the properties of the human tissues occur at the upper torso location, and (b) the upper torso area has tissues with higher fat content than the wrist, which causes higher WPT losses at the upper

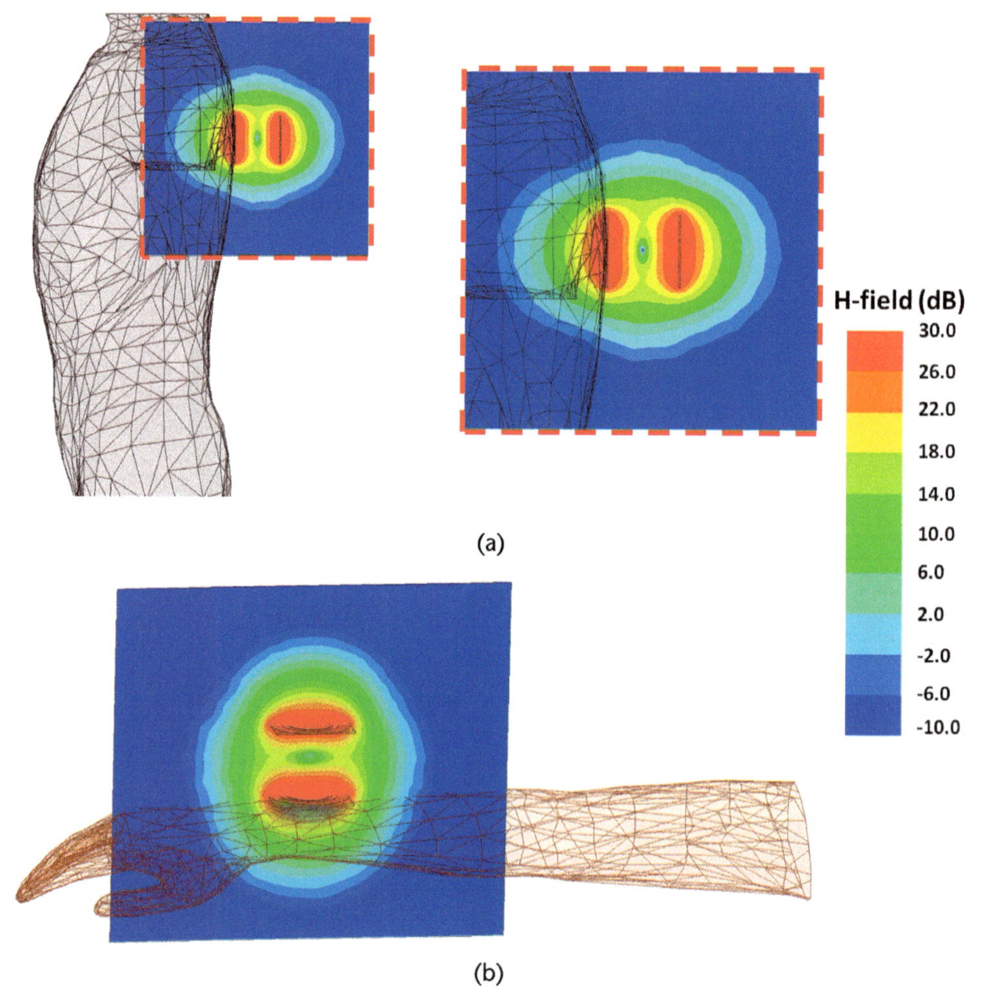

Figure 9.11 Magnetic field intensity of the CSCMR system with FR-4 substrate on (a) the upper back of the torso, and (b) top of the wrist [6].

torso. In fact, these conclusions and explanations are true for all of our placement locations; that is, locations that are similar to the torso provide lower PTEs compared to locations similar to the top of the wrist (e.g., since the inner thigh region has a larger area than the WPT system and similar tissue composition to that of the torso, it experiences similar losses). This is supported by our measured results that are shown in Table 9.1.

To mitigate the losses experienced by our WPT system on the human body, our CSCMR system's TX and RX are placed on a ferrite substrate instead of FR-4. Ferrites can isolate our CSCMR elements from the influence of the human body because of their high permeability. This high permeability confines the magnetic fields to the ferrite substrates, thereby not allowing them to reach and penetrate the human body. This reduces the intensity of the fields that interact with the human body, diminishing the losses caused by the human body. This is also supported by Figure 9.12, which compares the *H*-field distributions of the CSCMR systems with FR-4 and ferromagnetic substrates on the torso's upper back. This figure clearly illustrates that when a ferrite substrate is used, the strongest field intensity (shown in red) does not spread toward the bottom of the TX and RX substrates; thus the strongest fields are confined between the TX and RX. Hence, as shown in Figure 9.12(a), when the WPT system on the FR-4 substrate is placed on the upper back of the torso, strong magnetic field intensities penetrate the substrate and reach the torso, thereby causing a decrease in PTE due to the losses occurring in the surrounding human tissues. In contrast, the WPT system with the ferromagnetic substrate, as shown in Figure 9.12(b), confines the magnetic field in the area between the TX and RX, thus weakening the field intensities that reach the upper back torso, thereby reducing the losses caused by the surrounding human tissues and providing higher PTE than the WPT system on the FR-4 substrate.

9.3.4 Specific Absorption Rate

Safety considerations are always of the utmost importance when designing WPT systems, and SAR is one of the most important safety factors to consider. SAR is

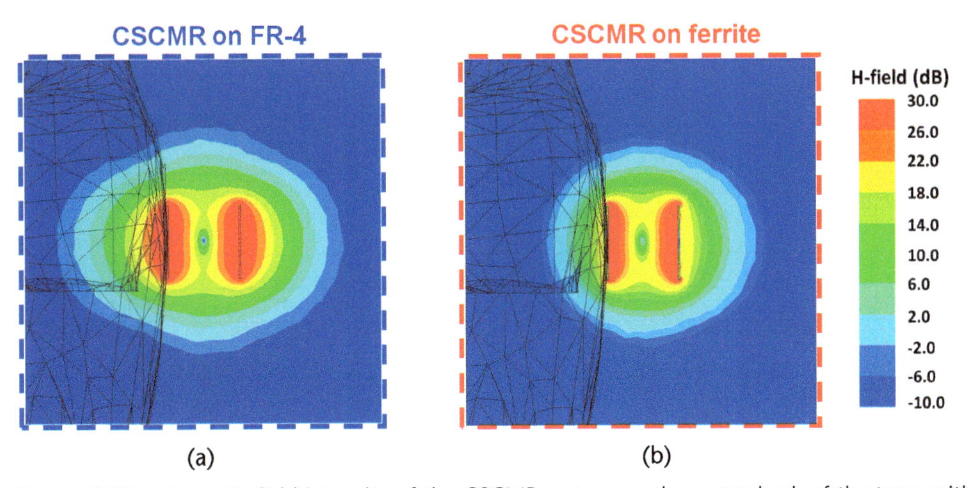

Figure 9.12 Magnetic field intensity of the CSCMR system on the upper back of the torso with (a) FR-4 substrate, and (b) ferrite substrate [6].

the measure of power absorbed per unit mass (W·kg-1) and has been the subject of extensive research efforts to ensure human safety. A high SAR may cause adverse effects on human health because biological tissues begin to absorb heat and exceedingly raise body core temperatures, especially RFs. The International Commission issues the standards on limiting RF exposure on non-ionizing radiation protection (ICNIRP 1998, ICNIRP 2010, ICNIRP 2020) [39] as does the Institute of Electrical and Electronics Engineers (IEEE C95.1) [40]. According to the ICNIRP and IEEE guidelines, local SAR is measured as the power absorbed per 10 g of cubical mass. Peak-spatial SAR (psSAR) is defined as the maximum SAR and it should be studied for our WPT system that radiates in a small portion of the human body; therefore, local exposures are of greater importance (i.e., our system will always comply with the whole body-average SAR established guidelines). The ICNIRP and IEEE impose different guidelines to different areas of the human body due to the temperature variations across the human body. The ICNIRP specifies the following body areas: (a) "head and torso," which consists of the head, eye, pinna, abdomen, back, thorax, and pelvis, and (b) "limbs," which consists of the upper arm, forearm, hand, thigh, leg, and foot. The safety guidelines require that the psSAR within the head and torso areas remain less than 10 W·kg-1 and 2 W·kg-1 for occupational exposure (OE) and general public exposure (GPE) scenarios, respectively. For the limb's region, the psSAR should be less than 20 W·kg-1 and 4 W·kg-1 for OE and GPE scenarios, respectively.

Table 9.2 tabulates the maximum psSAR for each body region, assuming an input power at the TX of 1W over 10 grams of cubical mass, and the maximum possible input power in watts based on the maximum psSAR. In our study, the head, neck/bicep, and torso body regions abide by the "head and torso" guidelines, whereas the arms and legs abide by the "limbs" guidelines. When the WPT system is placed flush to the skin (i.e., separation of 0 mm), the system with the ferrite substrate has a psSAR that, on average, is 2.5 times smaller than the corresponding psSAR of the system on FR 4. Also, compared to the case where the WPT system is placed flush to the skin, when there is a gap of 10 mm between the human body and the CSCMR systems, the average maximum psSAR for the body regions is 5 and 32 times smaller for the FR-4 and ferrite substrates, respectively. This is an important finding as it suggests that a gap between wearable WPT systems and the

Table 9.2 Maximum psSAR of CSCMR Systems on the Human Body and Maximum Possible Input Power

	Maximum psSAR (W·kg^{-1}) in Each Body Region				*Maximum Possible Input Power (W) for the Maximum psSAR*							
	CSCMR on FR-4		*CSCMR on Ferrite*		*CSCMR on FR-4*				*CSCMR on Ferrite*			
	0 mm	10 mm	0 mm	10 mm	0 mm		10 mm		0 mm		10 mm	
Body Region					GPE	OE	GPE	OE	GPE	OE	GPE	OE
Head	.7270	.1497	.4713	.0153	2.751	13.76	13.36	66.80	4.244	21.22	130.7	653.6
Neck/Bicep	.4730	.0844	.1260	.0113	4.230	21.14	23.70	118.5	15.87	79.37	177.0	885.9
Torso	.5547	.1371	.2223	.0274	3.610	18.03	14.59	72.94	9.000	44.99	73.00	364.9
Arms	.3025	.0662	.0932	.0108	13.22	66.12	60.42	302.1	42.92	214.6	370.4	1851
Legs	.5313	.0736	.4357	.0164	7.530	37.64	54.35	271.7	9.181	45.90	243.9	1219

human body and ferrite substrates is crucial for maintaining psSARs within the ICINRP and IEEE specifications, particularly for input power levels that are higher than 1 W.

In Figures 9.13(a) and 9.13(b), the SAR distributions are shown for the CSCMR system flush against the forehead (see location 1 in Figure 9.9) when it is on FR-4 and ferrite substrates, respectively. Notably, the simulated SAR on the forehead with FR-4 substrate (which is the highest recorded from all body locations) does not meet the ICNIRP specifications for GPE beyond an input power of 2.751W and 13.36W for the case of 0 and 10 mm, respectively. However, at the forehead location (placed flush to the skin) and assuming OE, our WPT systems on FR-4 and ferrite substrates meet the ICNIRP specifications as long as their input power does not exceed 13.76W and 21.22W, respectively. Figure 9.13(c) and (d) show the SAR distributions on top of the wrist (see location 12 in Figure 9.9.) for the CSCMR system with the FR-4 and ferrite substrate, respectively. According to our results, to meet the SAR ICNIRP guidelines for GPE at the wrist location, the maximum input powers of the WPT systems on FR-4 and ferrite substrates are 18.7W and 83.3W, respectively. Also, on top of the wrist (placed flush to the skin) and assuming OE, our WPT systems on FR-4 and ferrite substrates meet the ICNIRP specifications as long as their input power does not exceed 48.0W and 213.9W, respectively. These higher input power levels are justified since (a) the SAR is significantly smaller on top of the wrist compared to other locations (lowest recorded of all body locations), and (b) the SAR ICNIRP limits are higher for the limbs. Notably, the maximum transmitted power can be increased (if needed) by adding more ferrite layers,

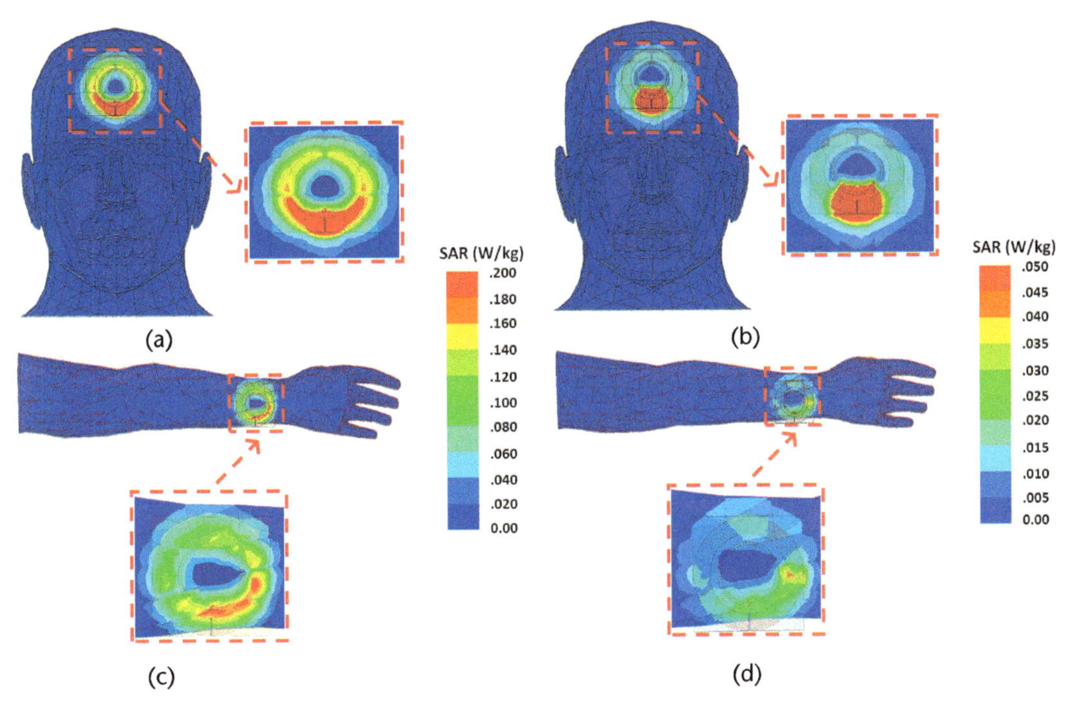

Figure 9.13 Simulated SAR of the CSCMR system: (a) with FR-4 substrate on the forehead, (b) with a ferrite substrate on the forehead, (c) with FR-4 substrate on the top of the wrist, and (d) with a ferrite substrate on the top of the wrist [6].

increasing the thickness of the ferrite substrate, or increasing the gap between the human body and the CSCMR system.

9.4 CSCMR Systems for Implantable Applications

In the previous sections, the effect of the human body was studied on the performance of WPT CSCMR systems for wearable applications. Here, we investigate the performance of implantable WPT CSCMR systems.

Implantable systems are very important for biomedical applications. Devices such as cochlear implants, gastric stimulator, pacemakers, foot drop implants, insulin pumps, and cardiac defibrillator implants could see significant improvements with the additions of WPT systems. These improvements include (a) size reduction due to batteryless operation, (b) increased lifetime, (c) elimination of undesirable and sometimes dangerous surgeries that are necessary to replace batteries of implantable units, and (d) elimination of risks due to chemical battery leakage. Therefore, in this section, we will examine the performance of WPT systems that are implanted in the human body through full-wave simulations.

Specifically, our simulation analysis here is performed using the EM SIM4LIFE software. This software utilizes a finite-difference time-domain (FDTD) EM solver. SIM4LIFE is often used to simulate biological and anatomical environments since it offers highly accurate models of the human body, ranging from different genders to different ages. Our studies are conducted using the SIM4LIFE Duke model. Duke is based on a 34-year-old man of 1.77m in height, with a weight of 70.2 kg, and his BMI is 22.4 kg/m2 [41] (see Figure 9.14). In this study, the areas of interest are the brain (the head), the heart (the chest), and the pelvic girdle (lower back).

To simplify the mesh and speed up our calculations, we designed the CSCMR system for this study with square loops. We also simulated our CSCMR elements as PEC, since our primary purpose is to compare the efficiency of implantable CSCMR systems to the efficiency of the same systems in free space. The geometry for our CSCMR TX/RX elements is shown in Figure 9.15 (R_S R_L= 15 mm, W_S = 0.5 mm, R_{TX} = 14 mm, W_{TX} = 2 mm). The dimensions of these elements are well suited for implantable devices (which in most cases are smaller than 40 mm × 40 mm). The proposed CSCMR system is encapsulated within the center of a 31 mm × 31 mm × 1 mm FR4 material for better isolation from the human body, thereby preventing a significant drop of the PTE.

The simulated efficiency is shown in Figures 9.16, 9.17, and 9.18 for varying implantation depths and locations [5], namely: (a) 10 mm for neurostimulator and insulin pump devices, and (b) 10, 15, and 20 mm for pacemaker devices. In free space, the CSCMR design achieves nearly 100% efficiency (this occurs as our elements are modeled as PEC; therefore, no conductive losses are considered here) at a transfer distance of 10 mm. The 10-mm distance is appropriate for an implantable device, as the implantation depth of such devices is typically in the range of 5 to 20 mm. When the WPT system is implanted within Duke's skull at an implantation depth of 10 mm (see Figure 9.16), the peak efficiency decreases by 10.51% compared to the free space case. Also, when the WPT system is implanted within Duke's chest at depths of 10, 15, and 20 mm (see Figure 9.17), the peak efficiency decreases by 7.37%, 22.12%, and 52.25%, respectively, compared to the

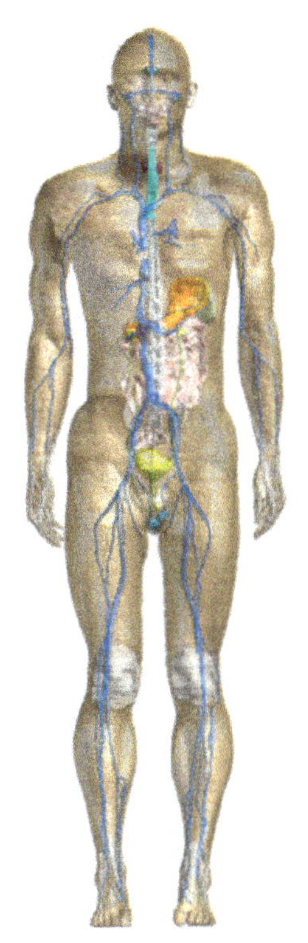

Figure 9.14 SIM4LIFE model "Duke" [41].

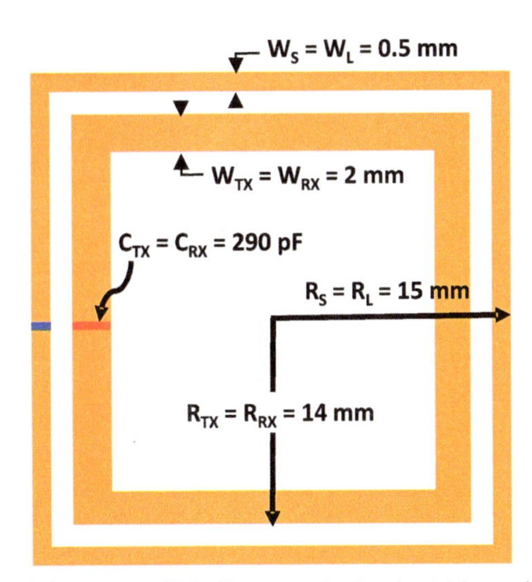

Figure 9.15 Geometry of the square CSCMR system for implantable applications.

free space case. Finally, when the WPT system is implanted within Duke's back at an implantation depth of 15 mm (see Figure 9.18), the peak efficiency decreases by 14.32% compared to the free space case. These important results are also summarized in Table 9.3, and they indicate that for reasonable implantation depths minimal losses occur in our CSCMR systems across different locations in the hu-

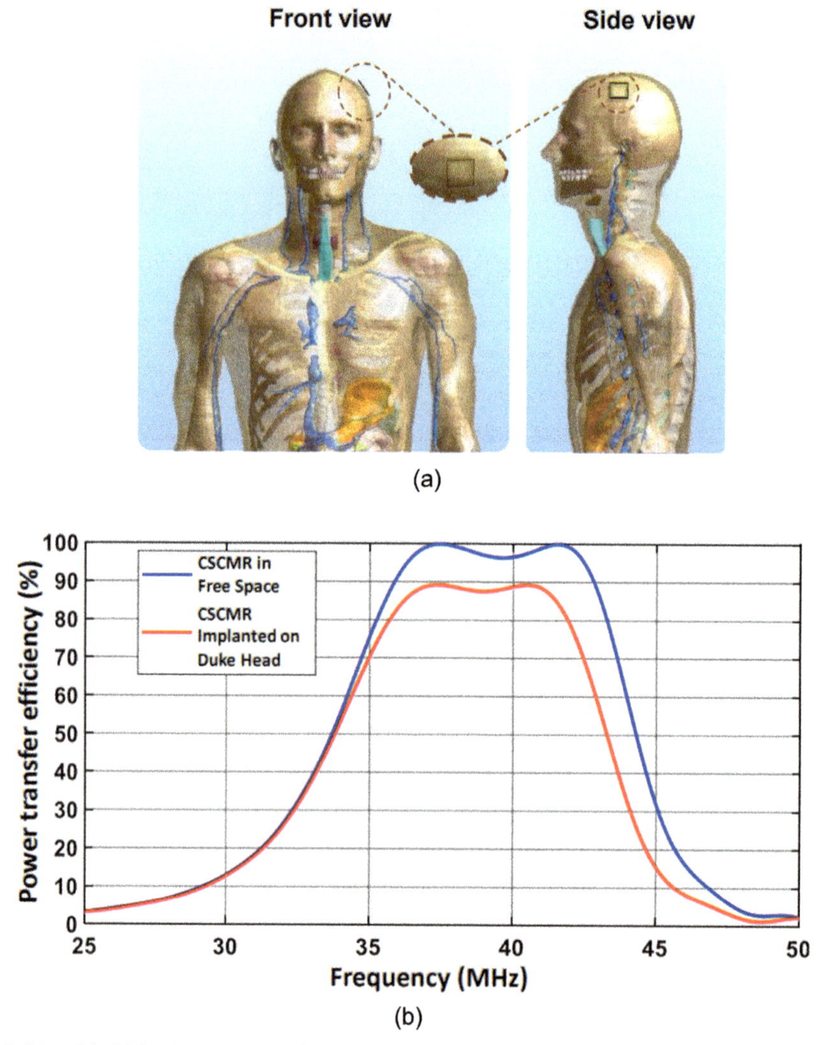

Figure 9.16 (a) CSCMR system implanted on Duke's head for neurostimulator applications, and (b) simulated PTEs of CSCMR system implanted at 10 mm inside Duke's head [5].

man body. These findings clearly prove that CSMCR systems are very well suited for implantable devices.

9.5 Conclusion

In this chapter, we focused on studying the performance of CSCMR systems on the human body for wearable applications. The SCMR method provides high efficiencies at midrange distances; however, it is hard to implement in applications that require compact designs. On the other hand, CSCMR systems are very well suited for wearable applications as they are highly efficient, compact, and conformal. The human body affects the performance of CSCMR systems. Therefore, we rigorously studied the performance of CSCMR systems at different locations on the human

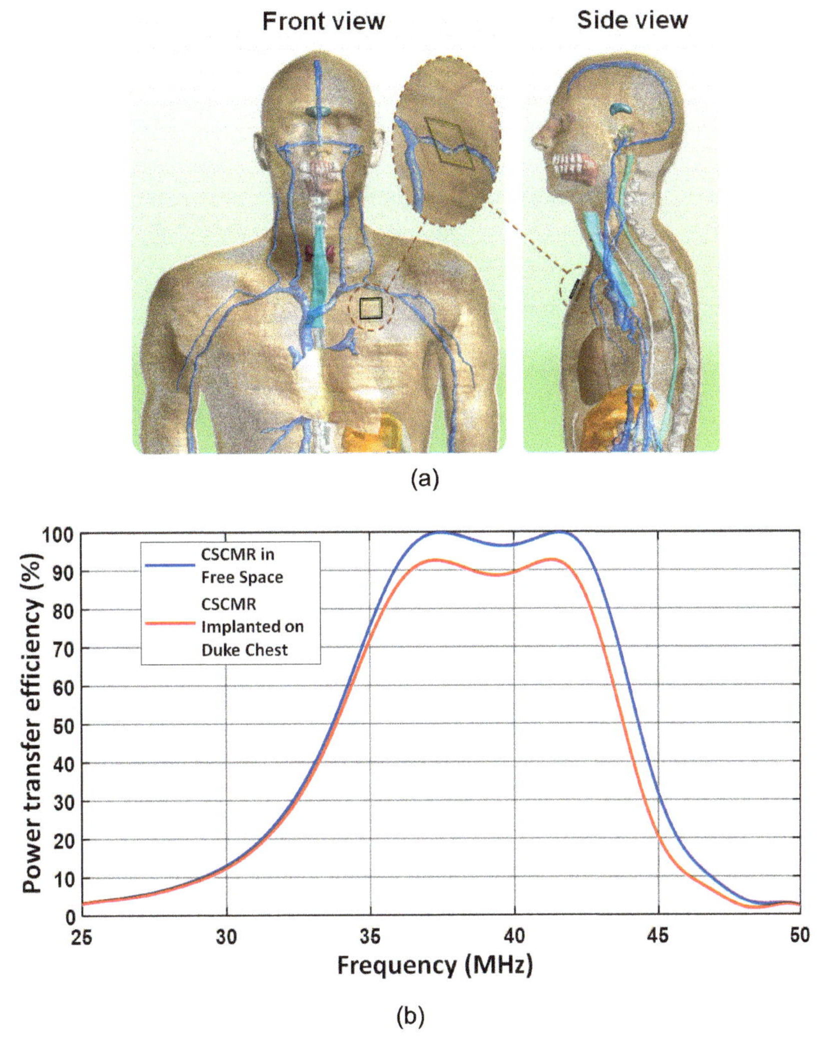

Figure 9.17 (a) CSCMR system implanted on Duke's chest for pacemaker applications, and (b) simulated PTEs of CSCMR system implanted at various depths in Duke's chest [5].

body. We also proposed a way to isolate the effects of the human body to WPT systems by using ferrite substrates that enable us to achieve higher efficiencies in wearable scenarios. As well, the efficiency of wearable WPT systems can be increased by introducing a gap (10 mm) between the CSCMR resonator and the human body. Since WPT wearable systems must meet ICNIRP and IEEE guidelines for safety, we calculated the maximum power that a CSCMR system could output before exceeding ICNIRP and IEEE guidelines for different placements on the human body. It was also shown that ferrites significantly reduce the amount of exposure of the human body, thereby allowing a larger amount of power to be transmitted. Finally, a simulation study of the effects of the human body on the performance of implantable CSCMR systems was presented. Results indicate that minimal losses are expected for reasonable implantation depths, thereby proving the potential applications of CSCMR systems in implantable devices.

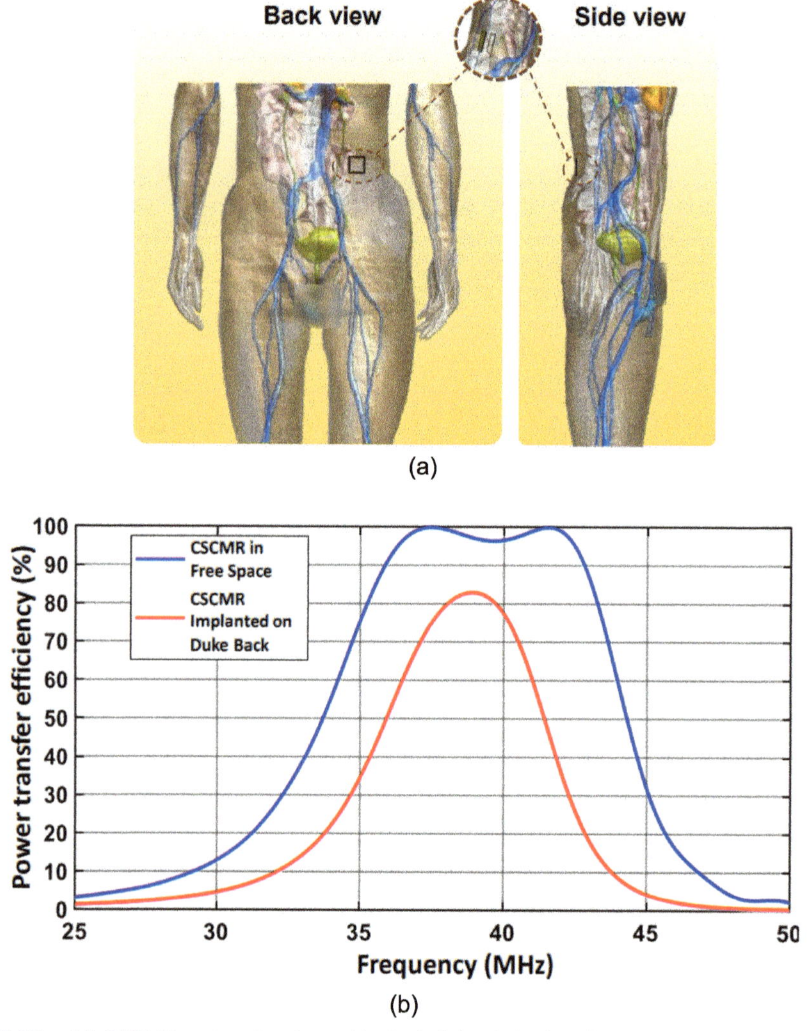

Figure 9.18 (a) CSCMR system implanted in Duke's back for insulin pump applications, and (b) simulated PTEs of CSCMR system implanted at 10 mm in Duke's back [5].

Table 9.3 PTEs of Implanted CSCMR System in the Human Body

	Frequency (MHz)	*PTE (%)*
Free space	37.44	99.7
Chest (10 mm)	37.29	92.33
Chest (15 mm)	38.65	77.25
Chest (20 mm)	38.65	47.45
Head (10 mm)	37.42	89
Back (15 mm)	38.9	82.69

References

[1] Market Research Future, "Global Wireless Power Transmission Market Research Report-Forecast 2022," Market Research Future, New York, 2016.

[2] Shin, J., S. Shin, Y. Kim, et al., "Design and Implementation of Shaped Magnetic Resonance Based Wireless Power Transfer System for Roadway Powered Moving Electric Vehicles," *IEEE Transactions on Industrial Electronics,* Vol. 61, No. 3, March 2014.

[3] Lee, J., and B. Han, "A Bidirectional Wireless Power Transfer EV Charger Using Self-Resonant PWM," *IEEE Transactions on Power Electronics,* Vol. 30, No. 4, April 2015, pp. 1784–1787.

[4] Ahn, D., and S. Hong, "Wireless Power Transmission with Self-Regulated Output Voltage for Biomedical Implant," *IEEE Transactions on Industrial Electronics,* Vol. 61, No. 5, May 2014, pp. 2225–2235.

[5] Barreto, J. C., A.- S. Kaddour, and S. V. Georgakopoulos, "Highly Efficient Wireless Power Transfer Systems for Wearable and Implantable Devices," *2020/2021 IEEE Wireless and Microwave Technology Conference (WAMICON),* Clearwater Beach, FL, April 28–29, 2021.

[6] Barreto, J., G. Perez, A.- S. Kaddour, and S. V. Georgakopoulos, "A Study of Wearable Wireless Power Transfer Systems on the Human Body," *IEEE Open Journal of Antennas and Propagation,* Vol. 2, 2021, pp. 86–94, doi: 10.1109/OJAP.2020.3043579.

[7] Barreto, J., A.- S. Kaddour, and S. V. Georgakopoulos, "A Wireless Power Transfer System on Clothes Using Conductive Threads," *2021 USNC-URSI National Radio Science Meeting (NRSM),* Boulder, CO, 2021, pp. 175–176, doi: 10.23919/USNC-URSINRSM51531.2021.9336480.

[8] Chang-Gyun, K., S. Dong-Hyun, Y. Jung-Sik, P. Jong-Hu, and B. H. Cho, "Design of a Contactless Battery Charger for Cellular Phone," *IEEE Transactions on Industrial Electronics,* Vol. 48, No. 6, December 2001, pp. 1238–1247.

[9] Liu, D., H. Hu, and S. V. Georgakopoulos, "Misalignment Sensitivity of Strongly Coupled Wireless Power Transfer Systems," *IEEE Transactions on Power Electronics,* Vol. 32, No. 7 July 2017, pp. 5509–5519, doi: 10.1109/TPEL.2016.2605698.

[10] Liu, D., and S. V. Georgakopoulos, "Cylindrical Misalignment Insensitive Wireless Power Transfer Systems," *IEEE Transactions on Power Electronics,* Vol. 33, No. 11, November 2018, pp. 9331–9343, doi: 10.1109/TPEL.2018.2791350.

[11] Lu, M., M. Bagheri, A. P. James and T. Phung, "Wireless Charging Techniques for UAVs: A Review, Reconceptualization, and Extension," *IEEE Access,* Vol. 6, 2018, pp. 29865–29884.

[12] Chen, J., R. Ghannam, M. Imran, and H. Heidari, "Wireless Power Transfer for 3D Printed Unmanned Aerial Vehicle (UAV) Systems," *IEEE Asia Pacific Conference on Postgraduate Research in Microelectronics and Electronics (PrimeAsia),* Chengdu, 2008.

[13] Na, W., J. Park, C. Lee, K. Park, J. Kim, and S. Cho, "Energy-Efficient Mobile Charging for Wireless Power Transfer in Internet of Things Networks," *IEEE Internet of Things Journal,* Vol. 5, No. 1, February 2018, pp. 79–92.

[14] Jiang, S., and S. V. Georgakopoulos, "Optimum Wireless Powering of Sensors Embedded in Concrete," *IEEE Transactions on Antennas and Propagation,* Vol. 60, No. 2, February 2012, pp. 1106–1113, doi: 10.1109/TAP.2011.2173147.

[15] Nikitin, P. V., K. V. S. Rao, and S. Lazar, "An Overview of Near Field UHF RFID," *IEEE International Conference on RFID,* March 2007, pp. 167–174.

[16] Finkenzeller, K., *RFID Handbook: Fundamentals and Applications in Contactless Smart Cards and Identification,* New York: Wiley, 2003, pp. 65–112.

[17] https://www.marketsandmarkets.com/Market-Reports/wearable-medical-device-market-81753973.html.

[18] Tesla, N., "High Frequency Oscillators for Electro-Therapeutic and Other Purposes," *Proceedings of the IEEE,* Vol. 87, July 1999, pp. 1282.

[19] Vandevoorde, G., and R. Puers, "Wireless Energy Transfer for Stand-Alone Systems: A Comparison Between Low and High Power Applicability," *Sensors and Actuators A: Physical,* Vol. 92, No. 1, August 2001, pp. 305–311.

[20] https://www.wirelesspowerconsortium.com/qi/.

[21] Kurs, A., A. Karalis, R. Moffatt, J. D. Joannopoulos, P. Fisher, and M. Soljaic, "Wireless Power Transfer Via Strongly Coupled Magnetic Resonances," *Science,* Vol. 317, No. 5834, July 2007, pp. 83–86.

[22] Barreto, J., A.- S Kaddour, and S. V. Georgakopoulos, "Conformal Strongly Coupled Magnetic Resonance Systems with Extended Range," *IEEE Open Journal of Antennas and Propagation*, Vol. 1, 2020, pp. 264–271, doi: 10.1109/OJAP.2020.2999447.

[23] Jonah, O., S. V. Georgakopoulos, and M. M. Tentzeris, "Optimal Design Parameters for Wireless Power Transfer by Resonance Magnetic," *IEEE Antennas and Wireless Propagation Letters,* Vol. 11, 2012, pp. 1390–1393.

[24] http://web.mit.edu/viz/EM/visualizations/coursenotes/modules/inductance.pdf.

[25] Balanis, C. A., *Antenna Theory: Analysis and Design,* Hoboken, NJ: Wiley, 2005.

[26] Lundin, R., "A Handbook Formula for the Inductance of a Single-Layer Circular Coil," *Proceedings of the IEEE,* Vol. 73, No. 9, September 1985 pp. 1428–1429.

[27] Fang, D., *Handbook of Electrical Calculations,* Jinan: Shandong Sci. Technol. Press, 1994.

[28] Volakis, J. L., *Antenna Engineering Handbook,* New York: McGraw-Hill, 2007.

[29] RamRakhyani, A. K., S. Mirabbasi, and C. Mu, "Design and Optimization of Resonance-Based Efficient Wireless Power Delivery Systems for Biomedical Implants," *IEEE Transactions on Biomedical Circuits and Systems,* Vol. 5, No. 1, February 2011, pp. 48–63.

[30] Wei, X., Z. Wang, and H. Dai, "A Critical Review of Wireless Power Transfer via Strongly Coupled Magnetic Resonances," *Energies,* Vol. 7, 2014, pp. 4316–4341, 10.3390/en7074316.

[31] Hu, H., K. Bao, J. Gibson, and S. V. Georgakopoulos, "Printable and Conformal Strongly Coupled Magnetic Resonant Systems for Wireless Powering," *WAMICON 2014,* 2014, pp. 1–4, doi: 10.1109/WAMICON.2014.6857762.

[32] Hu, H., and S. Georgakopoulos, "Multiband and Broadband Wireless Power Transfer Systems Using the Conformal Strongly Coupled Magnetic Resonance Method," in *IEEE Transactions on Industrial Electronics,* Vol. 64, No. 5, May 2017, pp. 3595–3607, doi: 10.1109/TIE.2016.2569459.

[33] Bao, K., C. L. Zekios, and S. V. Georgakopoulos, "Miniaturization of SCMR Systems Using Multilayer Resonators," in *IEEE Access*, Vol. 7, 2019, pp. 143445–143453, doi: 10.1109/ACCESS.2019.2945319.

[34] Barreto, J. C., A.- S. Kaddour, and S. V. Georgakopoulos, "Optimized and Miniaturized Conformal Strongly Coupled Magnetic Resonance Systems," *2020/2021 IEEE Wireless and Microwave Technology Conference (WAMICON),* Clearwater Beach, FL, April 28–29, 2021.

[35] Bao, K., C. L. Zekios, and S. V. Georgakopoulos, "A Wearable WPT System on Flexible Substrates," *IEEE Antennas and Wireless Propagation Letters*, Vol. 18, No. 5, May 2019, pp. 931–935, doi: 10.1109/LAWP.2019.2906069.

[36] Barreto, J., A.- S. Kaddour, and S. V. Georgakopoulos, "WPT Systems on Ferromagnetic Substrates with Enhanced Performance," *2020 IEEE International Symposium on Antennas and Propagation and North American Radio Science Meeting*, 2020, pp. 1337–1338, doi: 10.1109/IEEECONF35879.2020.9330258.

[37] https://www.fair-rite.com/flexible-ferrite/.

[38] Vogel, M., "Electromagnetic Safety in Wireless Communications and Bio-Medical Technologies," ANSYS white paper, ANSYS Inc., 2014.

[39] International Commission on Non-Ionizing Radiation Protection, "Guidelines for Limiting Exposure to Time-Varying Electric and Magnetic Fields (100 kHz–300 GHz)," *Health Physics,* Vol. 118, No. 5, March 2020, pp. 483–524.

[40] IEEE, C95.1-2019, "IEEE Standard for Safety Levels with Respect to Human Exposure to Electric, Magnetic, and Electromagnetic Fields, 0 Hz to 300 GHz," October 2019.

[41] Gosselin, M. C., E. Neufeld, H. Moser, et al., "Development of a New Generation of High-Resolution Anatomical Models for Medical Device Evaluation: The Virtual Population 3.0," *Physics in Medicine and Biology*, 2014, pp. 5287–5303.

About the Editors

Asimina Kiourti is an assistant professor in the Department of Electrical and Computer Engineering at The Ohio State University where she leads the Wearable and Implantable Technologies group. From 2013 to 2016, she served as a postdoctoral researcher and then a senior research associate at The Ohio State University ElectroScience Laboratory. Prior to that, she received a PhD in electrical and computer engineering from the National Technical University of Athens, Greece, in 2013 and an MSc from University College London, United Kingdom, in 2009.

Professor Kiourti's research interests are in bioelectromagnetics, wearable and implantable antennas, sensors for body area applications, and flexible e-textiles. Her work lies at the intersection of electromagnetics, sensors, materials, and medicine, and entails interdisciplinary collaborations with five colleges within The Ohio State University and beyond. Her team has developed (1) functionalized garments for seamless motion capture in real-world environments, (2) wireless and fully passive brain implants, (3) sensors for sensing the biomagnetic fields that are naturally emanated by the human body without the need for shielding, (4) into-body radiation antennas with unprecedented efficiency and bandwidth, useful for applications as diverse as radiometry and microwave imaging, and (5) conductive textile surfaces that exhibit radiofrequency performance similar to their copper equivalents while feeling and behaving like regular fabric.

As of August 2021, Professor Kiourti has mentored over 20 graduate and over 35 undergraduate researchers and has published with them 12 book chapters, 5 awarded patents, over 60 journal papers, and over 110 conference papers and abstracts. Her work has been recognized with over 40 international, national, and local awards, including the 2021 invitation to the National Academy of Engineering's US Frontiers of Engineering Symposium, the 2021 40 Under 40 recognition by Columbus Business First, the 2018 URSI Young Scientist Award, the 2014 IEEE Microwave Theory and Techniques Society Graduate Award for Medical Applications, and the 2011 IEEE Antennas and Propagation Society Doctoral Research Award. Her mentees have received another 40 awards at the international, national, and local levels. Her team's research contributions have also been featured by *TechCrunch, Gizmodo*, the *Times of India, Australia Network News,* and the *ALN Magazine,* among others.

Professor Kiourti is active in IEEE and USNC-URSI, where she serves in several elected and appointed roles, including Member of the IEEE Antennas and Propagation Society Meetings Committee, Member of the IEEE Antennas and Propagation

Society Young Professionals Committee, Secretary of USNC-URSI Commission K, and Founding Member of the USNC-URSI Women in Radio Science Chapter. She is the senior editor of the *IEEE Open Journal of Antennas and Propagation,* editor of the *IEEE Antennas and Propagation Magazine* "Bioelectromagnetics" column, and associate editor for the *IEEE Transactions on Antennas and Propagation,* the *IEEE Journal of Electromagnetics RF and Microwaves in Medicine and Biology,* and the *IEEE Antennas and Propagation Magazine.*

John L. Volakis is an IEEE, AAAS, NAI, URSI, and ACES Fellow. He currently serves as the dean of the College of Engineering and Computing at Florida International University (FIU), and a professor in the Electrical and Computer Engineering Department. Prior to joining FIU, he was the Roy and Lois Chope Chair in Engineering at Ohio State and a professor in the Electrical and Computer Engineering Department from 2003–2017. He continues to be an adjunct professor at Ohio State. He has also served as the director of the Ohio State University ElectroScience Laboratory for 14 years. His career spans 2 years at Boeing, 19 years on the faculty at the University of Michigan-Ann Arbor, and 15 years at Ohio State. At Michigan he also served as the director of the Radiation Laboratory from 1998–2000.

Professor Volakis has 39 years of engineering research experience, and has published over 435 journal papers, 900 conference papers, and over 30 book chapters. In 2004, he was listed by ISI Web of Science as one of the top 250 most referenced authors, and his google h-index=72 with over 26,500 citations, among the largest in engineering. He mentored nearly 100 PhD and postdoctoral students and has written with them 43 papers that received best paper awards. He is one of the most active researchers in electromagnetics, RF materials and metamaterials, antennas and phased array, RF transceivers, textile electronics, millimeter waves and terahertz, EMI/EMC, as well as EM diffraction and computational methods. He is also the author of eight books, including the *Antenna Engineering Handbook,* referred to as the "antenna bible." His research team is recognized for introducing and/or developing hybrid finite method for microwave engineering; now de facto methods in commercial RF design packages; novel composite materials for antennas and sensor miniaturization; a new class of wideband conformal antennas and arrays with over 30:1 of contiguous bandwidth, referred to as tightly coupled dipole antennas, already garnering over 9 million citations; textile surfaces for wearable electronics and sensors; batteryless and wireless medical implants for noninvasive brain signal collection; diffraction coefficients for material coated edges; and model-scaled radar scattering verification methods.

His service to professional societies includes 2004 President of the IEEE Antennas and Propagation Society, Chair of the International Union Radio Science-B (URSI-B) from 2020–2023, US URSI-B Chair from 2015–2017, twice the general chair of the IEEE Antennas and Propagation Symposium, 2019 cochair of the Applied Computational Electromagnetics Society (ACES) conference, and the 2019 chair of the International Workshop on Antenna Technologies (iWAT). He also served as an IEEE APS Distinguished Lecturer, IEEE APS Fellows Committee Chair, IEEE-wide Fellows committee member, and associate editor of several journals. Among his awards are University of Michigan College of Engineering Research Excellence award (1993), IEEE Tai Teaching Excellence award (2011), IEEE Henning Mentoring award (2013), IEEE APS Distinguished Achievement award (2015),

Ohio State University Distinguished Scholar Award (2016), Ohio State ElectroScience Lab Sinclair award (2016), and International Union of Radio Science Booker Gold Medal (2020).

About the Contributors

Abdulhameed Abdal joined the Bioelectronics Lab at Florida International University as a research associate in early 2020. His research contributed to several IEEE conferences, review articles, and proposals. The main focus of his research is on thermomechanical modeling and reliability engineering of thin flexible packages. He received his MS in mechanical engineering from Loyola Marymount University (2019), and BS from Drexel University (2017). Currently, he is at the University of California, San Diego, as a PhD student in the Stretchable Wearable Electronics Lab.

Juan Barreto received BS and MS degrees in electrical engineering from Florida International University, Miami, Florida, in 2019 and 2020, respectively. Since 2021, he has been a research associate with the Transforming Antennas Center at Florida International University. Currently, his main research interest is in wireless power transfer systems.

Shubhendu Bhardwaj finished his master's from UCLA (2012), PhD from Ohio State (2017), and joined Florida International University in August 2017 as an assistant professor. He completed his bachelor's degree from IIT Dhanbad (previously ISM-Dhanbad). His past research experience spans the areas of electromagnetics and applications, including computational modeling, terahertz and millimeter-wave antennas, and semiconductor devices. He has received multiple best-paper awards for his research work, including at the URSI-GASS-2017, iWat-2017, and AMTA-2014 and 2015 conferences, and awards at the Ohio State University, Presidential Fellowship and Louise B. Vetter Award.

Giulio Maria Bianco received a laurea degree in medical engineering at the University of Roma Tor Vergata, Italy, where he is currently pursuing a PhD in computer science, control, and geoinformation. He works within the Pervasive Electromagnetics Lab and the European Academy of Bolzano. His research is focused on wireless wearable devices based on the LoRa platform and radiofrequency identification. Mr. Bianco is a member of the Pontifical Council for Culture Youth Forum, where he is head of the Science thematic area. He won the Young Scientist Award from URSI with his work on radiofrequency finger augmentation devices for the tactile internet in 2020.

Balaji Dontha received a bachelor's degree in electronics and telecommunication engineering from the University of Mumbai, India, in 2018. He is currently pursuing a PhD degree in the Department of Electrical and Computer Engineering,

The Ohio State University, Columbus, Ohio, under the supervision of Dr. A. Kiourti. He is presently working with the ElectroScience Laboratory, The Ohio State University. His research interests include electromagnetics, wearable electronic sensors, and textile antennas for body-centric wireless communications.

Stavros V. Georgakopoulos received a diploma degree in electrical engineering from the University of Patras, Patras, Greece, in June 1996, and MS and PhD degrees in electrical engineering from Arizona State University (ASU), Tempe, Arizona, in 1998 and 2001, respectively. From 2001 to 2007, he held a position as the principal engineer at SV Microwave, Inc. Since 2007, he has been with the Department of Electrical and Computer Engineering, Florida International University, Miami, Florida, where he is currently a professor and director of the Transforming Antennas Center (a research center on foldable/origami, physically reconfigurable and deployable antennas).

Akeeb Hassan is an MS student at Florida International University in the Biomedical Engineering Department. He joined the Bioelectronics Lab as a research engineer, where his primary focus is on remateable implantable bioelectronics, and electromechanical characterization of flexible interconnects in embedded architectures. He also codeveloped the flex-embedded fan-out packages. He received his BS in biomedical science from the University of South Florida.

Umar Hasni received his undergraduate degree in electrical engineering at Virginia Commonwealth University (VCU). During his time at VCU, he researched electromagnetic wave interactions with the human body. He went on to study these effects culminating into the development of a novel antenna model for single resonance across multiple mediums in application of wearable antennas. After receiving his PhD, Umar went on to work in the aviation industry designing antennas for unmanned aerial autonomous vehicle platforms.

Abdul-Sattar Kaddour received a BS in electronics from the Lebanese University, Faculty of Science, Lebanon, in 2011, BS and MS degrees in electronics and embedded systems engineering from the Grenoble Institute of Technology, Grenoble, France, in 2012 and 2014, respectively, and a PhD in optics and radiofrequency from the Grenoble Alpes University, Grenoble, France in 2018. Since 2019, he has been a research fellow in the Transforming Antennas Center (TAC) at Florida International University, Miami, Florida. His main research interests include electrically small antennas, reconfigurable antennas, antenna arrays, reflectarrays, and wireless power transfer systems.

Jonathan Lundquist is a researcher at Virginia Commonwealth University's Wireless Communications Laboratory in the Department of Electrical and Computer Engineering. His research focus has been wearable electronics as well as subsurface communications. Jonathan served two enlistments in the United States Navy and worked as a technical specialist at Control Automation Technologies Laboratories prior to attending VCU. Jonathan is an active member of Richmond Virginia's Maker Community and occasionally volunteers his time teaching the skills he has learned in academia to other members of the community.

Gaetano Marrocco received a laurea in electronic engineering and PhD in applied electromagnetics at the University of L'Aquila, Italy. He is currently a full professor at the University of Roma Tor Vergata, where he chairs the Medical Engineering School and the Pervasive Electromagnetics Lab. His research has been recently focused on wireless-activated sensors, and in particular on wearable, epidermal, and implantable devices. Prof. Marrocco is chair of the Italian URSI Commission D Electronics and Photonics and cofounder and president of the university spin-off RADIO6ENSE, and is active in the areas of the Industrial Internet of Things, food, and biotech. He is listed in the *PLOS* ranking of Top 1E5 Scientist Worldwide.

McKenzie Piper is a biomedical engineering student at the VCU College of Engineering. She has been involved in research at the Medical Device Design and Prototyping Lab since August 2019 where she works to develop wearable antennas for a variety of proactive health and safety applications. She has published several conference papers and a journal paper in the *IEEE Open Journal of Antennas and Propagation*. She is also the recipient of the Dean's Scholarship and the Qimonda Endowed Scholarship awards. She is the cofounder and CEO of TekStyle.

Pulugurtha Markondeya Raj's expertise is in packaging of electronic and bioelectronic systems. He is an associate professor in BME and ECE at Florida International University. His research led to 340 publications, which include 8 patents. He received more than 25 best-paper awards. He coadvised more than 30 MS and PhD students who are current leaders and technology pioneers in the electronic packaging industry. He earned a PhD from Rutgers University in 1999 in ceramic engineering, a ME from the Indian Institute of Science, Bangalore, and a BS from the Indian Institute of Technology, Kanpur, in 1993.

Kelly Nair Rojas is an undergraduate biomedical engineering student at Florida International University. Her focus is on low-impedance microelectrode arrays to adapt a passive wireless neural recording system that will enable low-cost health monitoring. She also investigates electrode designs on flexible substrates to realize matched-impedance neural recording.

Erdem Topsakal is currently a tenured full professor and department chair at Virginia Commonwealth University. His research areas include antennas for medical and commercial applications including wearable and implantable antennas, numerical electromagnetics, and direct and inverse electromagnetic scattering. He has published over 200 journal and conference papers in these areas. He is a senior member of IEEE and an elected member of the URSI commissions B and K. He is the chair of the USNC URSI student paper competition. He serves on the steering committee for the *Journal of Electromagnetics, RF and Microwaves in Medicine and Biology* (IEEE J-ERM).

Satheesh Bojja Venkatakrishnan received his bachelor's degree in electronics and communication engineering from the National Institute of Technology, Tiruchirappalli, in 2009, and graduated with a PhD in electrical engineering from the Ohio State University (OSU), Columbus, Ohio, in 2017. He is currently a research assistant professor, electrical and computer engineering, at Florida International University. He has been a Phi Kappa Phi member since 2015 and is an associate member of USNC–URSI since 2021. He has won numerous awards including Young Scientist Award at EMTS 2019. He has authored more than 20 journal articles and 50 conference presentations.

Dieff Vital received a BSc degree (summa cum laude) in mechanical and industrial engineering from the Florida Polytechnic University, Florida, in 2017. He received a PhD in electrical and computer engineering at Florida International University (FIU), Miami, Florida, in 2021. His research interests include RF systems on-textile, wireless power transfer, and harvesting for Internet of Things applications and smart dressing solutions for electrochemical sensing and monitoring. He was a recipient of the McKnight Dissertation Fellowship. He is currently a postdoctoral associate in the Department of Electrical and Computer Engineering, University of Illinois Chicago.

Index

3-D digital image correlation (3-D DIC), 70
5G R-FADs, 198

A

Activated carbon (AC), 49
Actuators, textile-based, 36
Adaptive histogram equalization (AHE), 235
Ag/AgCl electrodes, 74–75
Amberstrand, 15
Analog-digital converters (ADCs), 43
Anchor-shaped antenna
  about, 146
  angular misalignments, 153
  E and H fields at resonant frequency, 154
  electric and magnetic field lines, 155
  frequency modulations, 152
  half-wavelength resonance, 147–48
  lateral misalignment, 153
  loop antenna comparison, 148
  magnetic versus electric coupling, lateral mismanagement, 157–59
  measurements, 159–60
  misalignment between transmitter and receiver, 155–57
  misalignment-resilient geometries, 163
  misalignment tests, 161
  near-field characteristics, 153–55, 159
  outer radius, 149, 151
  performance, 149
  power transfer characteristics, 155–57
  PTE, 148, 149, 150
  PTE, different degrees of misalignment, 157
  PTE comparison, 156
  self-capacitance, 150
  self-inductance, 149–50
  strip-line width, 150
  system integration, 146
  textile integration of, 163–67
  total capacitance, 151
  wave-impedance characteristics, 159
  wireless transfer system, 152
  See also Near-field integrated power transfer and harvesting
Angular misalignment
  about, 153
  in azimuth plane, 162
  in elevation plane, 162
  See also Anchor-shaped antenna
Anisotropic conductive adhesives (ACAs), 61–62, 85
Anisotropic conductive films (ACFs), 62, 64–65, 85
Anisotropic conductive pastes (ACPs), 85
Antennas, wearable
  about, 91–92
  anchor-shaped, 146–75
  applications of, 104–10
  biomatched, 109–10
  classification of, 92
  conductive fibers, 103
  conductive inks, 104
  conformal Archimedean spiral, 34
  corrugated crossed-dipole, 124
  design, 12
  dipole prototype, 20
  efficiency, crumpling cases, 31
  embroidered, 92–95
  e-textile, 5
  fabric coplanar keyhole, 105–6
  fabrics, 102–3
  fingertip, 191
  graphene, 107–8
  inkjet-printed, 100–102
  launderability, 31
  material considerations, 102
  nonfabric, fabrication techniques, 10
  patch, 134–38
  PIFA, 28, 29, 30

Antennas, wearable (continued)
RFID tag, 32, 34
screen-printed, 95–100
on-silicon sensors, 189
textile-based, 32–33
wrist, 189
Artificial intelligence/machine-learning (AI/ML) techniques, 115
Artificial magnetic conductors (AMC), 15
Augmented reality (AR) devices, 9

B

Back-end-of-the-line (BEOL), 64
Backprojection algorithm, 234
Backscattering imaging, 205–12
Bal Seal connectors, 80, 83
Batteries, 48–50
Bessel function, 210
Biological tissues, dielectric properties, 206–7
Biomarker extraction, 116–17
Biomatched antennas, 109–10
Biomedical sensing to wireless communication, 45
Biomedical wearable health-monitoring products, 41–43
Biosignal interfaces
about, 71
Ag/AgCl electrodes, 74–75
carbon-based electrodes, 76
dry electrodes, 75–76
electrochemical electrodes, 78
fractal gold electrodes, 77–78
long-term stability and, 73–74
low-impedance with skin and, 72
polymer-based electrodes, 76–77
skin compatibility and, 72–73
See also Electronics, wearable
Boundary shapes, phantoms with, 222

C

Capacitive micromachined ultrasound transducers (CMUTs)
about, 204
efficiency of, 223
integration of, 224
nozzle placement, 225
piezoelectric transducers versus, 223
probes, 223
Carbide-derived carbon (CDC), 49
Carbon-based electrodes, 76
Carbon nanotubes (CNTs), 48
Carotid intima-media (CIMT), 225
Ceramic-reinforced polymers, 21
Charge-coupled device (CCD), 230
Chemical vapor deposition (CVD), 106
Chip-embedded in flex, 64–65
Circuit formation
about, 55
flexographic, 57
gravure printing, 57
inkjet printing, 57
laser ablation patterning, 58–59
lithographic photopatterning, 58
nanoimprint lithography (NIL), 58
options and technology capabilities, 56
roll-to-roll (R2R) manufacturing, 59–60
screen-printing, 56–57
self-aligned imprint lithography (SAIL), 58
Cognitive remapping, 197
Cole-Cole dispersion, 216–17
Conducting elastomer inserts, 83
Conductive thread embroidery patch antenna, 134–38
Conductive threads, 14–16
Conformal Archimedean spiral antenna, 34
Conformal strongly coupled magnetic resonance (CSCMR)
about, 249
back implantation illustration, 262
chest implantation illustration, 260–62
efficiency, 258
head implantation illustration, 260
human body placement, 251
for implantable applications, 258–60
magnetic field distribution, 250–51, 254–55
magnetic field intensity, 251, 254, 255
maximum psSAR of, 256
measurement setup, 252
performance on human body, 251–54
PTE, 250, 253
specific absorption rate, 255–58

square, geometry, 259
substrates, 255
system design, 249–51
system geometry, 249
TX/RX elements, 258
for wearable applications, 249–58
Connectors, flat, 81–84
Coplanar waveguide (CPW) feeding, 15
Copper wire, e-threads versus, 24, 25
Corrugated crossed-dipole antenna, 124
Cross-correlation (CCR), 231
CST Design Studio, 217

D

Defense applications, 4
Deformation, effect of, 27–28
Device and component assembly
about, 60
anisotropic conductive adhesives (ACAs), 61–62
chip-embedded in flex, 64–65
dielectric ramp interconnections, 63–64
isotropic conductive adhesives or films (ICAs/ICFs), 62–63
low-temperature soldering, 60–61
See also Electronics, wearable
Dielectric ramp interconnections, 63–64
Dielectric-sensing R-FADs, 194–95
Digital fingerprints, 195
Direct flex transfer, 69–70
DOWSIL, 66
Drawing Interchange File (DXF), 13
Dry electrodes, 75–76

E

Elastomeric stamp, transfer with, 68
Electrical double-layer capacitance (EDLC), 48–49
Electrical impedance tomography, 220
Electrochemical electrodes, 78
Electrodes
Ag/AgCl, 74–75
carbon-based, 76
classification of, 71–74
dry, 75–76
electrochemical, 77–78
fractal gold, 77–78
long-term stability, 73–74
low impedance with skin and, 72
mini array, 79
polymer-based, 76–77
skin compatibility, 72–73
Electromagnetic (EM) fields, 124
Electromagnetic and circuit component applications, 33–35
Electromagnetic field (EMF) absorption, 242
Electromagnetic interference (EMI) shielding, 66
Electronics, wearable
active components, 46
biosignal interfaces, 71–78
circuit formation, 55–60
device and component assembly, 60–65
encapsulation, 66
energy storage, 48–50
functional building blocks, 43–50
inductive link constraints and, 46–47
passive components, 45–46
power and data telemetry, 46–48
remateable connectors, 78–85
system architecture and components, 43–46
technology building blocks, 50–66
technology drivers, 41–43
transferable on-skin electronics, 66–71
Elektrisola-7, 134–35, 171
Elektrisola, 15
Embroidered antennas
about, 92–93
advantages of, 93–94
construction of, 94–95
design of, 93, 94–95
disadvantages of, 94
fabric substrate, 93
illustrated, 94
simulation of, 95
See also Antennas, wearable
Embroidery
advanced aspects of, 17–20
antennas and, 92–93
basics of, 11–14
colorful prototypes, 19–20

Embroidery (continued)
density, grading, 18–19
elements of, 13–14
precision, improving, 17–18
process, operating principle, 11–14
prototypes, substrates used for, 15–16
See also e-textiles
Emergency applications, 4
Encapsulation
effect on interconnect stresses, 67
with PDMS, 64, 66
thin-film (TFE), 66
wearable electronics, 66
Energy storage, 48–50
e-textiles
antennas, 32–33
conductive threads, 14–16
digitization, 12–13
effect of deformation, 27–28
electromagnetic and circuit components, 33–35
embroidery, 11–20
example applications, 31–36
flex substrate embedding into, 70–71
nonconductive threads, 16–17
performance, 24–31
sensors and actuators, 36
testing and validation, 14
e-threads
in bobbin case, 20
copper surfaces versus, 24–25
copper wire versus, 24, 25
ground plane, effect of, 25–27
read range test, 35
transmission lines using, 26

F

Fabric-based sensors, 11, 12
Fabric coplanar keyhole antenna, 105–6
Fabrics, 5–6, 102–3
Far-field integrated power transfer and harvesting
conductive thread embroidery patch antenna, 134–38
introduction to, 133–34
power harvesting, 143–45
RF-power availability tests, 140–43
textile-based single-diode rectifier, 138–40
textile rectenna array, 140
Ferroelectrics, 47
Fibers, conductive, 103
Fidelity factor, radar-based, 217–20
Finger augmentation devices (FADs), 181
Fingertip antennas, 191
Fingertip sensors, 183, 189
Fingertip-wrist link
challenges, 183
system power gain, 184
two-port model, 183, 184
Finite-difference time domain (FDTD), 258
Fitness bands, 2
Flat connectors (FCs)
about, 81–82
conducting elastomer inserts, 83
magnetic snap-on, 82–83
z-elastomer interconnects, 83–84
See also Remateable connectors
Flexographic, 57
Flip-chip structure, 63
Fractal gold electrodes, 77–78
Fractional parameters, 212–16
Functional near-infrared spectroscopy (fNIRS), 204, 227

G

Gaussian pulse signal, 217, 218
Gauss-Newton (GN) method, 212, 215
Graphene antenna, 107–8
Graphene electrodes, 49–50
Gravure printing, 57
Green's function, 209, 211

H

Hankel function, 210
Harvesting
far-field, 133–45
near-field, 145–75

I

Imaging
backscattering, 205–12
domain, geometrical configuration of, 214

head, experimental setup of, 219
microwave (MWI), 204
performance, 213
photoacoustic, 203–5, 229–35
photoacoustic microscopy (PAM), 205
radar-based RF and THz, 205–23
simulated data from CST and, 218
ultrasound, 223–25
Implanted CSCMR
about, 258
in back, 262
on chest, 261
efficiency, 258
on head, 260
PTEs, 262
SIM4LIFE model, 258, 259
square, geometry, 259
See also Conformal strongly coupled magnetic resonance (CSCMR)
Inductive power transfer (IPT), 242, 244–45
Inkjet-printed antennas
about, 100
design and construction of, 101–2
illustrated, 101
inks, 100, 101
resolution and, 102
substrates and, 100
See also Antennas, wearable
Inkjet printing, 57
Inks, conductive, 104
Innovis VM5100 embroidery software, 134
Interferometric-based optical ultrasound detection, 229, 230
Internet of Things (IoT), 1
Isotropic conductive adhesives or films (ICAs/ICFs), 62–63

J

Jacobian matrix, 215, 221

L

Laser ablation patterning, 58–59
Laser-assisted release, 67–68
Lateral misalignment
about, 153
magnetic versus electric coupling, 157–59
PTE for, 155
SAR evaluation and, 168
wrapping over cylindrical surfaces and, 165, 166
See also Anchor-shaped antenna
Launderability, 31, 33
Lead zirconate titanate (PZT), 47
Liberator, 14
Linear frequency modulated (LFM) waveform, 232
Liquid crystal polymers (LCPs), 21, 52
Lithographic photopatterning, 58
Loop antenna
anchor-shaped antenna comparison, 148
E and H fields at resonant frequency, 154
evolution of anchor topology from, 164
PTE comparison, 156
Loss-compensated backpropagation (LC-BP) method, 209
Low-temperature soldering, 60–61

M

Mach-Zehnder interferometry (MZI), 229
Mag-Net connector, 81
Magnetic flux, 247
Magnetic resonance coupling (MRC), 242
Magnetic snap-on connectors, 82–83
Magneto-actuated prototypes, 23
Mechanical performance, 28–31
Medicine applications, 4–5
Metal foils, 54–55
Metallization, 55
Michelson interferometry (MI), 229
Microsupercapacitor (MSC) devices, 49
Microwave imaging (MWI), 204
Monostatic radar-based fidelity factor, 217–20

N

Nanoimprint lithography (NIL), 58
Near-field integrated power transfer and harvesting
anchor-shaped antenna, 146–63
conclusions, 175–76
introduction to, 145–46
RF-to-DC rectifier, 167–71
system design and tests, 172–75

Near-field integrated power transfer and harvesting (continued)
system integration illustration, 146
textile integration of anchor-shaped antenna, 163–67
transmitter and receiver antenna misalignment and, 147
Nonconductive threads, 16–17
Non-Ionizing Radiation Protection (ICNIRP), 242
Novacentrix inks, 96

O

Onion-like carbon (OLC), 49
Optical parametric oscillator (OPO), 204
Optical resolution photoacoustic microscopy (ORPAM), 235
Optical time-of-flight sampling, 226
Optical tomography (OT), 225–26
Organic-LED (OLED), 54
Organic surface protection (OSP) layer, 64–65
Organization, this book, 6–7
Organ monitoring sensor
comparison of values, 207, 208
electrodes, 206
functionality, validation of, 207
multistatic measurement scheme, 211
overview illustration, 206
placed around human torso, 210
placed on mannequin, 208

P

PAI imaging technique, 229
Paper substrates, 54
Patch antennas, 134–38
Perfect electrical conductor (PEC) model, 136
Performance
anchor-shaped antenna, 149
coil resistance and self-inductance results, 24
CSCM, 251–54
e-textiles, 24–31
imaging, 213
launderability and, 31, 33
mechanical, 28–31
power harvesting, 141–42
radio-frequency, 24–28
smart bandage, 122
strongly coupled magnetic resonance (SCMR), 248
Permittivities, 215
Phase shift migration (PSM) method, 232
Photoacoustic (PA) sensors, 235
Photoacoustic imagers, 231, 233
Photoacoustic imaging
about, 203–4, 229
illustrated, 231
implementation of, 230
Photoacoustic microscopy (PAM) imaging, 205
Photonic soldering, 62
Physical vapor deposition (PVD), 55
Piezo-magnetostrictive interfaces, 48
Pin-grid array (PGA), 81, 84–85
Pin-socket connectors, 80–81
Planar inverted F antenna (PIFA), 28, 29, 30
Polydimethylsiloxane (PDMS)
about, 16
bubble-free mixture, 23
encapsulation, 64, 66
polymer, 21, 22
stamp, 58
Polymer
integration, 21–23
magneto-actuated prototypes, 23
stretchable prototypes in, 21–23
substrates, 21, 51
Polymer-based electrodes, 76–77
Power harvesting (far-field)
measurements, 143–45
performance, 141–42
potential applications, 145
with textile rectenna arrays, 143–45
Power transfer efficiency (PTE)
anchor antenna compared to loop antenna, 156
anchor-shaped antenna, 145, 148–50, 155, 157
angular misalignment in azimuthal plane, 162
angular misalignment in elevation plane, 162
CSCMR system, 250, 253

under different degrees of misalignment, 157
effects of mechanical deformation and misalignment on, 163–67
implanted CSCMR, 262
lateral misalignment, 155, 160–62
near-field, 124, 145
performance of polygonal configurations, 160
potential applications, 163
Prototypes
colorful, 19–20
dipole antenna, 20
foldable, grading density for, 18–19
magneto-actuated, 23
stretchable, embedded in polymer, 21–23
substrates used for, 15–16
Pseudocapacitance, 48

Q

Q-factors, 246–47

R

Radar-based microwave imaging (MWI), 204
Radar-based RF and THz imaging
backscattering imaging and, 205–12
electrical impedance tomography, 220, 221–23
fractional parameters and, 212–16
monostatic fidelity factor, 217–20
reconstructed images, 220
time-domain approach, 216–17
Radiofrequency finger augmentation devices (R-FADs)
about, 181
application to cognitive remapping, 197
autotuning ICs, 186–87
body-worn, 185
conclusions, 197–98
constrained design of, 186–88
defined, 182
dielectric-sensing, 194–95
differential SCs returned by, 196
discrimination of materials, 194–96
fingertip sensors, 183
fingertip-wrist backscattering link, 183–86
5G, 198
hand movements and, 190
hypoesthesia and, 191
loop-matching fingertip antenna, 188
manufacturing, 189–91
multichannel, 190, 195–96
multichannel temperature sensing, 192
multisensor prototype, 186
plaster-finger sensors, 190
powering, 185
prototype testing, 185
self-tuning fingertip sensor, 187
sensorial impairment applications, 191–96
single-sensor, 195
temperature sensing, 192–94, 197, 198
training, 197
use of, 182
warm/cold sensing, 192
wrist antennas, 189
wrist-worn readers, 189
Radio frequency identification (RFID) sensing, 9, 31
Radio-frequency performance, 24–28
Radiometric-based optical ultrasound detection, 231
Redistribution layer (RDL) lamination, 64
Remateability, 78–79
Remateable connectors
about, 79
pin-socket, 80–81
reworkable adhesives, 84–85
technology classification, 80
Resonant inductive coupling (RIC), 245
Reworkable adhesives, 84–85
RF backscattering, 203
RFID tag antennas, 32, 34
RF-power availability tests, 140–43
RF-to-DC rectifier
about, 167
conjugate matching, 169, 170
conversion efficiencies, 170
design and optimization of, 167–71
fabric-based, 169
optimized parameters, 171
working principle of, 168–71
Roll-to-roll (R2R) manufacturing, 59–60

S

Scattering parameters, 212
Schottky diode, 168
Screen-printed antennas
  about, 95
  advantages of, 98
  coatings, 98
  composition of fabrics, 97
  design and construction of, 99–100
  disadvantages of, 98–99
  fabrication process, 99–100
  inflexibility of ink and, 99
  inks, 96
  patch illustration, 98
  performance, 95–96
  RFID, 107
  sheet resistance of fabrics, 97
  See also Antennas, wearable
Screen printing, 56–57, 95–96
Self-aligned imprint lithography (SAIL), 58
Sensorial impairment
  discrimination of materials, 194–96
  R-FAD applications, 191–96
  sensing an item's temperature, 192–94
Sensors, wearable
  about, 115–16
  for biomarker extraction, 116
  examples of, 3
  experimental setup, 129
  fabric, fabrication techniques, 11
  fingertip, 183, 189
  measurement setup, 127–29
  nonfabric, fabrication techniques, 10
  smart bandage integration, 123–24
  textile-based, 36
  textile-based VCO, 118–19
  wireless power telemetry link, 124–27
  for wound assessment with data modulation, 119–23
  for wound monitoring, 117–18
Shape memory polymers (SMPs), 16
SIM4LIFE, 258, 259
Single-diode rectifier, textile-based, 138–40
Smart bandage
  benchtop setup of, 121
  diagram illustration, 128
  performance of benchtop solution of, 122
  photos, 120
  for practical measurements, 123–24
  RFID, 117–18
  system components, 123
Space applications, 4–5
Specific absorption rate (SAR)
  about, 256
  adverse effects of, 256
  distributions, 257
  ICNIRP, 256–57
  lateral misalignment and, 168
  safety and, 255–56
  textile-based anchor-shaped antenna, 167
Strongly coupled magnetic resonance (SCMR)
  about, 245
  conformal (CSCMR), 249–60
  efficiency, 246, 247, 248
  equivalent circuit, 246
  performance, 248
  schematic, 245
  See also Wireless power transfer system
Substrates
  circuit transfer from, 69
  CSCMR, 255
  flexible, 53, 70
  inkjet-printed antennas and, 100
  metal foil, 54–55
  paper, 54
  polymer, 21
  for prototypes, 15–16
  thin, 51–55
  thin glass, 54
Supercapacitors, 48–50
System-on-chip (SoC) devices, 43
System power gain, 184

T

Tactile internet (TI), 181, 182
Taylor series expansion, 215
Temperature sensing, 192–94, 197
Temperature-sensing R-FADs, 192–94, 197, 198
Textile-based antennas
  about, 32–33
  anchor-shaped, 163–67

human body effects on, 138
single-diode rectifier, 138–40
types of, 32–33
See also specific antennas
Textile-based rectifier, 124–27
Textile-based VCO, 118–19
Textile integration (anchor-shaped antenna), 163–67
Textile rectenna array
design and optimization of, 140
efficiency comparison, 140
fabrication, 140
power-harvesting performance, 141–42
power harvesting with, 143
RF-power availability tests, 140–43
six-element prototype, 141
Thermal-assisted release, 67–68
Thermoplastic elastomers, 52–53
Thermoplastic nanocrystalline plastics, 52
Thermoplastic polyurethane (TPU), 53
Thin-film encapsulation (TFE), 66
Thin-film transistors (TFTs), 43–44
Thin glass substrates, 54
Thin substrates, 51–55
Time-domain approach, radar-based, 216–17
Tissue-mimicking gels, 104, 105
Transferable on-skin electronics
about, 66–67
broad classification of, 67
direct flex transfer, 69–70
flex substrate embedding, 70–71
laser or thermal-assisted release, 67–68
transfer with elastomeric stamp, 68
transfer with water-soluble tape, 68–69
See also Electronics, wearable
Transmission lines, embroidered, 26, 33

U

UHF RFID tag antenna, 32
Ultrasound imaging, 223–25, 226–27
Ultrasound transducers, 203
Ultra-thin glass, 54

V

Vector network analyzer (VNA), 211, 217, 219, 253
Virtual reality (VR) devices, 9
Voltage-controlled oscillator (VCO)
about, 116
DC-to-RF modulation, 129
in measurement setup, 127
output frequency variation, 122
textile-based, 118–19
for wound assessment with data modulation, 119–23
in wound monitoring RFID bandage, 117–18
Voltage standing wave ratio (VSWR), 32

W

Water-soluble tape, transfer with, 68–69
Wearable antennas. *See* Antennas, wearable
Wearable EIT, 221–23
Wearable electronics. *See* Electronics, wearable
Wearable imaging
about, 203–5
algorithms, 205–23
optical tomography, 225–29
photoacoustic, 229–35
radar-based RF and THz, 205–23
ultrasound imaging, 223–25
Wearable optical topography (WOT), 228
Wearables
applications of, 2–5
concurrent research, 4–5
defense applications, 4
emergency applications, 4
far-field power transfer and harvesting for, 133–45
future of, 5–6
history of, 1–2
market growth, 3
medicine applications, 4
milestones in evolution of, 2
near-field power transfer and harvesting for, 145–75
printing and weaving approaches, 9–36
space applications, 4–5
See also Antennas, wearable; Electronics, wearable; Sensors, wearable; *specific types of wearables*
Wearable sensors. *See* Sensors, wearable

Windows Metafile Format (WMF), 13
Wireless power telemetry link, 124–27
Wireless power transfer (WPT)
  applications of, 241
  human body with, 242
  introduction to, 241–42
  losses, 254–55
  methods, 242–49
  misalignments and, 46
  system diagram, 243
Wireless power transfer system
  conclusions, 260–62
  efficiency, 248
  illustrated, 172–73
  inductive power transfer (IPT), 242, 244–45
  measurement results, 174–75
  photo of measurement setup, 173
  quantitative power collection setup, 173–74
  resonant inductive coupling (RIC), 245
  RF power measurement, 174
  RF-to-DC rectifier, 167–71
  SAR evaluation, 167, 168
  strongly coupled magnetic resonance (SCMR), 242, 245–49
  system design and tests, 172–75
  textile integration, 163–67
  See also Anchor-shaped antenna
Wound assessment with data modulation, 119–23
Wound monitoring RFID bandage, 117–18
Wrist-worn readers, 189

X

X-steel, 14

Y

Young's moduli, 191

Z

Z-elastomer interconnects, 83–84

Recent Titles in the Artech House Antennas and Propagation Library

Christos Christodoulou, Series Editor

Adaptive Array Measurements in Communications, M. A. Halim

Antenna-on-Chip: Design, Challenges, and Opportunities, Hammad M. Cheema, Fatima Khalid, and Atif Shamim

Antenna Design for Cognitive Radio, Youssef Tawk, Joseph Costantine, and Christos Christodoulou

Antenna Design for CubeSats, Reyhan Baktur

Antenna Design with Fiber Optics, A. Kumar

Antenna Engineering Using Physical Optics: Practical CAD Techniques and Software, Leo Diaz and Thomas Milligan

Antenna Measurement Techniques, Gary E. Evans.

Antennas and Propagation for Body-Centric Wireless Communications, Second Edition, Peter S. Hall and Yang Hao, editors

Antennas and Site Engineering for Mobile Radio Networks, Bruno Delorme

Antennas for Small Mobile Terminals, Kyohei Fujimoto and Koichi Ito, editors

Analysis, Design and Measurement of Small and Low-profile Antennas, Kazuhiro Hirasawa

Analysis of Radome-Enclosed Antennas, Second Edition, Dennis J. Kozakoff

Applications of Geographic Information Systems for Wireless Network Planning, Francisco Saez de Adana, Josefa Gómez Pérez, Abdelhamid Tayebi Tayebi, and Juan Casado Ballesteros

The Art and Science of Ultrawideband Antennas, Second Edition, Hans G. Schantz

AWAS for Windows Version 2.0: Analysis of Wire Antennas and Scatterers, Antonije R. Djordjević, et al.

Broadband Microstrip Antennas, Girsh Kumar and K. P. Ray

Broadband Patch Antennas, Jean-François Zürcher and Fred E. Gardiol

Circularly Polarized Dielectric Resonator Antennas, Raghvendra Kumar Chaudhary, Rajkishor Kumar, and Rakesh Chowdhury

Design and Applications of Active Integrated Antennas, Mohammad S. Sharawi and Oualid Hammi

Designer Notes for Microwave Antennas, Richard C. Johnson

Dielectric Resonator Antenna Handbook, Aldo Petosa

Dielectric Resonator Antennas, Biswajeet Mukherjee and Monika Chauhan

Electromagnetics and Antenna Technology, Alan J. Fenn

Electromagnetics, Microwave Circuit and Antenna Design for Communications Engineering, Peter Russer

Engineering Applications of the Modulated Scatterer Technique, Jean-Charles Bolomey and Fred E. Gardiol

Fixed & Mobile Terminal Antennas, Akhileswar Kumar

Foundations of Antenna Engineering: A Unified Approach for Line-of -Sight and Multipath, Per-Simon Kildal

Four-Arm Spiral Antennas, Joseph A. Mosko

Frequency-Agile Antennas for Wireless Communications, Aldo Petosa

Fresnel Zones in Wireless Links, Zone Plate Lenses and Antennas, Hristo D. Hristov

Handbook of Antennas for EMC, Second Edition, Thereza Macnamara with John McAuley

Handbook of Reflector Antennas and Feed Systems, Volume I: Theory and Design of Reflectors, Satish Sharma, Sudhakar Rao, and Lotfollah Shafai, editors

Handbook of Reflector Antennas and Feed Systems, Volume II: Feed Systems, Lotfollah Shafai, Satish Sharma, and Sudhakar Rao, editors

Handbook of Reflector Antennas and Feed Systems, Volume III: Applications of Reflectors, Sudhakar Rao, Lotfollah Shafai, and Satish Sharma, editors

Interference Suppression Techniques for Microwave Antennas and Transmitters, Ernest Freeman

Introduction to Antenna Analysis Using EM Simulators, Hiroaki Kogure, Yoshie Kogure, and James C. Rautio

Introduction to Antennas and RF Propagation Analysis, Dean James Friesen

LONRS: Low-Noise Receiving Systems Performance and Analysis Toolkit, Charles T. Stelzried, Macgregor S. Reid, and Arthur J. Freiley

Measurement of Mobile Antenna Systems, Second Edition, Hiroyuki Arai

Microstrip Lines and Slotlines, Third Edition, Ramesh Garg Inder J. Bahl, Maurizio Bozzi

Microstrip Antenna Design, K.C Gupta and Abdelaziz Benalla

Microwave and Millimeter-Wave Remote Sensing for Security Applications, Jeffrey A. Nanzer

Microwave Cavity Antennas, Akhileswar Kumar and Hristo D. Hristov

Millimeter Wave Microstrip and Printed Circuit Antennas, Prakash Bhartia

Mobile Antenna Systems Handbook, Third Edition, Kyohei Fujimoto, editor

Moment Methods in Antennas and Scattering, Robert C. Hansen

Multiband Integrated Antennas for 4G Terminals, David A. Sánchez-Hernández, editor

Near-Field Antenna Measurements, Dan Slater

Noise Temperature Theory and Applications for Deep Space Communications Antenna Systems, Tom Y. Otoshi

Phased Array Antenna Handbook, Third Edition, Robert J. Mailloux

Phased Array Antennas, Arthur A. Oliner

Phased Array Antennas with Optimized Element Patterns, Sergei P. Skobelev

Plasma Antennas, Second Edition, Theodore Anderson

Polarization in Electromagnetic Systems, Second Edition, Warren Stutzman

Practical Antenna Design for Wireless Products. Henry Lau

Practical Microstrip and Printed Antenna Design, Anil Pandey

Practical Phased Array Antenna Systems, Eli Brookner

Practical Simulation of Radar Antennas and Radomes, Herbert L. Hirsch

Printed MIMO Antenna Engineering, Mohammad S. Sharawi

Radiowave Propagation and Antennas for Personal Communications, Third Edition, Kazimierz Siwiak

Reconfigurable Antenna Design and Analysis, Mohammod Ali

Reflectarray Antennas: Analysis, Design, Fabrication and Measurement, Jafar Shaker, Mohammad Reza Chaharmir, and Jonathan Ethier

Reflector Lens Antennas: Analysis and Design Using Personal Computers, Carlyle J. Sletten, editor

RF Coaxial Slot Radiators: Modeling, Measurements, and Applications, Kok Yeow You

Shipboard Antennas, Second Edition, Preston Law

Significant Phased Array Papers, Robert C. Hansen

Smart Antenna Engineering, Ahmed El Zooghby

Solid Dielectric Horn Antennas, Carlos Salema, Carlos Fernandes, and Rama Kant Jha

Ultrawideband Antennas for Microwave Imaging Systems, Tayeb A. Denidni and Gijo Augustin

Ultrawideband Short-Pulse Radio Systems, V. I. Koshelev, Yu. I. Buyanov, and V. P. Belichenko

Waveguide Components for Antenna Feed Systems: Theory and CAD, Jaroslaw Uher, Jens Bornemann, and Uwe Rosenberg

Wearable Antennas and Electronics, Asimina Kiourti and John L. Volakis, editors

For further information on these and other Artech House titles, including previously considered out-of-print books now available through our In-Print-Forever® (IPF®) program, contact:

Artech House
685 Canton Street
Norwood, MA 02062
Phone: 781-769-9750
Fax: 781-769-6334
e-mail: artech@artechhouse.com

Artech House
16 Sussex Street
London SW1V HRW UK
Phone: +44 (0)20 7596-8750
Fax: +44 (0)20 7630 0166
e-mail: artech-uk@artechhouse.com

Find us on the World Wide Web at: www.artechhouse.com